STEADY AIRCRAFT
FLIGHT AND PERFORMANCE

STEADY AIRCRAFT FLIGHT AND PERFORMANCE

N. Harris McClamroch

PRINCETON UNIVERSITY PRESS ✈ PRINCETON AND OXFORD

In the United Kingdom: Princeton University Press, 6 Oxford Street,
Woodstock, Oxfordshire OX20 1TW

Library of Congress Cataloging-in-Publication Data
McClamroch, N. H. (N. Harris)
Steady aircraft flight and performance / N. Harris McClamroch.
p. cm.
Includes bibliographical references and index.
ISBN 978-0-691-14719-2 (hardcover : alk. paper)
1. Airplanes—Performance.   2. Airplanes—Handling characteristics.
I. Title.
TL671.4.M38 2011
629.132—dc22
2010016334

British Library Cataloging-in-Publication Data is available

This book has been composed in Times New Roman
Printed on acid-free paper. ∞
press.princeton.edu

Interior designed by Marcella Engel Roberts
Typeset by S R Nova Pvt Ltd, Bangalore, India
Printed in the United States of America

1 3 5 7 9 10 8 6 4 2

CONTENTS

LIST OF ILLUSTRATIONS

LIST OF MATLAB M-FILES

PREFACE

This book treats the fundamental principles of steady aircraft flight and flight performance. It arose from notes that were developed for a sophomore-level University of Michigan course in aerospace engineering. This class covered most of the topics in this book, excluding chapters 12 and 13, in two credit hours. For most of the students, this course was their first exposure to the scientific principles of flight, presented within a mathematical framework.

There are a number of introductory textbooks (Anderson [1998, 2000], Asselin [1997], Eshelby [2000], Layton [1988], Lowry [1999], Mair and Birdsall [1996], Roskam and Lan [1997], Saarias [2006], Shevell [1988], Torenbeek and Wittenberg [2009], Yechout and Bossert [2003]) that treat topics related to aircraft flight performance. Almost all of these give extensive treatment to aerodynamics and propulsion systems issues. At the same time, most do not provide a thorough development of aircraft translational kinematics, nor do they provide a complete derivation of the fundamental equations of steady flight based on free-body diagrams and Newton's laws. In addition, most of the introductory books do not give a detailed analysis of steady flight in three dimensions and the performance issues that arise from this steady flight analysis.

The main objective of this book is to give a self-contained, detailed, and scientifically rigorous treatment of steady aircraft flight and associated aircraft flight performance topics; these are foundational aerospace engineering topics of importance in their own right. A secondary objective is to provide necessary background for further study of flight dynamics, flight control, and dynamic flight performance.

This book develops concepts and mathematical descriptions associated with steady aircraft flight and aircraft performance, emphasizing both the mathematical development and the engineering implications. Careful derivations and associated assumptions used in these derivations are emphasized. The basic steady flight relationships, viewed in mathematical terms, are derived, and numerous steady flight performance topics are studied. The presentation also emphasizes the concept of steady flight envelope that takes into account fundamental flight constraints associated with the aerodynamics, the propulsion system, and the structural limits of an aircraft. Another novel feature of the presentation is the treatment of flight maneuvers, concatenations of flight maneuvers and flight planning, and assessment of optimal and near optimal flight performance from the basic flight equations and the flight constraints.

The novel features of this book provide a substantial contrast with standard textbooks (Anderson [1998], Asselin [1997], Eshelby [2000], Layton [1988], Lowry [1999], Mair and Birdsall [1996], Shevell [1988], Torenbeek and Wittenberg [2009], Roskam and Lan [1997], Saarias [2006], Yechout and Bossert [2003]). These books cover a number of topics in detail that are only briefly introduced herein. An introductory treatment of the history of flight is given in Anderson (1998) and Torenbeek and Wittenberg (2009). Anderson (1998, 2000), Asselin (1997), Mair and Birdsall (1996), McCormick (1994), Roskam and Lan (1997), Torenbeek and Wittenberg (2009), and Yechout and Bossert (2003) provide extensive treatments of aerodynamics and propulsion topics that are only briefly covered here.

In this book, the coverage of the concepts of flight envelope, flight maneuverability, and flight performance in three dimensions is significantly more detailed than what is covered in any other introductory textbook on aircraft performance. Results for steady turning flight in three dimensions are developed in considerable detail; these turning flight results follow most closely the development in Vinh (1995), but the presentation here is restricted to steady turning flight. Another novel feature of this book is that the basic steady flight concepts are illustrated through detailed studies of various steady flight conditions for a high-speed executive jet aircraft and for a low-speed, small, single-engine propeller-driven general aviation aircraft; these detailed case studies are constructed around computations and performance graphs developed in the Matlab programming language. In particular, these two types of aircraft are described and used to illustrate application of the various steady flight and flight performance concepts that are introduced. These two running case studies are developed at the end of chapters 5–13 to provide logical continuity to the development of the fundamental concepts. A thorough and unified treatment of each of the aircraft flight properties also illustrates the integration of the theoretical and computational approaches and the physical interpretation of results as part of a complete investigation of steady flight analysis and flight performance of an aircraft. A third case study example of an uninhabited aerial vehicle (UAV) is introduced and used to define problems that are given at the end of chapters 5–13.

Although the two case studies illustrate conventional jet-propelled flight and propeller-driven flight for fixed-wing aircraft, many of the conceptual ideas and approaches are applicable to other categories of flight vehicles. Although they are not explicitly treated in this book, the methods of steady flight and performance analysis are applicable, with changes in the assumptions about aerodynamics and propulsion, to microflight vehicles, supersonic flight vehicles, and hypersonic flight vehicles. The same methods are applicable to piloted flight vehicles, remote-controlled flight vehicles, and autonomous flight vehicles.

The book is intended for persons who have a good background in geometry in two and three dimensions, calculus, and classical physics. Chapters 1–12 should be accessible to readers with little or no background in differential equations. The main additional requirement is a willingness to follow careful mathematical developments that use both engineering and mathematical arguments. Although students with limited mathematical and science backgrounds are the intended audience for this book, it is hoped that more advanced students and engineering and science professionals may also benefit from study of the book's treatment of aircraft steady flight and performance.

In chapters 1–12 there is no explicit use of theoretical material on differential equations, although certain differential equations are solved in analytical form using elementary methods of calculus. Differential equations are solved numerically using Matlab only in the analysis of climbing or descending flight; this approach is taken when analytical solutions using elementary methods of calculus are not easily obtained. Further, chapter 13 summarizes the fundamental differential equations that can be used to treat aircraft translational flight dynamics and dynamic flight performance; aerodynamics and propulsion assumptions are made that are consistent with the assumptions made in studying steady flight performance. Chapter 13 provides a bridge from the analysis of steady flight and performance to dynamic flight and peformance.

The treatment of optimal aircraft flight performance requires the application of optimization theory. A few comments about fundamental optimization concepts are necessary as background for the subsequent consideration of optimal flight performance. The flight optimization problems considered are formulated in terms of a performance function of one or more flight variables that is to be maximized (or minimized). If there are no constraints on the variables then standard results from calculus imply that the derivative of the performance function with respect to each of the variables should be zero at the optimal flight condition; this property can be used to determine the optimal values of the flight variables and hence the optimal value of the performance function. On the other hand, for many flight optimization problems equality constraints and inequality constraints must be taken into account. For a small number of variables and a small number of constraints, the optimal solution may be determined by simplification of the optimization problem by enforcing the equality constraints and by assuming that the optimal solution occurs with only a few active inequality constraints. In the relatively simple cases studied in this book, geometry, calculus, and logic can often be used to determine the optimal flight conditions and the corresponding optimal performance. Flight optimization problems that involve more complicated equations with many variables and constraints require more advanced formal knowledge of optimization theory and associated numerical optimization methods.

Basic knowledge of the Matlab programming language is required to follow the development of the computational results in the two aircraft case studies that run throughout the book. Matlab, in this book, is primarily used to solve certain algebraic equations, and its graphics capabilities are used to visualize various geometric features of steady flight. Matlab is used in only a few cases to numerically solve optimization problems and to numerically integrate differential equations in chapter 13. Matlab is shown to be a useful computational tool for aircraft flight and performance analysis.

The organization of the steady flight and flight performance topics in chapters 6–12 flows from the simplest to the more complicated. After presentation of background material on aerodynamics, translational kinematics, and propulsion, steady gliding flight is first studied assuming there is no thrust force on the aircraft. The subsequent chapters present material on powered flight: steady level longitudinal flight, steady climbing or descending flight, steady level turns, and steady climbing or descending turns. Subsequent chapters cover material on aircraft range and endurance and on aircraft flight maneuvers and flight planning. The last chapter provides an introduction to translational flight dynamics. Concrete illustrations of many of the general flight concepts are provided by evaluating specific flight conditions and flight performance measures for an executive jet aircraft and for a propeller-driven general aviation aircraft. The terminology and notation are consistent throughout the book, and they have been chosen to reflect what is most common in the flight literature.

There is considerable redundancy in the development of steady gliding flight, steady longitudinal flight, and steady turning flight. This redundancy is intentional, keeping in mind that the intended audience is not assumed to have prior experience with the mathematical aspects of aircraft flight. The organizational flow of the book from steady gliding flight to steady turning and climbing or descending flight enables the reader to digest the material in an incremental way. However, readers who prefer to move directly to the most general form of steady climbing or descending and turning flight can skip the chapters on steady gliding flight, steady longitudinal flight, and steady level turning flight, viewing these topics as special cases of this most general form of steady flight that is treated in chapter 10.

Acknowledgments

This book has been developed with the help of Dr. Taeyoung Lee, who has contributed to the writing of several sections and to the computations for the case studies that are important illustrations of the fundamental flight concepts. He developed most of the figures in the book and provided many suggestions in the use of LaTeX throughout the development of the book. His contributions and support of this endeavor are gratefully acknowledged.

Many of my former students and colleagues have provided comments and suggestions; this feedback has been crucial in the development of the material in its present form. I am grateful for their comments and suggestions, and I acknowledge all of my past students for providing the inspiration for this effort.

NHM

STEADY AIRCRAFT FLIGHT AND PERFORMANCE

Aircraft Components and Subsystems

This chapter deals with the fundamental physical components and properties of conventional fixed-wing aircraft. This material is covered in detail in many textbooks (see Anderson [1998, 2000], Asselin [1997], Eshelby [2000], Layton [1988], Lowry [1999], Mair and Birdsall [1996], Roskam and Lan [1997], Saarias [2006], Shevell [1988], Torenbeek and Wittenberg [2009], and Yechout and Bossert [2003]); see Collinson (1996) for an overview that also describes flight instruments and avionics. The treatment here is brief and emphasis is given to the aspects of conventional aircraft that are most related to their flight characteristics.

1.1 Aircraft Subsystems for Conventional Fixed-Wing Aircraft

Figure 1.1 illustrates a conventional fixed-wing aircraft that is the basic flight vehicle of interest in this book. The key physical components, or subsystems, that define the aircraft are the fuselage, the wings, the horizontal tail, the vertical tail, and the propulsion system. The fuselage provides working volume for passengers, cargo, and aircraft subsystems that are internal to the aircraft. The fuselage is important in terms of achieving particular flight missions, but it is not especially important from a flight performance perspective. The two wings are crucial for flight, since their main purpose is to generate lift. The aircraft illustrated in Figure 1.1, and all aircraft considered hereafter, are fixed-wing aircraft, since the wings are rigidly attached to the fuselage. This is in contrast with helicopters or other rotary wing flight vehicles that generate lift using rotating blades.

Other important flight subsystems, illustrated in Figure 1.1, are the horizontal tail, the vertical tail, and the engines. The horizontal and vertical tails are rigidly attached to the fuselage as indicated. The horizontal tail provides longitudinal stability and control capability, while the vertical tail provides directional stability and control capability. The engines are crucial flight subsystems, since they generate the thrust force that acts on the aircraft. Note that gliding flight, studied in chapter 6, occurs if the engines are turned off so that they do not generate thrust; gliders have no propulsion system.

The above descriptions imply that the aircraft can be viewed as a rigid body, and this is the perspective that is taken throughout. That is, there is no relative motion between the physical aircraft subsystems such as the fuselage, the wings, and the vertical and horizontal

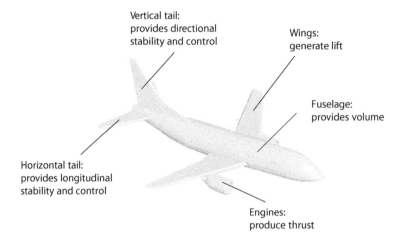

Figure 1.1. Aircraft subsystems.

tails. Since many forces act on these physical subsystems, the rigid body assumption is only a crude approximation. In fact, the aircraft physical structures deform under the applied forces that occur during flight. Issues of structural design and analysis are important to guarantee that the rigid body assumption is justified.

This is the appropriate point to mention another important assumption that holds throughout the analysis presented subsequently. The complete aircraft, consisting of the fuselage, the wings, the horizontal and vertical tails, and all other flight subsystems, has a plane of mass symmetry that exactly bisects the aircraft. This assumption is a consequence of the design of conventional fixed-wing aircraft where, in particular, engines mounted on the fuselage or the wings are balanced to satisfy this mass symmetry assumption.

1.2 Aerodynamic Control Surfaces

Figure 1.2 illustrates three types of aerodynamic control surfaces: the elevator, the ailerons, and the rudder. The elevator is one (or more than one) movable flap, located on the trailing edge of the horizontal tail. Deflection of the elevator changes the air flow over the horizontal tail in such a way that a pitch moment on the aircraft is generated. The ailerons consist of a pair of movable flaps, located on the trailing edge of each wing; ailerons usually operate in differential mode so that if one flap is deflected up the other flap is deflected down by the same amount or vice versa. Differential deflection of the ailerons changes the air flow over the wings in such a way that a roll moment on the aircraft is generated. The rudder is one (or more than one) movable flap, located on the trailing edge of the vertical tail. Deflection of the rudder changes the air flow over the vertical tail in such a way that a yaw moment on the aircraft is generated. The elevator deflection, rudder deflection, and the differential deflection of the ailerons are typically viewed as angles measured from some reference values.

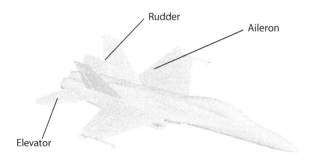

Figure 1.2. Aerodynamic control surfaces.

These movable flaps are referred to as aerodynamic control surfaces; they generate moments on the aircraft according to the principles of aerodynamics. The precise meanings of pitch, roll, and yaw moments are described later. These moments are used to maneuver and control the flight of the aircraft.

Some modern aircraft have unconventional elevators, ailerons, and rudders, as well as additional flaps on the fuselage referred to as canards. Although many aerodynamic control surface designs are possible, they all are intended to generate pitch, roll, and yaw moments. The subsequent development in this book is based on the assumption that conventional elevators, ailerons, and rudders are utilized.

1.3 Aircraft Propulsion Systems

The aircraft engines, together with associated fuel tanks and related hardware, are referred to as propulsion systems. The purpose of the propulsion system is to generate a thrust force that propels the aircraft in flight. Although engines and propulsion systems are extremely complicated, detailed knowledge of the engine specifications is not required to analyze flight properties of an aircraft. Rather, the key features for the study of steady flight are the maximum thrust (or the maximum power) that the engine can produce and the rate at which fuel is burned to produce a given thrust level (or power level).

From the earliest days of powered flight, propulsion systems have consisted of an internal combustion engine that causes rotation of a propeller. The blades of the propeller are designed so that they generate a thrust force. Such propulsion systems remain in common use today, especially for low-speed aircraft for which cost and durability considerations are primary. The important specifications for this type of propulsion system are the maximum power that the internal combustion engine can produce, the rate of fuel burned to provide a specified power level, and the efficiency of the propeller.

An important technological outcome of World War II was the development of jet engine technology. Such propulsion systems make high-speed flight possible. Most jet engines consist of a compressor, a turbine, and a combustor that are used to accelerate the flow of air through the engine, thereby producing a thrust force on the aircraft. These turbojet engines are extremely complicated, but the important specifications are the maximum thrust that the turbojet engine can produce and the rate of fuel burned to provide a specified thrust level.

Turboprop engines use turbine and jet engine technologies to turn a propeller that generates a thrust force on the aircraft. These propulsion systems exhibit some of the features of conventional jet engines, but the efficiency of the propeller is also important.

Finally, rocket engines have been used to generate propulsive thrust forces on certain experimental aircraft. A rocket engine consists of fuel and an oxidizer that are stored internally in the aircraft; as the fuel is burned the combustion products are exhausted out a nozzle to produce a thrust force on the aircraft. Rocket engines can generate extremely large levels of thrust, but they have limited duration of operation.

Our focus here is on conventional propulsion systems: either a propeller driven by an internal combustion engine or a turbojet engine. Many of the subsequent developments can be modified to handle other types of propulsion systems.

The aircraft propulsion system produces a thrust force vector that has a fixed direction with respect to the aircraft. This property arises since the engines are typically fixed in the aircraft, either on the wings or on the fuselage. The direction of the thrust vector is also assumed to lie in the plane of mass symmetry of the aircraft. Finally, it is assumed that the engine can be throttled or adjusted so that any thrust level (or power level) between zero thrust (or zero power) and maximum thrust (or maximum power) can be achieved using a throttle setting between 0 and 1.

In some modern fighter aircraft, the direction of the thrust vector, with respect to the aircraft, can be adjusted within limits; this feature is referred to as thrust vectoring. The advantage of thrust vectoring is that control moments due to the thrust can be generated; this extra control capability can add to the maneuvering capability that can be achieved using conventional aerodynamic control surfaces. In certain flight conditions, such as post-stall flight, the aerodynamic control surfaces may be ineffective; in such cases thrust vectoring is essential to maintain maneuvering capability. Thrust-vectored aircraft are not explicitly treated in this presentation, although the methods that are subsequently developed can be modified to treat thrust-vectored aircraft.

1.4 Aircraft Structural Systems

As mentioned previously, fixed-wing aircraft consist of many structural subsystems such as the wings, the fuselage, horizontal and vertical tails, and aerodynamic control surfaces, as well as additional structural subsystems for attachment of the engines and for landing gear. Each of these structural subsystems must be designed to withstand forces and moments that are due to gravity, propulsive, and aerodynamics loadings, the most important being the aerodynamic forces and moments. In particular, these structural subsystems must maintain their structural integrity and exhibit limited deformation in the presence of expected aerodynamic forces and moments. Structural issues are crucial to the design of any aircraft.

Although the details of aircraft structural design are not central to the subsequent development of aircraft steady flight and performance that are addressed here, it is important to keep them in mind. In particular, they provide the justification for the rigid body aircraft assumption. In addition, aircraft performance is significantly limited by the aerodynamic forces for which the aircraft structure is designed. This structural limitation arises most importantly in the allowed wing loading, which, as will be shown later, limits the tightness of a turn that an aircraft can safely achieve. These issues are developed in some detail in subsequent chapters on steady turning flight.

1.5 Air Data and Flight Instrumentation

Modern aircraft are complex vehicles that contain many instruments and devices that support flight. These flight instruments use extensive electronics and are referred to as the aircraft avionics. Although these instruments do not directly influence the flight performance of an aircraft, they are important instruments for measuring flight properties and flight performance. They provide useful measurements that enable the pilot to effectively fly the aircraft. The following is a brief summary of typical avionics found on many aircraft:

- An altimeter provides a pressure measurement that approximately indicates the flight altitude above sea level, based on properties of the standard atmospheric model, as developed in chapter 2.

- The airspeed indicator provides a measurement that is used to estimate the speed of the aircraft with respect to the surrounding air.

- Engine gauges provide information about the engine status, including the engine speed and the engine temperature.

- The tachometer measures either the speed of rotation of the propeller or the turbine speed in a jet engine.

- An artificial horizon, based on gyroscopic instruments, provides measurements that can be used to estimate the roll angle and pitch angle of the aircraft (see chapter 3); this provides the pilot with an Earth-fixed reference.

1.6 Guidance, Navigation, and Control

Many aircraft also include avionics that are used for guidance, navigation, and control functions. Several types of instruments are common:

- Distance measuring equipment provides an estimate of the distance of the aircraft from a ground station based on a signal received from the ground station.

- A magnetic compass is used to estimate the heading angle of an aircraft relative to magnetic north.

- Gyroscopes are used to estimate the angular velocity of the aircraft; they are often used as components of more complicated instrument packages.

- Inertial navigation systems consist of accelerometers, gyroscopes, and associated electronics that provide estimates of the attitude of the aircraft and estimates of the velocity vector and the position vector of the aircraft in an Earth-fixed reference frame; the estimates are obtained through numerical integration of the gyroscopic and the accelerometer data.

- Global Positioning System (GPS) is a satellite-based navigation system that provides extremely accurate estimates of the inertial position of an aircraft. GPS operates by determining precise measurements of the distances of the aircraft from several satellites, located at exactly known positions, using signals broadcast by those satellites; these distances are processed to estimate the position vector of the aircraft in an Earth-fixed reference frame. In some versions, GPS can also be used to estimate the aircraft attitude.

- Instrument landing systems are microwave-based or GPS-based systems that provide information to the aircraft that is used for semiautomatic or automatic landing.

1.7 Flight Control Computers

Computer systems are now a common and essential part of many modern flight vehicles. Flight computations can be carried out by embedded computer systems that are part of an integrated avionics package that performs guidance, navigation, and control functions. A flight control computer is often available that carries out computations associated with high-level planning, including routing, automatic pilots, and flight management tasks. The flight control computer often provides an interface with the flight avionics, and it is responsible for flight data displays on the flight deck in the cockpit. The flight computer system consists of both hardware and software. Flight software in many modern flight vehicles is amazingly complex. Flight computers constitute the brains of many flight vehicles, and they are often essential for flight operations.

1.8 Communication Systems

Communication functions are important for many aircraft. As indicated previously, communications are necessary for navigation functions such as distance measurement, instrument landing, and global positioning. A transponder is a radar-based system that provides positive identification of the aircraft to an air traffic controller. In addition, VHF voice radios and omnidirectional radios allow the aircraft pilot to communicate directly with the ground and with other aircraft.

1.9 Aircraft Pilots

The pilot is an important aircraft subsystem, and a complete analysis of aircraft flight characteristics requires attention to pilot operations. Depending on the type of aircraft, human pilots can have varying degrees of flight authority. The most common situation is that the pilot directly controls the aircraft flight conditions using manual flight control inputs, namely the throttle, the elevator deflection, the rudder deflection, and the differential deflection of the ailerons. In more advanced aircraft, the flight conditions are controlled by the pilot but with aid from an automatic control system, often referred to as an autopilot. In some of the most advanced aircraft, autonomous operation is possible where the flight conditions are completely controlled by the autopilots, with little or no supervision by a human.

Pilots operate aircraft according to flight operational protocols. These are an extensive set of flight rules that pilots must follow within the common airspace. The two main flight protocols are visual flight rules and instrument flight rules. Visual flight rules are most restrictive; they are required for low-speed general aviation aircraft with few flight instruments. Instrument flight rules apply to high-speed aircraft that are suitably equipped with flight instrumentation and flown by certified pilots.

In the subsequent development, it is sufficient to consider the pilot, or equivalently the autopilot, as providing control input commands that are used to keep the aircraft in a desired steady flight condition or to maneuver the aircraft from one steady flight condition to another. The four types of flight control inputs are the elevator, the ailerons, the rudder, and the throttle. All flight conditions and all flight maneuvers can be obtained by appropriate adjustment of these four flight control inputs. Subsequently, a complete analysis is developed that describes how a given steady flight condition is achieved by appropriate selection of the values of the flight control inputs. Flight maneuvers are also analyzed and expressed in terms of the required changes for these four flight control inputs.

1.10 Autonomous Aircraft

Some aircraft, such as small hobbyist radio-controlled aircraft or uninhabited aerial vehicles (UAVs), can be controlled through a remote operator. In this case, the remote operator is effectively the pilot in the sense that the operator flies the aircraft by adjusting the elevator, ailerons, rudder, and throttle.

It is possible, within limited environments, to develop aircraft that fly with complete or nearly complete autonomy. These autonomous flight vehicles require the development of advanced computer systems, often based on advanced control and artificial intelligence theories, that can make automatic flight decisions without direct intervention from a human. This is a flight technology that is growing rapidly; the fundamental flight principles described herein form the basis for many of these new developments.

1.11 Interconnection and Integration of Flight Systems

The flight subsystems of a single aircraft have been briefly described. The complete aircraft should be viewed as an interconnection of all of these flight subsystems. That is, aircraft flight operations depend on all of the flight subsystems working together. The concept of a system interconnection means that the outputs of one subsystem can be viewed as inputs to other subsystems. This flight interconnection perspective is very powerful, and it is essential to understand the flight operations of a single complex aircraft.

A topic of current study in the research and development community is the coordination of multiple aircraft, referred to as an aircraft team, to achieve some overall cooperative mission or flight formation. This requires a high level of integration of all of the aircraft that constitute the aircraft team. The flight properties of the aircraft team depend on the flight properties of the individual aircraft and the coordination strategy that is employed to control the team. This topic is not discussed further here, but it clearly requires, as background, a detailed knowledge of the flight properties of a single aircraft.

Fluid Mechanics and Aerodynamics

Aircraft flight depends on the physical properties of the atmosphere, which are characterized by the pressure, temperature, and density of air. Aircraft flight requires relative motion of the aircraft with respect to the surrounding air; this relative motion gives rise to aerodynamic forces and aerodynamic moments on the aircraft. It is essential to quantify the aerodynamic forces and moments on an aircraft in flight in order to develop a theory of steady aircraft flight and performance. This chapter treats the fundamental properties of air, both when it is stationary and when it has relative motion with respect to the aircraft. The latter case is referred to as aerodynamics. This background is subsequently used in chapter 3 to develop mathematical models for the aerodynamic forces and moments that act on an aircraft in flight.

These models of the aerodynamics are critical ingredients in the subsequent steady flight and performance analyses. Our treatment of aerodynamics in this chapter is limited to the aspects of aerodynamics that are required for the subsequent steady flight and performance analyses. (For more comprehensive treatments of aerodynamics, see Anderson [1998, 2000], Mair and Birdsall [1996], McCormick [1994], Roskam and Lan [1997], and Torenbeek and Wittenberg [2009].)

2.1 Fundamental Properties of Air

There are three fundamental characteristics of air that are important for understanding aircraft flight in the atmosphere: pressure, temperature, and density. Conceptually, the air pressure denotes the force per unit area exerted by an element of air. The temperature is a measure of the thermal intensity of an element of air, and the density is mass per unit volume of an element of air. The pressure, temperature, and density can be defined at any location in the atmosphere, and each can vary as a function of location and time. The pressure, temperature, and density have real values and are examples of scalar fields. In this book, we use the British or U.S. system of units. The pressure has the units of pounds per square foot; the temperature can be measured according to the common relative temperature scale in degrees F or according to the absolute temperature scale in degrees R, wherein absolute zero corresponds to zero motion of the air molecules; and the density of air has the units of slugs per cubic foot.

The other important property of an element of air is its flow velocity in three dimensions. As usual, the velocity is a vector that defines the magnitude, sometimes referred to as the airspeed, and the direction of the flow. The velocity vector has three components, or equivalently a magnitude and direction, that can vary as a function of location and time. The flow velocity is an example of a vector field. If the air has a zero velocity vector at some location at some instant, that element of air is not in motion. If the flow velocity vector is nonzero at some location and some instant, that element of air is in motion. Each scalar component of a velocity field has units feet per second (ft/s).

The ideal gas law is a general relation between the pressure, temperature, and density of an ideal gas such as air. It is often referred to as an equation of state. Let p denote the air pressure, T denote the absolute temperature, and ρ denote the density of an element of air at a particular location and time. Pressure, temperature, and density of that element of air are related by the ideal gas law

$$p = \rho R T, \tag{2.1}$$

where R is a constant appropriate to air. The value for this constant for air is $1,716 \text{ ft}^2/\text{s}^2 \text{°R}$.

Since R is a constant, the ideal gas law shows that if the values of any two of the three pressure, temperature, and density variables are known then the value of the third can be determined from the ideal gas law. It also implies many other relationships. For example, if the temperature does not change but the pressure is increased (decreased), then the density must increase (decrease) accordingly.

2.2 Standard Atmosphere Model

In this section, theoretical properties of the atmosphere are described. These properties demonstrate how the pressure, temperature, and density in the atmosphere vary with altitude. By assuming that the atmosphere is stationary with respect to the Earth, the properties of the standard atmospheric model are obtained; this model is widely used in many engineering applications. In particular, it plays a critical role in the study of aircraft steady flight and performance.

Properties of the standard atmospheric model are derived by combining the ideal gas law and the hydrostatic equation. Empirical data about the temperature variation with altitude are also introduced. The final form of the standard atmospheric model is developed by using the methods of calculus.

Assume the atmosphere is stationary with respect to the Earth; that is, the atmospheric velocity field is identically zero. The standard atmospheric model is based on the assumption that the air temperature, pressure, and density of an element of air do not vary within the horizontal plane at a fixed altitude. The air temperature, pressure, and density do vary with altitude, and these variations are now studied.

Let h denote the altitude measured from sea level; then the hydrostatic equation describes the vertical air pressure gradient as

$$\frac{dp}{dh} = -\rho g. \tag{2.2}$$

In this equation, g is the acceleration of gravity. Although the acceleration of gravity is smaller at higher altitudes, this effect is quite small and is omitted in this analysis. The acceleration of gravity at sea level has the value 32.2 ft/s^2.

The hydrostatic equation is easily derived by examining the forces acting on an air element and using the fact that the sum of the vertical forces and the sum of the horizontal forces acting on the element of air must be zero since the air is stationary.

Empirical results obtained from averaged atmospheric measurements are introduced to model the variation of temperature in the atmosphere as it depends on altitude. The important part of the atmosphere for flight can be naturally divided into several horizontal layers; only the troposphere, the tropopause, and the stratosphere are considered here.

- The lowest atmospheric layer is the *troposphere*, where the temperature decreases linearly with altitude with an assumed sea level atmospheric temperature of $518.67°R = 59°F$; this layer extends from sea level to an altitude of 36,089.2 ft.

- In the second atmospheric layer, referred to as the *tropopause*, the temperature does not change with altitude, having the constant value $389.97°R = -69.7°F$; the tropopause is the layer from an altitude of 36,089.2 ft up to an altitude of 65,616.2 ft.

- The next highest atmospheric layer, the *stratosphere*, extends from an altitude of 65,616.2 ft up to an altitude of 104,986.9 ft; in this region the temperature increases linearly with altitude.

Although it is possible to characterize the temperature variation above the stratosphere, it is not necessary to consider this region for purposes of flight analysis. In fact, the most common flight conditions occur within the troposphere.

An analytical formula for the empirical atmospheric temperature variation in the troposphere is given by

$$T = 518.67 - 0.003567h°\text{R}, \tag{2.3}$$

where h is expressed in ft.

In the tropopause, the empirical atmospheric temperature variation is given by

$$T = 389.97°\text{R}. \tag{2.4}$$

In the stratosphere, the empirical atmospheric temperature variation is given by

$$T = 389.97 + 0.0005494\,(h - 65,616.2)\,°\text{R}, \tag{2.5}$$

where h is expressed in ft.

The hydrostatic equation can be rewritten by substituting the expression for the air density expressed in terms of temperature and pressure from the ideal gas law (2.1) into equation (2.2). The resulting expression gives the vertical pressure gradient in terms of the pressure and temperature as

$$\frac{dp}{dh} = -\frac{p}{RT}g. \tag{2.6}$$

In the troposphere, the temperature variation is linear with altitude. This allows integration of equation (2.6), using the value for the sea level atmospheric pressure, to determine the analytical expression for the pressure within the troposphere as

$$p = p_0 \left(\frac{T}{T_0} \right)^{-\frac{g}{a_0 R}}, \tag{2.7}$$

where p_0 and T_0 are the pressure and the temperature at sea level, respectively, and a_0 is the constant slope of the temperature variation with altitude in the troposphere:

$$p_0 = 2,116.2 \, \text{lb/ft}^2,$$
$$T_0 = 518.67°\text{R},$$
$$a_0 = -3.567 \times 10^{-3}°\text{R/ft}.$$

This then allows determination of the atmospheric pressure at the highest altitude in the troposphere or, equivalently, the lowest altitude in the tropopause.

We now integrate equation (2.6), using the constant value for the atmospheric temperature in the tropopause, to determine the analytical expression for the pressure in the tropopause as

$$p = p_1 \exp \left(-\frac{g}{R T_1} (h - h_1) \right), \tag{2.8}$$

where p_1 and T_1 are the pressure and temperature at the lowest altitude in the tropopause, respectively, and h_1 is the altitude at which the tropopause begins:

$$p_1 = 472.7 \, \text{lb/ft}^2,$$
$$T_1 = 389.97°\text{R},$$
$$h_1 = 36,089.2 \, \text{ft}.$$

This allows determination of the atmospheric pressure at the highest altitude in the tropopause or, equivalently, at the lowest altitude in the stratosphere.

We now integrate equation (2.6), using the value for the atmospheric pressure at the lowest altitude of the stratosphere, to determine the analytical expression for the pressure in the stratosphere as

$$p = p_2 \left(\frac{T}{T_2} \right)^{-\frac{g}{a_2 R}}, \tag{2.9}$$

where p_2 and T_2 are the pressure and the temperature at the lowest altitude of the stratosphere, respectively, and a_2 is the constant slope of the temperature variation with altitude in the stratosphere:

$$p_2 = 114.3 \, \text{lb/ft}^2,$$

$$T_2 = 389.97°\text{R},$$

$$a_2 = 5.494 \times 10^{-4}°\text{R/ft}.$$

It is now possible to determine the variation of the atmospheric density from expressions (2.3)–(2.9) using the ideal gas law. This gives the air density expression in the troposphere as

$$\rho = \rho_0 \left(\frac{T}{T_0} \right)^{-\frac{g}{a_0 R} - 1}, \tag{2.10}$$

where ρ_0 is the air density at sea level:

$$\rho_0 = 2.38 \times 10^{-3} \, \text{slugs/ft}^3.$$

The air density expression in the tropopause is

$$\rho = \rho_1 \exp \left(-\frac{g}{RT_1} (h - h_1) \right), \tag{2.11}$$

where ρ_1 is the density at the lowest altitude of the tropopause:

$$\rho_1 = 7.061 \times 10^{-4} \, \text{slugs/ft}^3.$$

The air density expression in the stratosphere is

$$\rho = \rho_2 \left(\frac{T}{T_2} \right)^{-\frac{g}{a_2 R} - 1}, \tag{2.12}$$

where ρ_2 is the air density at the lowest altitude of the stratosphere:

$$\rho_2 = 1.708 \times 10^{-4} \, \text{slugs/ft}^3.$$

The above expressions for the temperature, pressure, and density of the atmosphere are shown in graphical form in Figure 2.1. Note that for altitudes exceeding 90,000 ft, the pressure and density of the atmosphere are very small, which is why conventional aircraft flight is not possible at these high altitudes.

A table for the standard atmospheric model is given in appendix A. This table is convenient for our subsequent flight analysis. It includes data for atmospheric temperature, pressure, and density, as they each depend on the altitude.

It is important to keep in mind that the standard atmospheric model describes an idealization of the atmosphere that is stationary. This standard atmosphere is not intended to represent any real atmospheric properties at any specific location near the Earth or at any particular instant. Rather, the standard atmosphere is a model that is widely adopted and useful for analysis of aircraft flight.

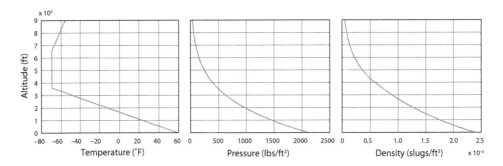

Figure 2.1. Standard atmosphere.

The following Matlab function computes the temperature, pressure, and density at a specified altitude less than or equal to 90,000 ft.

Matlab function 2.1 `StdAtpUS.m`

```
function [T p rho]=StdAtpUS(h)
%Standard Atmosphere
%Input : h altitude (ft)
%Output : T temperature (F), p pressure (lbs/ft^2),
rho density (slug/ft^3)
h1=3.6089e4; h2=6.5616e4; h3=9.0e4; a0=-3.567e-3;
a2=5.494e-4; g=32.2;
R=1716;
T0=518.67; p0=2116.2; rho0=2.3769e-3; T1=T0+a0*h1;
p1=p0*(T1/T0)^(-g/a0/R); rho1=rho0*(T1/T0)^(-g/a0/R-1);
T2=T1;
p2=p1*exp(-g/R/T2*(h2-h1)); rho2=rho1*exp(-g/R/T2*
(h2-h1));
if h <= h1
    disp('Troposphere');
    T=T0+a0*h;
    p=p0*(T/T0)^(-g/a0/R);
    rho=rho0*(T/T0)^(-g/a0/R-1);
elseif h <= h2
    disp('Tropopause');
    T=T1;
    p=p1*exp(-g/R/T*(h-h1));
    rho=rho1*exp(-g/R/T*(h-h1));
elseif h <= h3
    disp('Stratosphere');
    T=T2+a2*(h-h2);
    p=p2*(T/T2)^(-g/a2/R);
    rho=rho2*(T/T2)^(-g/a2/R-1);
else
    disp('Error: the altitute should be less then 90000 ft');
end
```

Figure 2.1 is obtained using the Matlab function StdAtpUS for computing the dependence of temperature, pressure, and density as a function of altitude and the following Matlab m-file.

Matlab function 2.2 FigStdAtpUS.m

```
h=0:100:90000;
for k=1:size(h,2)
    [T(k) p(k) rho(k)]=StdAtpUS(h(k));
end
figure;
plot(T-459.67,h);
xlabel('Temperature (^{o}F)');
ylabel('Altitude (ft)');
grid on;
figure;
plot(p,h);
xlabel('Pressure (lb/ft^2)');
ylabel('Altitude (ft)');
grid on;
figure; plot(rho,h);
xlabel('Density (slugs/ft^3)');
ylabel('Altitude (ft)');
grid on;
```

2.3 Aerodynamics Fundamentals

The assumption in section 2.2 is that the atmosphere is stationary, namely that the velocity field is identically zero. In this section, air is assumed to be in motion with a nontrivial velocity field. Steady flow properties are briefly summarized, where steady flow means that the velocity field may depend on location but is time independent.

Figure 2.2 shows a schematic drawing of the flow of air over a fixed airfoil in a wind tunnel. The two-dimensional schematic drawing illustrates the flow characteristics that are exposed by injection of smoke particles into the flow. The smoke particles define streamlines that clearly separate the flow around the airfoil. There is no flow across the streamlines. Equivalently, the velocity field must be always tangent to the streamlines. Properties of the flow based on the streamlines shown in Figure 2.2 can be determined, assuming the flow is two-dimensional. For more complex three-dimensional steady flows, stream tubes can be defined analogously to streamlines.

Incompressible flow means that the air density field is constant. Conversely, if the air density field is not constant, the result is compressible flow. Incompressible flow is a reasonable approximation so long as the magnitude of the velocity field is substantially smaller than the speed of sound in air. When the velocity field is sufficiently high, compressible flow, for which the density is not constant, must be considered. Most of our subsequent analysis assumes incompressible flow.

The continuity equation, equivalent to conservation of mass, is a fundamental physical relationship. The continuity equation can be written in a general form for air flow. It is

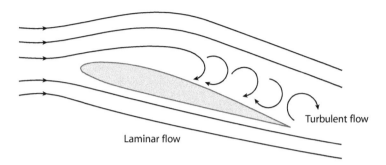

Figure 2.2. Flow over an airfoil.

sufficient to consider the steady flow of air between two cross sections of a stream tube. Suppose the cross-sectional area of a stream tube at the first location is A_1 and let V_1 and ρ_1 denote the speed of the flow and the air density at this location. Similarly, at some other location, suppose the cross-sectional area of the stream tube is A_2 and the speed of the flow and the air density at this location are V_2 and ρ_2. The flow is assumed steady; that is, it is time independent. Continuity requires that the mass of air within the stream tube between the two cross sections remains constant, since there is no flow across the stream tube boundary. It follows that the continuity equation for steady flow is

$$\rho_1 A_1 V_1 = \rho_2 A_2 V_2. \tag{2.13}$$

If the flow is assumed to be incompressible, then the air densities are identical at the two cross sections and the continuity equation for incompressible flow gives

$$A_1 V_1 = A_2 V_2. \tag{2.14}$$

This relation demonstrates the general property that if the cross-sectional area is decreased, for example by requiring that the flow go through a nozzle, then the flow velocity must increase. This general fact is also true for compressible flow, although the change in air density must be taken into account.

Flow over an aircraft wing, or flow over an airfoil in a wind tunnel, provides another illustration of the continuity equation. As seen in Figure 2.2, the stream tubes are deformed by the presence of the airfoil in the flow. Consequently, changes in the cross-sectional area of the stream tube correspond to changes in the velocity of the flow.

Another fundamental physical principle is conservation of momentum for a steady flow. Suppose that the flow properties vary only along the axis of a stream tube. Conservation of momentum along that axis can be shown to lead to the following equation:

$$\frac{dp}{dx} = -\rho V \frac{dV}{dx}. \tag{2.15}$$

At the location denoted by x along the stream tube axis, p is the pressure, referred to as the static air pressure, ρ is the air density, and V is the speed of the flow. This equation relates the pressure gradient of the flow to the velocity gradient of the flow. Note that the pressure gradient and the velocity gradient have opposite signs. Along the stream tube, if the pressure increases (decreases) then the flow speed necessarily decreases (increases).

Assuming incompressible flow, equation (2.15) can be integrated along the axis of the stream tube to obtain Bernoulli's equation

$$p + \frac{1}{2}\rho V^2 = C, \tag{2.16}$$

where C is an integration constant; this equation provides a flow function that is constant anywhere within a given stream tube. That is, the sum of the static pressure p and the dynamic pressure $\frac{1}{2}\rho V^2$ is a constant along the stream tube. The dynamic pressure term

$$q = \frac{1}{2}\rho V^2 \tag{2.17}$$

is an important variable in aerodynamics and aircraft flight. It has the same units as pressure.

Bernoulli's equation is extremely useful. Using both the continuity equation and Bernoulli's equation for flow along a stream tube, it can be shown that if the cross-sectional area of the stream tube is decreased the flow velocity must increase and the static pressure must decrease. Bernoulli's equation also forms the theoretical basis for various flow instruments. For example, if one knows the constant C then direct measurement of the static pressure at some location allows determination of the flow velocity at that location; this assumes knowledge of the air density, such as might be obtained from the standard atmospheric model. A device for measuring flow velocities using Bernoulli's equation is called a pitot tube.

Although equation (2.16) is given only for incompressible flow, it is possible to obtain a modified form of the equation that includes the effects of compressible flow. The key conclusion is that the dynamic pressure is an important quantity in characterizing both incompressible flow and compressible flow.

There are important features of the flow illustrated in Figure 2.2. For a part of the flow, the streamlines (or stream tubes) are clearly defined and regular. This type of flow is referred to as laminar flow and is the easiest to analyze. The streamlines are clearly defined and regular over the leading edge of the airfoil, but as the flow proceeds over the trailing edge, the streamlines become irregular and are no longer clearly defined. This latter type of flow regime is said to be turbulent. Turbulent flow is quite difficult to model accurately. In our subsequent development, the aerodynamic models employed are based on the assumption of laminar flow. However, the onset of turbulence is important since it defines the aerodynamic limits of conventional aircraft flight. The onset of turbulence is primarily influenced by the Reynolds number and the deflection of the streamlines, due to the presence of the airfoil, from its free stream flow direction.

More complex models of fluid flow that take into account non-steady three-dimensional effects, as well as thermodynamic effects, can be developed, and they are important for more advanced studies of aerodynamics and non-steady aircraft flight. Such models are not required for the analysis of steady flow properties and associated steady flight characteristics of aircraft.

An important factor in determining whether compressibility effects of air are important is the Mach number of a flow. The Mach number is the ratio of the airspeed to the speed of sound, or acoustic propagation speed, in air. The Mach number M of a flow is

$$M = \frac{V}{a},\qquad(2.18)$$

where V is the airspeed and a is the speed of sound. The speed of sound depends on the air temperature according to

$$a = \sqrt{\gamma R T},\qquad(2.19)$$

where $\gamma = 1.4$ is the specific heat of air, R is the universal gas constant given previously, and T is the absolute temperature. The speed of sound depends on the air temperature, which depends on the altitude according to the standard atmospheric model. Values of the speed of sound at different altitudes are tabulated in appendix A. Note that the Mach number is a dimensionless quantity.

If the air flow Mach number satisfies $M < 0.4$, the flow is effectively incompressible. If the air flow Mach number satisfies $0.4 < M < 0.85$, compressible flow effects may be important. If $0.85 < M$, then compressibility effects are significant. If $M < 0.85$, the flow is said to be subsonic; if $0.85 < M < 1$, the flow is said to be transonic; if $1 < M < 5$, the flow is said to be supersonic; if $M > 5$, the flow is said to be hypersonic. Our subsequent flight analysis is based on the assumption of subsonic aerodynamics. The general approach for analysis of steady flight is, however, applicable to transonic, supersonic, and even hypersonic flight if suitable aerodynamics models are utilized.

2.4 Aerodynamics of Flow over a Wing

In section 2.3, some fundamental physical properties of steady flow were presented, emphasizing the concept of flow past an airfoil. Many of the concepts introduced in section 2.3 are directly applicable to steady aircraft flight. In particular, steady aircraft flight corresponds to steady flow of air past the aircraft. The wings and other aircraft components deform the stream tubes that define the flow past the aircraft. This deformation of the stream tubes is associated with changes in the airspeed and the static pressure of the flow past the aircraft. The change in the static pressure is important, since the aerodynamic forces and moments on the aircraft arise directly from these pressure variations. In particular, the net aerodynamic forces on the wing (or any other aircraft component) can be obtained by integrating the static pressure over the total wing surface area (over the surface area of that aircraft component).

The airspeed of the aircraft, namely its speed with respect to the surrounding air, can be used to define the Mach number of the aircraft. In this way, we can refer to aircraft subsonic flight, aircraft supersonic flight, or aircraft hypersonic flight. We only treat the case of subsonic aircraft flight. Note that the Mach number of the aircraft is only a crude indicator of the Mach number of the flow of air past a particular point on the aircraft.

In our subsequent flight analysis, the aerodynamic forces and moments that act on an aircraft in flight are fundamental ingredients in characterizing the steady flight properties of an aircraft. Models of these aerodynamic forces and moments are essential to understand the principles of flight. The most important aerodynamic forces act on the aircraft wing, which is designed to generate a large lift force.

Consider an aircraft in flight or, equivalently, an aircraft wing with air flowing past the wing. This perspective allows the development of the basic aerodynamics required for our subsequent analysis. A fixed wing in a moving flow field is exactly the situation encountered in a wind tunnel, which explains the importance of wind tunnels in experimental studies of aerodynamics. In fact, Figure 2.2 can be thought of as a two-dimensional exposure that shows the flow of smoke particles injected into the flow past a stationary wing in a wind tunnel.

As seen previously, from Bernoulli's equation, the dynamic pressure on the wing has a strong influence on the static pressure field, and hence the dynamic pressure influences the aerodynamic forces that are exerted on the wing by the air flow. Although the pressure distribution may vary in a complex way over the wing surface, the magnitude of the net aerodynamic force vector on the wing is proportional to the product of the free stream dynamic pressure and the wing surface area. The free stream dynamic pressure is determined by the steady flow speed before the air flow is influenced by the presence of the wing.

It is convenient to decompose the net aerodynamic force vector into the sum of two component vectors: one component that is along the free stream velocity vector of the flow past the wing and the other component that is perpendicular to the free stream velocity vector and lies in the plane of mass symmetry of the aircraft. The aerodynamic force vector along the free stream velocity vector can be further decomposed into the sum of a component that lies in the plane of mass symmetry, referred to as the drag, and a component that is perpendicular to the plane of mass symmetry, referred to as the side force. The component of the aerodynamic force vector that is perpendicular to the free stream velocity vector is referred to as the lift force. In many important flight conditions the side force is small and can be ignored; the most important aerodynamic forces are the drag and the lift. The aerodynamic force vector has a magnitude that is proportional to the dynamic pressure and the wing surface area, and it depends on the geometry of the wing.

These aerodynamic forces also give rise to an aerodynamic moment vector with respect to an aircraft frame whose origin is fixed at the center of mass of the aircraft (see chapter 3). It is convenient to decompose the aerodynamic moment into components along the axes of the aircraft-fixed frame; these are referred to as the roll moment, the pitch moment, and the yaw moment. This aerodynamic moment vector has a magnitude that is proportional to the dynamic pressure and the wing surface area. Mathematical models for these aerodynamic forces and moments that act on an aircraft in steady flight are given in chapter 3.

2.5 Wing Geometry

Figure 2.3 shows a typical wing cross section or airfoil. Figure 2.4 shows a top view of the wings as part of the aircraft. These figures illustrate important geometric features of a conventional aircraft wing. The most important wing parameters are the wing span, the camber, the mean wing chord, and the wing surface area. These geometric features of the wing are important in determining the aerodynamic forces and the aerodynamic moments that act on the wing and hence on the aircraft.

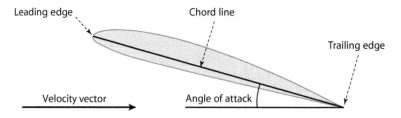

Figure 2.3. *Wing cross section, airfoil (NACA2411).*

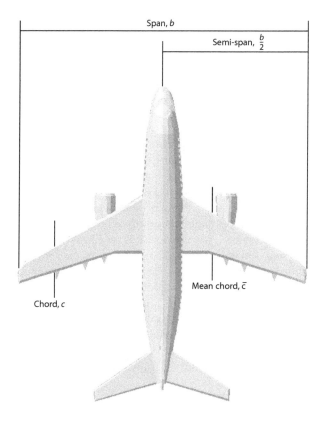

Figure 2.4. *Wing geometry.*

There are many types of aircraft wings with a variety of design features. Figure 2.5 illustrates many different wing designs. Each wing design has its strengths and weaknesses. The history of flight is, in part, a story of creative wing designs, and this is easily recognized in the variety of designs shown.

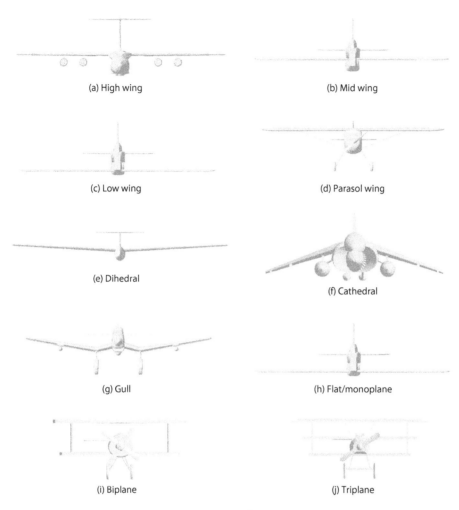

Figure 2.5. Aircraft wing designs.

Each of these designs is characterized by the same basic fundamental aerodynamics properties. The specific wing design influences the values of parameters that appear in the aerodynamics models, but the basic wing theory principles apply to all types of wings shown.

2.6 Problems

2.1. What are the temperature (in degrees C), the pressure (in newtons/m^2), and the density (in kg/m^3) of air according to the standard atmospheric model at the following altitudes: sea level, 5,000 m, 10,000 m, and 25,000 m?

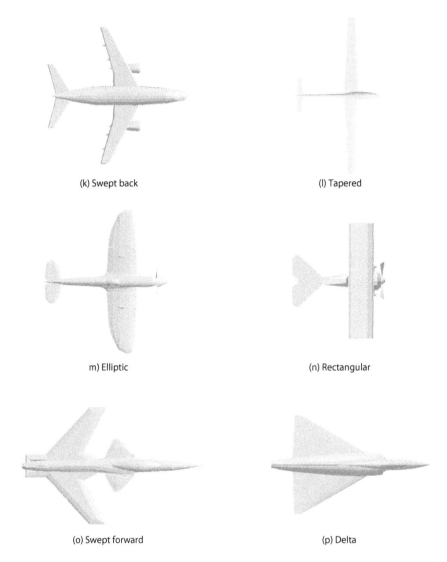

(k) Swept back   (l) Tapered

m) Elliptic   (n) Rectangular

(o) Swept forward   (p) Delta

Figure 2.5. continued.

2.2. In the development of the standard atmospheric model, the variation of
     the acceleration of gravity with altitude is ignored; that is, the acceleration
     of gravity is approximated as constant with altitude. Using the Newtonian
     gravitational model, compute the acceleration of gravity at sea level, 50,000 ft,
     and 100,000 ft. Based on these computations, describe the accuracy of this
     approximation.

2.3. Based on the standard atmospheric model, what are the temperature, pressure, and air density at an altitude of 8,000 ft above sea level? What are the temperature, pressure, and air density at an altitude of 18,000 ft?

2.4. At a particular sea level location at a particular time the atmospheric pressure is 14.8 lbs/in^2 and the temperature is 80 degrees F. Estimate the temperature, pressure, and air density at an altitude of 8,000 ft. Estimate the temperature, pressure, and air density at an altitude of 18,000 ft above sea level. Note: You must modify the standard atmosphere model using the given values for sea level pressure and temperature.

2.5. An executive jet aircraft is in a rapid climb in a standard atmosphere. Its time rate of change of altitude (its climb rate) is 15 ft/s.
  (a) At the instant it passes through an altitude of 20,000 ft above sea level, what are the temperature, pressure, and density of the surrounding air?
  (b) At the instant it passes through an altitude of 20,000 ft above sea level, what are the time rate of change of the temperature, pressure, and density of the surrounding air?

2.6. An executive jet aircraft is in a rapid descent in a standard atmosphere. Its time rate of change of altitude (its climb rate) is -15 ft/s.
  (a) At the instant it passes through an altitude of 20,000 ft above sea level, what are the temperature, pressure, and density of the surrounding air?
  (b) At the instant it passes through an altitude of 20,000 ft above sea level, what are the time rate of change of the temperature, pressure, and density of the surrounding air?

2.7. A single stream tube is selected to characterize the air flow past the wing of an aircraft in flight at an altitude of 20,000 ft at an airspeed of 275 ft/s. The free stream cross-sectional area of the stream tube is 0.07 ft^2.
  (a) What is the air flow speed in the stream tube at a location near the wing where the cross-sectional area of the stream tube is 0.05 ft^2? What is the static pressure at this location?
  (b) What is the air flow speed in the stream tube at a location near the wing where the cross-sectional area of the stream tube is 0.04 ft^2? What is the static pressure at this location?

2.8. A high altitude aircraft has an airspeed of 750 ft/s at an altitude of 70,000 ft above sea level. What is the Mach number of the aircraft in this flight condition?

2.9. A hypersonic flight vehicle is in flight at Mach number of 5.2 at an altitude of 98,000 ft above sea level. What is the airspeed of the flight vehicle in this flight condition?

3

Aircraft Translational Kinematics, Attitude, Aerodynamic Forces and Moments

This chapter presents fundamental relationships for aircraft translational kinematics in three dimensions. The aircraft velocity vector can be decomposed into three components in a ground-fixed (or Earth-fixed) frame or into three components in an aircraft-fixed frame. The aircraft velocity vector can be characterized by its magnitude and its direction with respect to an aircraft-fixed frame; the direction in this case is described by two angles, namely the angle of attack and the side-slip angle. The aircraft velocity vector can also be characterized by its magnitude and its direction with respect to a ground-fixed frame; the direction in this case is described by two angles, namely the flight path angle and the heading angle.

We also introduce the concept of the aircraft attitude, which is defined by three angles that describe the orientation of the aircraft-fixed frame with respect to the ground-fixed frame. These three attitude angles are referred to as the yaw angle, the pitch angle, and the roll angle. Several translational kinematics relationships are expressed in terms of the attitude angles. Additional material on aircraft kinematics and aircraft attitude can be found in Stengel (2004) and other textbooks on flight dynamics.

Mathematical models for the aerodynamic forces and moments that act on an aircraft in flight are presented. The aerodynamic forces and moments depend on the relative velocity of air as it flows past the aircraft in flight. If the atmosphere is assumed to be stationary, the relative velocity vector of the air with respect to the aircraft is the negative of the relative velocity vector of the aircraft with respect to the atmosphere; the magnitude of the velocity vector is the airspeed of the aircraft. This implies that the velocity vector of the aircraft is an important factor that determines the aerodynamic forces and moments that act on an aircraft in flight. Throughout this book, we most often express the airspeed of an aircraft in units of ft/s. In a few cases, airspeed is given in units of miles per hour. Although airspeed units of knots (nautical miles per hour where 1 knot is 1.15 miles per hour) are common, this is not used in the book.

Our treatment of aerodynamic forces and moments in this chapter is limited to the features that are required for aircraft steady flight and performance analyses. (See Anderson [1998, 2000], Mair and Birdsall [1996], McCormick [1994], Roskam and Lan [1997], and Torenbeek and Wittenberg [2009] for more comprehensive treatments of aerodynamic forces and moments.)

3.1 Cartesian Frames

Our subsequent analysis assumes the aircraft to be a rigid body with constant mass. A rigid body does not allow relative motion between any of its mass elements. In fact, real aircraft do contain moving parts, including the elevator, ailerons, and rudder, as well as engine components and possibly occupants. For the purposes of steady flight analysis over relatively short time periods (on the order of several minutes), it is reasonable to assume that an aircraft is a rigid body with constant mass (equivalently constant weight). For the purposes of steady flight over relatively long time periods (more than twenty minutes), as considered in chapter 11, the mass (or weight) of the aircraft is assumed to change slowly as fuel is burned.

It is necessary to introduce several different Euclidean or Cartesian frames. These reference frames are defined by the location of their origin and by three mutually orthogonal (perpendicular) axes. The axes are ordered (as the x-axis, the y-axis, and the z-axis) so that if the positive x-axis is rotated about the positive z-axis, it coincides with the positive y-axis after a 90-degree rotation; this is referred to as a right-hand Cartesian frame.

The first frame of interest is one that is fixed with respect to the Earth (assumed flat); the center of this frame is often located at sea level but this is not essential. The x_E-axis and the y_E-axis are assumed to lie in a horizontal plane and the z_E-axis is assumed to be vertical; for convenience the positive z_E-axis is often assumed to point downward. This frame is referred to as a ground-fixed frame or an Earth-fixed frame; the subscript denotes the Earth-fixed frame. It is an important frame, since aircraft motion with respect to this frame defines aircraft motion with respect to the Earth. This frame is often assumed to be an inertial frame; that is, Newton's laws, in their simplest form, are valid in this frame.

Assuming the aircraft mass is constant (or only slowly varying), the center of mass of the aircraft is defined as usual for a rigid body. The next reference frame of interest is one that is rigidly fixed to the aircraft and translates and rotates with the aircraft; the origin of the frame is located at the center of mass of the aircraft. The positive x_A-axis is assumed to lie in the direction of the nose of the aircraft, the positive y_A-axis is assumed to lie in the direction of the right wing (viewed from above the aircraft), and the positive z_A-axis is assumed to be perpendicular to the plane formed by the x_A-axis and the y_A-axis pointing to the bottom of the aircraft. These aircraft-fixed axes are selected so that the x_A and z_A axes define the plane of mass symmetry of the aircraft. This frame is referred to as a body-fixed frame or an aircraft-fixed frame; the subscript denotes the aircraft-fixed frame. It is an important frame, since this frame represents the perspective of instruments or humans in the aircraft. This frame is, in general, not an inertial frame.

A three-dimensional perspective of a ground-fixed frame and an aircraft-fixed frame, for a typical aircraft attitude, is shown in Figure 3.1. The origin of the ground-fixed frame is located at an arbitrary point on the ground. In subsequent developments, it is sometimes desirable to express the aircraft velocity vector (or some other vector) in each of these frames; in such cases it is convenient to visualize the origin of the ground-fixed frame as located at the aircraft center of mass. This is allowed for visual clarity, since it is only the directions of the three axes of the ground-fixed frame and not the location of its origin that are required by such expressions.

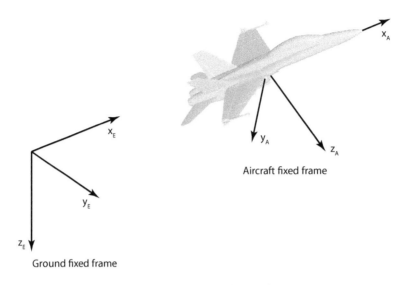

Aircraft fixed frame

Ground fixed frame

Figure 3.1. Ground- and aircraft-fixed frames in 3D.

Other reference frames can be introduced, but for the present these two frames are sufficient. They are each a Euclidean, right-hand frame. Any vector can be represented in terms of a basis of three unit vectors in the directions of the three frame axes in either of these frames. The subsequent development studies the aircraft velocity vector and shows how it can be represented in terms of components in the ground frame or in terms of components in the aircraft-fixed frame; further, knowledge of the components in one of the frames allows determination of components of the velocity vector in the other frame. This subject is referred to as aircraft translational kinematics.

3.2 Aircraft Translational Kinematics

The general motion of an aircraft occurs in three-dimensional space. There are three degrees of freedom that describe the translational motion of the aircraft and three degrees of freedom that describe the rotational motion of the aircraft. The three translational degrees of freedom are consistent with the fact that the translational velocity vector of the aircraft center of mass has three components with respect to each of the frames. The three rotational degrees of freedom are consistent with the fact that the aircraft can rotate about each of the three aircraft-fixed axes.

We show that the aircraft translational velocity vector, subsequently referred to as the aircraft velocity vector, can be expressed in several different forms. We first introduce important notation. Let u, v, and w denote the components of the velocity vector with respect to the aircraft-fixed frame; let V_x, V_y, and V_z denote the components of the velocity vector with respect to the ground-fixed frame. Let V denote the magnitude of the velocity vector.

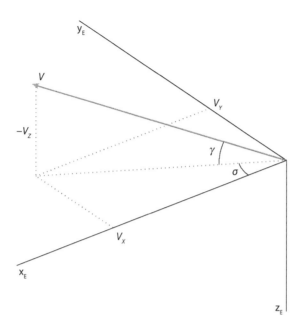

Figure 3.2. Aircraft velocity vector in ground-fixed frame in 3D.

Assume that γ denotes the angle between the velocity vector and the Earth-fixed $x_E - y_E$ (horizontal) plane and σ denotes the angle between the projection of the velocity vector onto the ground-fixed $x_E - y_E$ (horizontal) plane and the ground-fixed x_E-axis. The angle γ is the flight path angle; the angle σ is the heading angle. These two angles define the direction of the velocity vector with respect to the ground; hence they are important in characterizing the aircraft motion with respect to the ground. A three-dimensional perspective of the velocity vector in the ground-fixed frame is shown in Figure 3.2. This figure helps us visualize the flight path angle and the heading angle, as well as the components of the velocity vector in the ground-fixed frame.

Assume that α denotes the angle between the velocity vector and the aircraft-fixed $x_A - y_A$ plane and β denotes the angle between the projection of the velocity vector onto the aircraft-fixed $x_A - y_A$ plane and the aircraft-fixed x_A-axis. The angle α is the aircraft angle of attack or the angle of incidence; the angle β is the side-slip angle. These two angles define the direction of the velocity vector with respect to the aircraft; hence they are important in characterizing the aerodynamic forces and moments on the aircraft. A three-dimensional perspective of the velocity vector in the aircraft-fixed frame is shown in Figure 3.3. This figure helps us visualize the angle of attack and the side-slip angle, as well as the components of the velocity vector in the aircraft-fixed frame.

Careful attention to the three-dimensional geometry and the definitions given above allows us to express the ground-fixed components of the aircraft velocity vector in terms

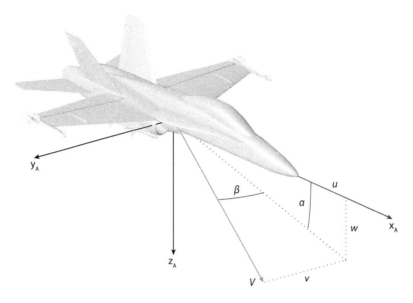

Figure 3.3. Aircraft velocity vector in aircraft-fixed frame in 3D.

of the flight path angle and the heading angle as

$$V_x = V \cos \gamma \cos \sigma, \tag{3.1}$$

$$V_y = V \cos \gamma \sin \sigma, \tag{3.2}$$

$$V_z = -V \sin \gamma. \tag{3.3}$$

Similarly, expressions for the aircraft-fixed components of the aircraft velocity vector in terms of the angle of attack and the side-slip angle are

$$u = V \cos \alpha \cos \beta, \tag{3.4}$$

$$v = V \cos \alpha \sin \beta, \tag{3.5}$$

$$w = V \sin \alpha. \tag{3.6}$$

It is easy to show that the magnitude of the aircraft velocity vector can be expressed in terms of the ground-fixed velocity components as

$$V = \sqrt{V_x^2 + V_y^2 + V_z^2}, \tag{3.7}$$

and it can also be expressed in terms of the aircraft-fixed velocity components as

$$V = \sqrt{u^2 + v^2 + w^2}. \tag{3.8}$$

It is also easy to show that the flight path angle and the heading angle satisfy

$$\tan \gamma = -\frac{V_z}{\sqrt{V_x^2 + V_y^2}}, \tag{3.9}$$

$$\sin \sigma = \frac{V_y}{\sqrt{V_x^2 + V_y^2}}. \tag{3.10}$$

The flight path angle can be either positive or negative. A positive flight path angle means that the aircraft is climbing. A negative flight path angle means that the aircraft is descending. A zero flight path angle means that the aircraft is in level flight.

It is also easy to show that the angle of attack and the side-slip angle satisfy

$$\tan \alpha = \frac{w}{\sqrt{u^2 + v^2}}, \tag{3.11}$$

$$\sin \beta = \frac{v}{\sqrt{u^2 + v^2}}. \tag{3.12}$$

The angle of attack can be either positive or negative, although most common steady flight conditions correspond to positive values of the angle of attack.

3.3 Aircraft Attitude and the Translational Kinematics

In this section, the concept of aircraft attitude is introduced and used to directly relate velocity components in the aircraft-fixed frame to velocity components in the ground-fixed frame and vice versa.

Since the aircraft is assumed to be a rigid body, the attitude of the aircraft can be viewed as the attitude of the aircraft-fixed frame with respect to the ground-fixed frame. Since each frame has three axes, the attitude can be represented by the nine real numbers that are the cosines of the angles formed by one axis from one frame and one axis from the other frame. This is a useful but redundant method for describing the aircraft attitude. Since the aircraft has three rotational degrees of freedom, the aircraft attitude can also be described by three angles, called Euler angles, that are commonly denoted by yaw angle ψ, pitch angle θ, and bank (or roll) angle ϕ.

An attitude can be achieved by three successive simple rotations: the first rotation is about the aircraft-fixed z_A-axis by the yaw angle, followed by a second rotation about the resulting aircraft-fixed y_A-axis by the pitch angle, followed by a third rotation about the resulting aircraft-fixed x_A-axis by the bank angle. The order of these simple rotations is important. This description is limited to only certain aircraft attitudes, namely the pitch angle must remain between -90 and $+90$ degrees. More details about aircraft attitude are given in Stengel (2004).

The components of the aircraft velocity vector in the ground-fixed frame can be expressed in terms of the components of the aircraft velocity vector in the aircraft-fixed frame and the Euler angles. These expressions are:

$$V_x = u \cos\theta \cos\psi + v(-\cos\phi \sin\psi + \sin\phi \sin\theta \cos\psi)$$
$$+ w(\sin\phi \sin\psi + \cos\phi \sin\theta \cos\psi), \tag{3.13}$$

$$V_y = u \cos\theta \sin\psi + v(\cos\phi \cos\psi + \sin\phi \sin\theta \sin\psi)$$
$$+ w(-\sin\phi \cos\psi + \cos\phi \sin\theta \sin\psi), \tag{3.14}$$

$$V_z = -u \sin\theta + v \sin\phi \cos\theta + w \cos\phi \cos\theta. \tag{3.15}$$

Similarly, the components of the aircraft velocity vector in the aircraft-fixed frame can be expressed in terms of the components of the aircraft velocity vector in the ground-fixed frame and the Euler angles. These expressions are:

$$u = V_x \cos\theta \cos\psi + V_y \cos\theta \sin\psi - V_z \sin\theta, \tag{3.16}$$

$$v = V_x(-\cos\phi \sin\psi + \sin\phi \sin\theta \cos\psi)$$
$$+ V_y(\cos\phi \cos\psi + \sin\phi \sin\theta \sin\psi) + V_z \sin\phi \cos\theta, \tag{3.17}$$

$$w = V_x(\sin\phi \sin\psi + \cos\phi \sin\theta \cos\psi)$$
$$+ V_y(-\sin\phi \cos\psi + \cos\phi \sin\theta \sin\psi) + V_z \cos\phi \cos\theta. \tag{3.18}$$

These are complicated expressions that can be simplified if additional assumptions are made.

3.4 Translational Kinematics for Flight in a Fixed Vertical Plane

Longitudinal flight corresponds to aircraft flight in a fixed vertical plane; that is, the aircraft is not turning. This special type of flight in two dimensions occurs when the heading angle, the side-slip angle, and the bank angle are identically zero. The previously presented equations for three-dimensional flight can be considerably simplified, and a schematic drawing of longitudinal flight is helpful in understanding the flight geometry. This longitudinal flight geometry is shown in Figure 3.4.

The expressions presented above can be simplified to describe longitudinal flight translational kinematics, assuming the vertical flight plane is the plane defined by $y_A = y_E = 0$ in both the aircraft-fixed frame and the ground-fixed frame. The longitudinal flight velocity components with respect to the ground-fixed frame are:

$$V_x = V \cos\gamma, \tag{3.19}$$

$$V_y = 0, \tag{3.20}$$

$$V_z = -V \sin\gamma. \tag{3.21}$$

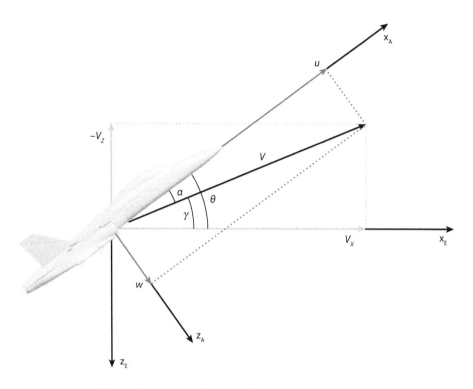

Figure 3.4. Longitudinal aircraft translational kinematics.

Similarly, the longitudinal flight velocity components with respect to the aircraft-fixed frame are:

$$u = V \cos \alpha, \tag{3.22}$$

$$v = 0, \tag{3.23}$$

$$w = V \sin \alpha. \tag{3.24}$$

The flight path angle in longitudinal flight satisfies

$$\tan \gamma = -\frac{V_z}{V_x}. \tag{3.25}$$

The angle of attack in longitudinal flight satisfies

$$\tan \alpha = \frac{w}{u}. \tag{3.26}$$

The aircraft velocity components in the aircraft-fixed frame and in the ground-fixed frame are related by

$$u = V_x \cos\theta - V_z \sin\theta, \tag{3.27}$$

$$w = V_x \sin\theta + V_z \cos\theta \tag{3.28}$$

or, equivalently,

$$V_x = u \cos\theta + w \sin\theta, \tag{3.29}$$

$$V_z = -u \sin\theta + w \cos\theta. \tag{3.30}$$

According to the geometry in Figure 3.4, the pitch angle, the angle of attack, and the flight path angle satisfy the equation

$$\theta = \alpha + \gamma. \tag{3.31}$$

The results in this section hold only for longitudinal flight, that is, flight in a fixed vertical plane. Since these results are based on two-dimensional flight, the geometry is simple and the results are easily derived.

3.5 Translational Kinematics for Flight in a Fixed Horizontal Plane

Aircraft flight in a fixed horizontal plane is often referred to as lateral flight; that is, the aircraft is not climbing or descending. This special type of flight in two dimensions occurs when the flight path angle is identically zero. The previously presented equations for three-dimensional flight can be considerably simplified, and a schematic drawing that illustrates lateral flight is helpful in understanding the flight geometry. This lateral flight geometry is shown in Figure 3.5.

The expressions presented above can be simplified to describe lateral flight translational kinematics, assuming the horizontal plane is the plane defined by constant altitude in the ground-fixed frame. The lateral flight velocity components with respect to the ground-fixed frame are

$$V_x = V \cos\sigma, \tag{3.32}$$

$$V_y = V \sin\sigma, \tag{3.33}$$

$$V_z = 0. \tag{3.34}$$

The heading angle in lateral flight satisfies

$$\tan\sigma = \frac{V_y}{V_x}. \tag{3.35}$$

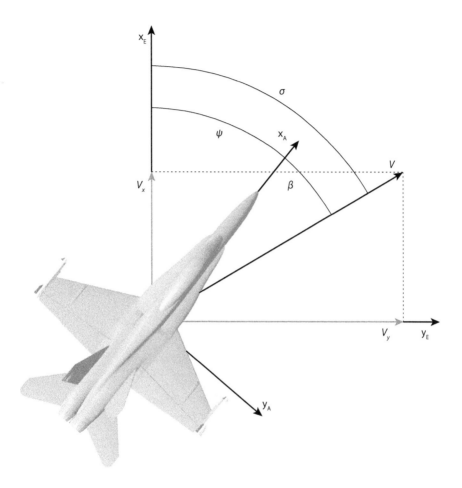

Figure 3.5. Lateral aircraft translational kinematics.

The side-slip angle in lateral flight satisfies

$$\tan \sigma = \frac{V}{u}.$$ (3.36)

According to the geometry in Figure 3.5, the yaw angle, the side-slip angle, and the heading angle satisfy the equation

$$\sigma = \beta + \psi.$$ (3.37)

The results in this section hold only for lateral flight, that is, flight in a fixed horizontal plane. Since these results are based on two-dimensional flight, the geometry is simple.

3.6 Small Angle Approximations

Aircraft flight often occurs with small flight path angle, small heading angle, small angle of attack, small side-slip angle, and small bank, pitch, and yaw angles. A small angle assumption means that the angle, when expressed in terms of radian measure, is much less than 1. In this case, the equations in sections 3.4 and 3.5 can be simplified by using the small angle approximation that the sine and tangent of the angle are approximated by the angle and the cosine of the angle is approximated by 1. This gives the relations that approximate the flight path angle, the heading angle, the angle of attack, and the side-slip angle:

$$\gamma = -\frac{V_z}{V_x}, \tag{3.38}$$

$$\sigma = \frac{V_y}{V_x}, \tag{3.39}$$

$$\alpha = \frac{w}{u}, \tag{3.40}$$

$$\beta = \frac{v}{u}. \tag{3.41}$$

A small angle of attack and small side-slip angle imply that v and w are small compared with u. The approximate aircraft translational kinematics are

$$V_x = u, \tag{3.42}$$
$$V_y = u\psi + v, \tag{3.43}$$
$$V_z = -u\theta + w. \tag{3.44}$$

A small flight path angle and small heading angle imply that V_y and V_z are small compared with V_x. The approximate aircraft translational kinematics can also be expressed as

$$u = V_x, \tag{3.45}$$
$$v = -V_x\psi + V_y, \tag{3.46}$$
$$w = V_x\theta + V_z. \tag{3.47}$$

3.7 Coordinated Flight

For most conventional aircraft flight it is desired to maintain the side-slip velocity v near zero or, equivalently, to maintain the side-slip angle β near zero. Coordinated flight occurs when the side-slip angle is exactly zero. Most propulsion systems are designed to operate efficiently in coordinated flight, and coordinated flight is preferred by the pilot and passengers. Much of the subsequent development in this book, including the analysis of steady turning flight, assumes coordinated flight, that is, zero side-slip velocity.

3.8 Clarification of Bank Angles

It is important to clarify two different notions of aircraft bank angle or roll angle. As indicated above, the bank attitude angle ϕ is the rotation angle of the aircraft about its aircraft-fixed x_A-axis. It is also useful to introduce a slightly different notion, namely the bank angle about the aircraft velocity vector; this bank angle is denoted by μ. This bank angle μ is the angle between the aircraft fixed y_A-axis and the horizontal plane, based on a simple rotation about the aircraft velocity vector.

In general the two bank angles are different. In the special case for which the angle of attack $\alpha = 0$ and the side-slip angle $\beta = 0$ (coordinated flight), the two bank angles are identical; that is, $\phi = \mu$.

An aircraft is usually banked to achieve turning flight. Consequently, banked aircraft flight is studied only in the chapters that treat turning flight. In these chapters, it is the bank angle about the velocity vector μ that plays the most important role in the flight analysis.

3.9 Aerodynamic Forces

Expressions for the lift and drag forces on an aircraft in steady flight are presented. The aircraft is assumed to be in steady flight through a stationary atmosphere; that is, there is no wind. Thus, the magnitude of the aircraft velocity vector is identical with the airspeed past the aircraft. The lift and drag forces are primarily due to the flow of air over the surface of the wing. This allows us to introduce the lift coefficient and the drag coefficient according to the wing geometry; it should be kept in mind that the lift force and drag force are assumed to characterize the total lift and total drag on the aircraft.

The air density, at a given flight altitude, is denoted by ρ, whose value is determined according to the standard atmospheric model; V is the airspeed of the aircraft; S is the wing surface area; c is the mean wing chord; and b is the wing span.

According to basic aerodynamics, the magnitude of the lift force vector on the aircraft is given by

$$L = \frac{1}{2}\rho V^2 S C_L, \tag{3.48}$$

where $\frac{1}{2}\rho V^2$ is the aircraft dynamic pressure. The direction of the lift force vector is perpendicular to the velocity vector of the aircraft and it lies in the plane of mass symmetry of the aircraft. In this expression, C_L is the dimensionless lift coefficient. The lift coefficient has a strong dependence on the aircraft angle of attack according to

$$C_L = C_{L_0} + C_{L_\alpha}\alpha. \tag{3.49}$$

This linear dependence on the angle of attack is a good approximation, so long as the aircraft wing is not in stall. The values of the lift coefficient parameters C_{L_0} and $C_{L_\alpha} > 0$ that appear in the lift coefficient expression above are empirical parameters that depend on the specific geometry of the wing.

There is a maximum lift coefficient that arises due to the fact that the flow of air over the wing at a sufficiently high angle of attack becomes turbulent with flow detachment from the wing. This maximum value of the lift coefficient is denoted by $C_{L_{max}}$. For the purposes of steady flight analysis, the stall constraint can be represented by the inequality

$$C_L \leq C_{L_{max}}, \tag{3.50}$$

or, equivalently, by the inequality

$$\alpha \leq \alpha_{max}, \tag{3.51}$$

where α_{max} is the maximum angle of attack at stall and the constants satisfy

$$C_{L_{max}} = C_{L_0} + C_{L_\alpha} \alpha_{max}. \tag{3.52}$$

The following Matlab commands produce a plot that illustrates the linear dependence of the lift coefficient on the angle of attack for a typical aircraft.

Matlab function 3.1 CLvsAj.m
```
a=linspace (0, 23. 2, 500);
CL0=0.02;  CLa=0.12;
CL=CL0+CLa*a;
plot (a, CL);
xlabel('Angle of attack (degrees)');
ylabel ('Lift coefficient');
grid on;
```

This dependence is shown in Figure 3.6. The maximum value of the lift coefficient, 2.804, is also indicated in Figure 3.6; this defines the point at which the aircraft stalls.

According to basic aerodynamics, the magnitude of the drag force vector on the aircraft is given by

$$D = \frac{1}{2}\rho V^2 S C_D, \tag{3.53}$$

where $\frac{1}{2}\rho V^2$ is the aircraft dynamic pressure. The direction of the drag force vector is opposite to the velocity vector of the aircraft. In this expression, C_D is the dimensionless drag coefficient. The drag coefficient has a strong dependence on the angle of attack. For typical subsonic flight conditions, the drag coefficient is most often expressed in terms of the lift coefficient. The most common dependence is a quadratic dependence given by

$$C_D = C_{D_0} + \frac{C_L^2}{\pi e AR}. \tag{3.54}$$

This expression, referred to as a quadratic drag polar, indicates that there are two sources for the drag force. The second term indicates that there is a part of the drag that is due to the fact that the wing generates lift; this is often referred to as the induced drag term. The first term above indicates that there is a part of the drag that is independent of the lift; this is

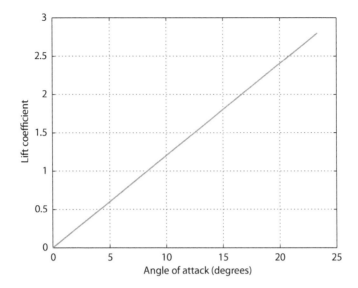

Figure 3.6. Dependence of lift coefficient on angle of attack.

aircraft friction drag due to viscous flow effects. The parameter $AR = \frac{b^2}{S}$ is the wing aspect ratio, where b is the wing span and S is the wing surface area; e is an empirical parameter referred to as the Oswald efficiency factor, and C_{D_0} is the zero-lift drag coefficient, that is, the drag coefficient when the lift coefficient is zero. The values of the drag polar parameters C_{D_0} and $K = \frac{1}{\pi e AR}$ in the drag polar expression are empirical parameters that depend on the specific geometry of the wing.

The drag coefficient can also be expressed in terms of the angle of attack; equation (3.54) for the quadratic drag polar and equation (3.49) for the linear dependence of lift coefficient on the angle of attack imply that the drag coefficient is a quadratic function of the angle of attack. A plot of the drag coefficient as a function of the angle of attack for a typical aircraft is developed in the following Matlab m-file.

Matlab function 3.2 CDvsAj.m
```
CL0=0.02;  CLa=0.12;  CD0=0.015;  K=0.05;
a=linspace (0, 23. 2, 500);
CL=CL0+CLa*a;
CD=CD0+K*CL.*CL;
plot (a, CD);
xlabel ('Angle of attack (degrees)');
ylabel ('Drag coefficient');
grid on;
```

and shown in Figure 3.7.

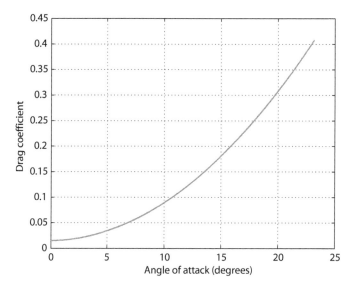

Figure 3.7. Dependence of drag coefficient on angle of attack.

Another aerodynamic quantity that is often introduced is the lift to drag ratio L/D. It is clear that the lift to drag ratio is the ratio of the lift coefficient to the drag coefficient, namely

$$\frac{L}{D} = \frac{C_L}{C_D},\tag{3.55}$$

which can also be expressed in terms of the lift coefficient only using the drag polar expression.

The aerodynamic side force is usually given by

$$S = \frac{1}{2}\rho V^2 S C_S,\tag{3.56}$$

where C_S is the dimensionless side force coefficient that typically depends on the side-slip angle β according to

$$C_S = C_{S_\beta}\beta.\tag{3.57}$$

Note that in coordinated flight when the side-slip angle is zero, the aerodynamic side force is zero.

The lift, drag, and side force coefficients also depend on the characteristics of the flow over the wings through the Mach number: for subsonic flight the parameters in the lift coefficient expression, the side force coefficient, and the parameters in the drag polar expression can be considered constant; for transonic flight there can be a substantial increase in the drag compared with subsonic flight and a substantial reduction in the maximum lift coefficient; and for supersonic flight the drag polar parameters and the maximum lift coefficient have substantial dependence on the aircraft Mach number. The additional aerodynamics complications that arise in transonic flight or supersonic flight are not considered.

3.10 Aerodynamic Moments

The aerodynamic forces give rise to aerodynamic moments about the aircraft-fixed x_A, y_A, and z_A axes. These aerodynamic moments are referred to as the roll moment, the pitch moment, and the yaw moment.

Each moment is expressed as a product of the dynamic pressure, wing surface area, a moment arm or length, and a dimensionless roll moment coefficient, a dimensionless pitch moment coefficient, or a dimensionless yaw moment coefficient, respectively.

The aerodynamic roll moment is described by

$$\mathcal{L} = \frac{1}{2}\rho V^2 SbC_{\mathcal{L}}, \tag{3.58}$$

where $C_{\mathcal{L}}$ denotes the dimensionless roll moment coefficient. By convention, the moment arm for the roll moment is taken as the wing span b. The roll moment coefficient depends strongly on the side-slip angle β and the differential ailerons deflection δ_a, and it depends weakly on the rudder deflection δ_r. The roll moment coefficient is most commonly described by

$$C_{\mathcal{L}} = C_{\mathcal{L}_\beta}\beta + C_{\mathcal{L}_{\delta_a}}\delta_a + C_{\mathcal{L}_{\delta_r}}\delta_r, \tag{3.59}$$

which indicates a linear dependence on the side-slip angle, the ailerons deflection, and the rudder deflection. This linear dependence is a good approximation, so long as the flight condition is such that the aircraft ailerons are not stalled and they remain within their deflection limits. The aerodynamic roll moment vector acts along the aircraft-fixed x_A-axis.

Flight with zero aerodynamic roll moment occurs when $\mathcal{L} = 0$, that is, when the roll moment coefficient $C_{\mathcal{L}} = 0$. The ailerons and rudder should be adjusted to satisfy

$$0 = C_{\mathcal{L}_\beta}\beta + C_{\mathcal{L}_{\delta_a}}\delta_a + C_{\mathcal{L}_{\delta_r}}\delta_r \tag{3.60}$$

to achieve zero aerodynamic roll moment for flight with an aircraft side-slip angle β.

The pitch moment on the aircraft is given by

$$M = \frac{1}{2}\rho V^2 ScC_M, \tag{3.61}$$

where C_M denotes the dimensionless pitch moment coefficient. By convention, the moment arm for the pitch moment is taken as the mean wing chord c. The pitch moment coefficient depends strongly on the angle of attack and the elevator deflection δ_e. The pitch moment coefficient is most commonly described by

$$C_M = C_{M_0} + C_{M_\alpha}\alpha + C_{M_{\delta_e}}\delta_e, \tag{3.62}$$

which indicates a linear dependence on the angle of attack and the elevator deflection. This linear dependence is a good approximation, so long as the flight condition is such that the aircraft elevator is not stalled and it remains within its deflection limits. The aerodynamic pitch moment acts along the aircraft-fixed y_A-axis.

Flight with zero pitch moment occurs when $M = 0$, that is, when the pitch moment coefficient $C_M = 0$. The elevator deflection should satisfy

$$0 = C_{M_0} + C_{M_\alpha}\alpha + C_{M_{\delta_e}}\delta_e \tag{3.63}$$

to achieve zero aerodynamic pitch moment for flight at an aircraft angle of attack α.

The aerodynamic yaw moment N is given by

$$N = \frac{1}{2}\rho V^2 SbC_N, \tag{3.64}$$

where C_N denotes the dimensionless yaw moment coefficient. By convention, the moment arm for the yaw moment is taken as the wing span b. The yaw moment coefficient depends strongly on the side-slip angle β and the rudder deflection δ_r, and it depends weakly on the differential ailerons deflection δ_a. Ideally, the yaw moment should not depend on the ailerons deflection. In practice this dependence is often present and is referred to as adverse yaw; that is, a change in the ailerons results in a yaw moment. The yaw moment coefficient is most commonly described by

$$C_N = C_{N_\beta}\beta + C_{N_{\delta_a}}\delta_a + C_{N_{\delta_r}}\delta_r, \tag{3.65}$$

which indicates a linear dependence on the side-slip angle, the ailerons deflection, and the rudder deflection. This linear dependence is a good approximation, so long as the flight condition is such that the aircraft rudder is not close to stall and remains within its deflection limits. The aerodynamic yaw moment vector acts along the aircraft-fixed z_A-axis.

Flight with zero aerodynamic yaw moment occurs when $N = 0$, that is, when the yaw moment coefficient $C_N = 0$. The ailerons and rudder should be adjusted to satisfy

$$0 = C_{N_\beta}\beta + C_{N_{\delta_a}}\delta_a + C_{N_{\delta_r}}\delta_r \tag{3.66}$$

to achieve zero aerodynamic yaw moment for flight at side-slip angle β. Adverse yaw can be countered by proper adjustment of the rudder.

If the side-slip velocity (the side-slip angle) is zero, then the side force is zero, resulting in coordinated flight. Further, if the rudder deflection and the differential ailerons deflection are zero, then the roll moment and the yaw moment are zero. If this flight condition is maintained, it is referred to as steady coordinated flight. Thus the aerodynamic forces and moments in steady coordinated flight depend essentially on the elevator deflection and the associated angle of attack; the elevator deflection can be selected so that the pitch moment coefficient is zero.

It should be kept in mind throughout the subsequent development that the aerodynamics expressions introduced in this chapter are only valid within a certain range of steady subsonic flight conditions.

3.11 Problems

3.1. Suppose an executive jet aircraft is in steady longitudinal flight in a vertical plane. The translational velocity vector of the aircraft, when expressed in an aircraft-fixed frame, has constant velocity components 250 ft/s along its aircraft-fixed x_A-axis and 30 ft/s along its aircraft-fixed z_A-axis. The constant pitch attitude of the aircraft is given by an angle of 3 degrees. Assume a stationary atmosphere. The aerodynamic drag polar for the aircraft is given by

$$C_D = 0.015 + 0.05C_L^2.$$

The lift coefficient, expressed in terms of the angle of attack, is

$$C_L = 0.02 + 0.12\alpha,$$

where the angle of attack is measured in degrees.
   (a) What is the aircraft airspeed?
   (b) What are the horizontal ground speed and the rate of climb of the aircraft?
   (c) What is the aircraft angle of attack?
   (d) What is the aircraft flight path angle?
   (e) Draw a side view of the aircraft indicating the relevant aircraft-fixed and ground axes with origins at the aircraft center of mass; also draw the velocity vector and indicate the pitch angle, the angle of attack, and the flight path angle. Verify that the pitch angle is the sum of the angle of attack and the flight path angle.
   (f) What is the value of the lift coefficient at this flight condition?
   (g) What is the value of the drag coefficient at this flight condition?
   (h) What should be the elevator deflection at this flight condition so that the pitch moment is zero?

3.2. Suppose a general aviation aircraft is in steady longitudinal flight in a vertical plane. The translational velocity vector of the aircraft, when expressed in an aircraft-fixed frame, has constant velocity components 150 ft/s along its aircraft-fixed x_A-axis and 0 ft/s along its aircraft-fixed z_A-axis. The constant pitch attitude of the aircraft is given by an angle of 2 degrees. Assume a stationary atmosphere. The aerodynamic drag polar of the aircraft is given by

$$C_D = 0.026 + 0.054C_L^2.$$

The lift coefficient, expressed in terms of the angle of attack, is

$$C_L = 0.02 + 0.12\alpha,$$

where the angle of attack is measured in degrees.
   (a) What is the aircraft airspeed?
   (b) What are the horizontal ground speed and the rate of climb of the aircraft?
   (c) What is the aircraft angle of attack?
   (d) What is the aircraft flight path angle?
   (e) Draw a side view of the aircraft indicating the relevant aircraft-fixed and ground axes with origins at the aircraft center of mass; also draw the velocity vector and indicate the pitch angle, the angle of attack, and the flight path angle. Verify that the pitch angle is the sum of the angle of attack and the flight path angle.
   (f) What is the value of the lift coefficient at this flight condition?
   (g) What is the value of the drag coefficient at this flight condition?
   (h) What should be the elevator deflection at this flight condition so that the pitch moment is zero?

3.3. Suppose a small remote-controlled aircraft is in steady longitudinal flight in a vertical plane. The translational velocity vector of the aircraft, when expressed in an aircraft-fixed frame, has constant velocity components 90 ft/s along its aircraft-fixed x_A-axis and -4 ft/s along its aircraft-fixed z_A-axis. The constant pitch attitude of the aircraft is given by an angle of 3 degrees. Assume a stationary atmosphere. The aerodynamic drag polar of the aircraft is given by

$$C_D = 0.025 + 0.05 \, (C_L - 0.7)^2 \, .$$

The lift coefficient, expressed in terms of the angle of attack, is

$$C_L = 0.18 + 0.06\alpha,$$

where the angle of attack is measured in degrees.
   (a) What is the aircraft airspeed?
   (b) What are the horizontal ground speed and the rate of climb of the aircraft?
   (c) What is the aircraft angle of attack?
   (d) What is the aircraft flight path angle?
   (e) Draw a side view of the aircraft indicating the relevant aircraft-fixed and ground axes with origins at the aircraft center of mass; also draw the velocity vector and indicate the pitch angle, the angle of attack, and the flight path angle. Verify that the pitch angle is the sum of the angle of attack and the flight path angle.
   (f) What is the value of the lift coefficient at this flight condition?
   (g) What is the value of the drag coefficient at this flight condition?
   (h) What should be the elevator deflection at this flight condition so that the pitch moment is zero?

3.4. Suppose an executive jet aircraft is in steady longitudinal flight in a vertical plane. The translational velocity vector of the aircraft with respect to a stationary atmosphere, expressed in a ground-fixed coordinate frame, has a constant ground speed of 260 ft/s and a constant rate of climb of -15 ft/s; the constant pitch attitude of the aircraft is given by 4 degrees. The aerodynamic drag polar for the aircraft is given by

$$C_D = 0.015 + 0.05 C_L^2.$$

The lift coefficient, expressed in terms of the angle of attack, is

$$C_L = 0.02 + 0.12\alpha,$$

where the angle of attack is measured in degrees.
   (a) What is the aircraft airspeed?
   (b) What are the velocity components of the aircraft when expressed in the aircraft-fixed frame?
   (c) What is the aircraft angle of attack?
   (d) What is the aircraft flight path angle?
   (e) Draw a side view of the aircraft indicating the relevant aircraft-fixed and ground axes with origins at the aircraft center of mass; draw the velocity vector and indicate the pitch angle, the angle of attack, and the flight path angle. Verify that the pitch angle is the sum of the angle of attack and the flight path angle.
   (f) What is the value of the lift coefficient at this flight condition?
   (g) What is the value of the drag coefficient at this flight condition?
   (h) What should be the elevator deflection at this flight condition so that the pitch moment is zero?

3.5. Suppose a general aviation aircraft is in steady longitudinal flight in a vertical plane. The translational velocity vector of the aircraft with respect to a stationary atmosphere, expressed in a ground-fixed frame, has a constant ground speed of 160 ft/s and a constant rate of climb of 5 ft/s; the constant pitch attitude of the aircraft is given by 4 degrees. The aerodynamic drag polar of the aircraft is given by

$$C_D = 0.026 + 0.054 C_L^2.$$

The lift coefficient, expressed in terms of the angle of attack, is

$$C_L = 0.02 + 0.12\alpha,$$

where the angle of attack is measured in degrees.
   (a) What is the aircraft airspeed?
   (b) What are the velocity components of the aircraft when expressed in the aircraft-fixed frame?

(c) What is the aircraft angle of attack?

(d) What is the aircraft flight path angle?

(e) Draw a side view of the aircraft indicating the relevant aircraft-fixed and ground axes with origins at the aircraft center of mass; draw the velocity vector and indicate the pitch angle, the angle of attack, and the flight path angle. Verify that the pitch angle is the sum of the angle of attack and the flight path angle.

(f) What is the value of the lift coefficient at this flight condition?

(g) What is the value of the drag coefficient at this flight condition?

(h) What should be the elevator deflection at this flight condition so that the pitch moment is zero?

3.6. Suppose a small remote-controlled aircraft is in steady longitudinal flight in a vertical plane. The translational velocity vector of the aircraft with respect to a stationary atmosphere, expressed in a ground-fixed frame, has a constant ground speed of 80 ft/s and a constant rate of climb of −4 ft/s; the constant pitch attitude of the aircraft is given by 4 degrees. The aerodynamic drag polar of the aircraft is given by

$$C_D = 0.025 + 0.05 \, (C_L - 0.7)^2 \,.$$

The lift coefficient, expressed in terms of the angle of attack, is

$$C_L = 0.18 + 0.06\alpha,$$

where the angle of attack is measured in degrees.

(a) What is the aircraft airspeed?

(b) What are the velocity components of the aircraft when expressed in the aircraft-fixed frame?

(c) What is the aircraft angle of attack?

(d) What is the aircraft flight path angle?

(e) Draw a side view of the aircraft indicating the relevant aircraft-fixed and ground axes with origins at the aircraft center of mass; draw the velocity vector and indicate the pitch angle, the angle of attack, and the flight path angle. Verify that the pitch angle is the sum of the angle of attack and the flight path angle.

(f) What is the value of the lift coefficient at this flight condition?

(g) What is the value of the drag coefficient at this flight condition?

(h) What should be the elevator deflection at this flight condition so that the pitch moment is zero?

3.7. Suppose an executive jet aircraft is in steady lateral flight in a horizontal plane with zero side-slip angle and a constant angle of attack of 10 degrees. The constant aircraft speed with respect to a stationary atmosphere is 500 ft/s and the constant turn rate is 0.01 rad/s. Assume the heading angle is initially zero degrees.

(a) What are the initial velocity components of the aircraft when expressed in the ground-fixed frame?

(b) What are the initial velocity components of the aircraft when expressed in the aircraft-fixed frame?

(c) After 200 seconds, what are the velocity components of the aircraft when expressed in the ground-fixed frame?

(d) After 200 seconds, what are the velocity components of the aircraft when expressed in the aircraft-fixed frame?

(e) The path of the aircraft is circular in the horizontal plane. What is the period of this circular motion?

3.8. Suppose a general aviation aircraft is in steady lateral flight in a horizontal plane with zero side-slip angle and a constant angle of attack of 5 degrees. The constant aircraft speed with respect to a stationary atmosphere is 250 ft/s and the constant turn rate is 0.015 rad/s. Assume the heading angle is initially zero degrees.

(a) What are the initial velocity components of the aircraft when expressed in the ground-fixed frame?

(b) What are the initial velocity components of the aircraft when expressed in the aircraft-fixed frame?

(c) After 150 seconds, what are the velocity components of the aircraft when expressed in the ground-fixed frame?

(d) After 150 seconds, what are the velocity components of the aircraft when expressed in the aircraft-fixed frame?

(e) The path of the aircraft is circular in the horizontal plane. What is the period of this circular motion?

3.9. Suppose that the flight of an executive jet aircraft is characterized by the following drag polar expression:

$$C_D = 0.015 + 0.05 C_L^2.$$

(a) What is the maximum value of the lift to drag ratio $\frac{C_L}{C_D}$; what are the corresponding values of the lift coefficient and the drag coefficient?

(b) What is the maximum value of the lift to drag ratio $\frac{C_L^{\frac{1}{2}}}{C_D}$; what are the corresponding values of the lift coefficient and the drag coefficient?

(c) What is the maximum value of the lift to drag ratio $\frac{C_L^{\frac{3}{2}}}{C_D}$; what are the corresponding values of the lift coefficient and the drag coefficient?

3.10. Suppose that the flight of a general aviation propeller-driven aircraft is characterized by the following drag polar expression:

$$C_D = 0.026 + 0.054 C_L^2.$$

(a) What is the maximum value of the lift to drag ratio $\frac{C_L}{C_D}$; what are the corresponding values of the lift coefficient and the drag coefficient?

(b) What is the maximum value of the lift to drag ratio $\frac{C_L^{\frac{1}{2}}}{C_D}$; what are the corresponding values of the lift coefficient and the drag coefficient?

(c) What is the maximum value of the lift to drag ratio $\frac{C_L^{\frac{3}{2}}}{C_D}$; what are the corresponding values of the lift coefficient and the drag coefficient?

3.11. Suppose that the flight of a small remote-controlled aircraft is characterized by the following drag polar expression:

$$C_D = 0.025 + 0.05 \, (C_L - 0.7)^2 \,.$$

(a) What is the maximum value of the lift to drag ratio $\frac{C_L}{C_D}$; what are the corresponding values of the lift coefficient and the drag coefficient?

(b) What is the maximum value of the lift to drag ratio $\frac{C_L^{\frac{1}{2}}}{C_D}$; what are the corresponding values of the lift coefficient and the drag coefficient?

(c) What is the maximum value of the lift to drag ratio $\frac{C_L^{\frac{3}{2}}}{C_D}$; what are the corresponding values of the lift coefficient and the drag coefficient?

Propulsion Systems

This chapter summarizes the basic properties of aircraft propulsion systems that directly impact the aircraft steady flight characteristics. It is not necessary to go into the details of aircraft engines and their associated flow and combustion physics in order to understand the properties of propulsion systems that affect steady aircraft flight and performance. It is sufficient to describe aircraft propulsion systems in terms of the steady thrust force or steady power that the engine produces and the steady rate at which it burns fuel. These are the topics that constitute the main focus of this chapter. (See Anderson [1998, 2000], Mair and Birdsall [1996], McCormick [1994], Roskam and Lan [1997], and Torenbeek and Wittenberg [2009] for more comprehensive treatments of aircraft propulsion systems.)

4.1 Steady Thrust and Power Relations

In our analysis of steady flight, it is assumed that the propulsion system is operated to produce a constant level of thrust or power while the aircraft is in steady flight at a constant airspeed. The thrust T produced by an engine (or the thrust required for flight) and the power P produced by that engine (or the power required for flight), in steady operation, are related by

$$P = TV. \tag{4.1}$$

Here V is the flow speed of air through the engine, which is assumed to be identical to the airspeed of the aircraft. Equivalently, the thrust is related to the power by

$$T = \frac{P}{V}. \tag{4.2}$$

If the airspeed has units ft/s and the thrust has units lbs, then the power has units $\mathrm{ft-lbs/s}$. It is common to express engine power in terms of horsepower where $1\,\mathrm{hp} = 550\,\mathrm{ft-lbs/s}$.

4.2 Jet Engines

Many modern aircraft have a propulsion system based on jet engine technology that operates in combination with a turbine. It is not necessary to go into detail about the operation of jet engines; it suffices to describe the overall properties of a typical jet engine.

A jet engine or jet engines produce a net thrust force along a direction that is fixed in the aircraft plane of mass symmetry. It is often assumed that the net thrust is along the aircraft-fixed x_A-axis. A propulsion system that consists of one or more jet engines is typically described by the maximum rated thrust that can be produced at sea level. Although our subsequent terminology in reference to such a propulsion system may refer to a jet engine, it should be kept in mind that the development also applies to a propulsion system that consists of multiple jet engines so long as they are operated by a single throttle.

For a given flight condition, the actual thrust that a jet engine produces depends on the product of the throttle setting, the ratio of the air density at the flight altitude to the air density at sea level, and the maximum rated thrust of the engine at sea level. If T_{max}^s denotes the maximum rated thrust that an engine can produce at sea level, then the actual thrust T that the engine produces at altitude is given by

$$T = \delta_t \left(\frac{\rho}{\rho^s}\right)^m T_{max}^s. \tag{4.3}$$

Here δ_t denotes the throttle setting, which is assumed to denote a value between 0 and 1; a throttle setting of zero corresponds to engine off while a throttle setting of 1 corresponds to maximum thrust of the jet engine at that altitude. Since the minimum thrust occurs at zero throttle and the maximum rated thrust occurs at full throttle, the constraints on the thrust available for flight can be written as

$$0 \leq T \leq \left(\frac{\rho}{\rho^s}\right)^m T_{max}^s. \tag{4.4}$$

In expressions (4.3) and (4.4)

$$\left(\frac{\rho}{\rho^s}\right)$$

is the ratio of the air density at the flight altitude to the air density at sea level. Recall that the air density depends on the altitude according to the standard atmospheric model; thus, this factor expresses the fact that the maximum thrust that an engine can produce decreases as the flight altitude increases. The dimensionless air density exponent $m > 0$ in expressions (4.3) and (4.4) is a characteristic of the jet engine. The throttle setting that produces an engine thrust T for an aircraft in steady level flight is given by

$$\delta_t = \frac{T}{T_{max}^s} \left(\frac{\rho^s}{\rho}\right)^m. \tag{4.5}$$

The power produced by a jet engine in steady flight is

$$P = \delta_t \left(\frac{\rho}{\rho^s}\right)^m T_{max}^s V. \tag{4.6}$$

Another important property is the fuel consumed by a jet engine. The rate of fuel flow (the weight of fuel burned per unit of time) is proportional to the thrust produced by the engine according to

$$\frac{dW}{dt} = -cT. \tag{4.7}$$

Here the variable W is the weight of the aircraft, including fuel, and c is the constant thrust specific fuel consumption rate for the jet engine. The negative sign appears in equation (4.7) to express the fact that the time rate of change of the aircraft weight is negative. The fuel consumption rate for a jet engine can also be expressed in terms of the throttle, the air density ratio, and the maximum rated thrust produced by the engine as

$$\frac{dW}{dt} = -c\,\delta_t \left(\frac{\rho}{\rho^s}\right)^m T^s_{max}. \tag{4.8}$$

An ideal jet engine is described by the thrust, power, and fuel consumption formulas in expressions (4.3), (4.6), and (4.8). The maximum rated engine thrust at sea level, the air density exponent, and the thrust specific fuel consumption rate are typically viewed as constants, independent of throttle, altitude, and airspeed. For many jet engines, the ideal engine is a good approximation. In some cases, non-ideal jet engines are described in graphical form by providing tables or graphs that describe how the thrust produced by the engine that is available for flight and the fuel consumption rate depend on throttle, altitude, and airspeed.

Plots for a typical jet engine that describe the maximum thrust produced by the engine as it depends on the flight altitude and the maximum fuel flow rate as it depends on the flight altitude are developed in the following Matlab m-file.

Matlab function 4.1 FigJetEng.m
```
alt=linspace (0,65000,500);
Tsmax=12500; m=0.6; c=-.69;
[Temp Pr  rho]=StdAtpUS (alt);
T=((rho/0.0023769).^m)*Tsmax;
FR=((rho/0.0023769).^m)*Tsmax*c;
figure;
plot (alt,T);
xlabel ('Altitude (ft)');
ylabel ('Thrust (lbs)');
grid on;
figure;
plot (alt,FR);
xlabel ('Altitude (ft)');
ylabel ('Fuel flow rate (lbs/hr)');
grid on;
```

Figures 4.1(a) and 4.1(b) show curves based on the assumption of full throttle.

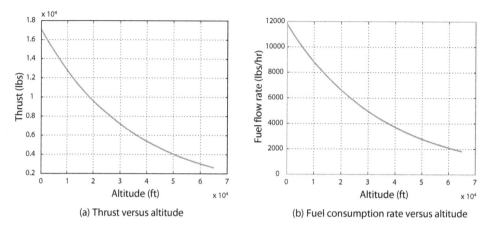

Figure 4.1. *Typical full throttle jet engine characteristics.*

4.3 Propeller Driven by Internal Combustion Engine

Many general aviation aircraft are propelled by one or more internal combustion engines, each of which rotates a shaft to turn a propeller. Each propeller produces a thrust force along a direction that is fixed in the aircraft plane of mass symmetry. It is often assumed that the net thrust is along the aircraft-fixed x_A-axis. Internal combustion engines are typically described by the maximum rated power that they can produce at sea level. Propellers are described by their efficiency in converting engine power into power available for flight. A propulsion system that consists of one or more engine and propeller combinations is typically described by the maximum rated power that the engine or engines can produce at sea level and by the propeller efficiency. Although our subsequent terminology with regard to such a propulsion system may refer to an internal combustion engine and propeller, it should be kept in mind that the development also applies to a propulsion system that consists of multiple engines and propellers so long as they are operated by a single throttle.

For a given flight condition, the actual power that an internal combustion engine produces depends on the product of the throttle setting, the ratio of the air density at the flight altitude to the air density at sea level, and the maximum rated power of the engine. If $P^s_{\max}$ denotes the rated maximum power that an engine can produce at sea level, then the actual power that the engine produces at altitude is given by

$$\delta_t \left(\frac{\rho}{\rho^s} \right)^m P^s_{\max}.$$

Here δ_t denotes the throttle setting, which is a value between 0 and 1; a throttle setting of zero corresponds to engine off while a throttle setting of 1 corresponds to maximum power at that altitude. The factor

$$\left(\frac{\rho}{\rho^s}\right)$$

is the ratio of the air density at the flight altitude to the air density at the sea level. Recall that the air density depends on the altitude according to the standard atmospheric model; thus, this factor expresses the fact that the maximum power that the engine can produce decreases as the altitude increases. The dimensionless air density exponent $m > 0$ is a characteristic of the internal combustion engine.

The power available for flight, denoted by P, is the power generated by the engine multiplied by the efficiency η of the propeller. The propeller efficiency is assumed to satisfy $0 \le \eta \le 1$. Thus, the power available for flight is

$$P = \eta \delta_t \left(\frac{\rho}{\rho^s}\right)^m P_{\max}^s. \tag{4.9}$$

Since the minimum power occurs at zero throttle and the maximum rated power occurs at full throttle, the power constraints on the power available for flight can be written as

$$0 \le P \le \eta \left(\frac{\rho}{\rho^s}\right)^m P_{\max}^s. \tag{4.10}$$

The thrust on an aircraft in steady flight produced by a propeller driven by an internal combustion engine is

$$T = \eta \delta_t \left(\frac{\rho}{\rho^s}\right)^m \frac{P_{\max}^s}{V}. \tag{4.11}$$

The fuel flow rate (the weight of fuel burned per unit of time) is proportional to the power produced by the engine according to

$$\frac{dW}{dt} = -\frac{c}{\eta} P. \tag{4.12}$$

Here c is the power specific fuel consumption rate for the engine. The fuel flow rate can also be expressed in terms of the throttle, the air density ratio, and the maximum rated power produced by the engine as

$$\frac{dW}{dt} = -c \, \delta_t \left(\frac{\rho}{\rho^s}\right)^m P_{\max}^s. \tag{4.13}$$

An ideal internal combustion engine is described by the thrust, power, and fuel consumption formulas in expressions (4.11), (4.9), and (4.12). The maximum rated engine power at sea level, the air density exponent, the engine power specific fuel consumption rate, and the propeller efficiency are viewed as constants, independent of throttle, altitude,

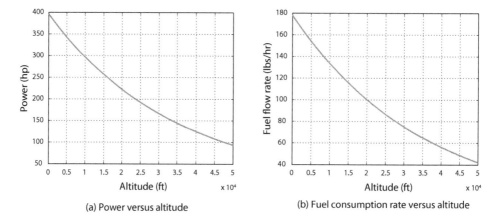

(a) Power versus altitude

(b) Fuel consumption rate versus altitude

Figure 4.2. Typical full throttle internal combustion engine characteristics.

and airspeed. For many internal combustion engines and propellers, the ideal engine and propeller models are good approximations. In some cases, non-ideal engines and propellers are described by providing tables or graphs that describe how the power produced by the engine that is available for flight and the fuel consumption rate depend on throttle, altitude, and airspeed.

Plots for a typical internal combustion engine that describe the maximum power produced by the engine as it depends on the flight altitude and the maximum fuel flow rate as it depends on the flight altitude are developed in the following Matlab m-file.

Matlab function 4.2 FigPropEng.m
```
alt=linspace (0,50000,500);
Tsmax=290; m=0.6; c=0.45;
[Temp Pr  rho]=StdAtpUS (alt);
P=((rho/0.0023769) .^m)*Psmax;
FR=((rho/0.0023769) .^m)*Psmax*c;
figure;
plot (alt,P);
xlabel ('Altitude (ft)');
ylabel ('Power (hp)');
grid on;
figure;
plot (alt,FR);
xlabel ('Altitude (ft)');
ylabel ('Fuel flow rate (lbs/hr)');
grid on;
```

Figures 4.2(a) and 4.2(b) are based on the assumption of full throttle.

4.4 Turboprop Engines

Turboprop engines and turbofan engines use a turbine to turn a propeller that produces the thrust on the aircraft. Consequently, these engines combine some of the features of jet engines and internal combustion engines but with a propeller.

It is most common to describe turboprop engine performance using tables or graphs that express thrust or power produced by the engine and the fuel consumption rate in terms of throttle, altitude, and airspeed.

Steady flight analysis of aircraft with these types of propulsion systems are not explicitly studied in the subsequent development. However, the analysis methods that are presented can be modified to handle these cases.

4.5 Throttle as a Pilot Input

The pilot of an aircraft typically controls the thrust or power that the engine provides for flight by adjustment of the throttle setting. In particular, the pilot is able to select a throttle setting between 0 and 1. Depending on the type of engine, this means that the pilot has limits on the thrust or power that can be used to carry out a flight maneuver or to change flight conditions.

In order to conserve fuel, most pilots would prefer to carry out flight maneuvers using only the aerodynamic control surfaces without changes in the throttle, if possible. As will be seen in subsequent chapters, some flight maneuvers can be accomplished without using the throttle; however, some flight maneuvers require use of the throttle.

4.6 Problems

4.1. Consider an executive jet aircraft, assuming it is powered by two jet engines, each of which can provide a maximum sea level thrust of 6,250 lbs. The thrust produced by the propulsion system depends on the flight altitude and the throttle setting according to

$$ T = \delta_t \left(\frac{\rho}{2.3769 \times 10^{-3}} \right)^{0.6} 12,500 \, \text{lbs}, $$

where δ_t is the throttle setting and ρ is the air density in slug/ft^3 at the flight altitude. The thrust specific fuel consumption rate for the jet engines is 0.69 lb fuel/hr/lb.

   (a) What is the maximum available thrust provided by the jet engine when it is operating at a flight altitude of 25,000 ft? What is the fuel flow rate at this operating condition?

   (b) What is the thrust provided by the jet engine when it is operating at a flight altitude of 25,000 ft and 70% throttle? What is the fuel flow rate at this operating condition?

   (c) Suppose the jet aircraft is in steady level flight at an altitude of 25,000 ft with an airspeed of 520 ft/s. Assuming 70% throttle, what is the power produced by the jet engine?

(d) What is the throttle setting required for the jet engine to produce 9,250 lbs of thrust at an altitude of 25,000 ft? What is the fuel flow rate at this operating condition?

4.2. Consider a general aviation aircraft, assuming a single propeller generates the thrust. The propeller is driven by an internal combustion engine that produces a maximum of 290 hp at sea level. The power produced by the propulsion system that is available for flight depends on the flight altitude and the throttle setting according to

$$P = \eta \delta_t \left(\frac{\rho}{2.3769 \times 10^{-3}} \right)^{0.6} 290 \text{ hp},$$

where $\eta = 0.80$ is the propeller efficiency, δ_t is the throttle setting, and ρ is the air density in slug/ft^3 at the flight altitude. The power specific fuel consumption rate for the internal combustion engine is 0.45 lb fuel/hr/hp.

(a) What is the maximum available power provided by the engine and propeller when it is operating at a flight altitude of 20,000 ft? What is the fuel flow rate at this operating condition?

(b) What is the available power provided by the engine and propeller when it is operating at a flight altitude of 20,000 ft and 70% throttle? What is the fuel flow rate at this operating condition?

(c) Suppose the general aviation aircraft is in steady level flight at an altitude of 20,000 ft with an airspeed of 410 ft/s. Assuming 70% throttle what is the thrust on the aircraft produced by the engine and propeller?

(d) What is the throttle setting required for the engine and propeller to produce 210 hp at an altitude of 20,000 ft? What is the fuel flow rate at this operating condition?

4.3. Consider a small remote-controlled aircraft, assuming the aircraft is powered by a single propeller. The net thrust produced by the propulsion system depends on the airspeed, flight altitude, and the throttle setting according to

$$T = \delta_t \left(\frac{\rho}{2.3769 \times 10^{-3}} \right)^{0.6} (22.0 - 0.126V) \text{ lbs},$$

where δ_t is the throttle setting, V is the airspeed in ft/s, and ρ is the air density in slug/ft^3 at the flight altitude. The propeller losses are included. This expression for the thrust, which depends explicitly on the airspeed, is valid for airspeeds up to a maximum of 170 ft/s. The thrust specific fuel consumption rate for the engine is 0.6 lb fuel/hr/lb.

(a) What is the maximum available thrust provided by the engine and propeller when it is operating at a flight altitude of 5,000 ft and an airspeed of 95 ft/s? What is the fuel flow rate at this operating condition?

(b) What is the maximum available thrust provided by the engine and propeller when it is operating at a flight altitude of 5,000 ft? Give an expression for the maximum available thrust as a function of the airspeed. Give an expression for the corresponding fuel flow rate as a function of airspeed.

(c) What is the maximum available power, in hp, provided by the engine and propeller when it is operating at a flight altitude of 5,000 ft and an airspeed of 95 ft/s? What is the fuel flow rate at this operating condition?

(d) What is the maximum available power, in hp, provided by the engine and propeller when it is operating at a flight altitude of 5,000 ft? Give an expression for the maximum available power as a function of the airspeed. Give an expression for the corresponding fuel flow rate as a function of airspeed?

(e) What is the throttle setting required for the engine and propeller to produce 210 hp at an altitude of 5,000 ft and airspeed of 95 ft/s? What is the fuel flow rate at this operating condition?

5

Prelude to Steady Flight Analysis

In this chapter, the basic physical principles for steady aircraft flight through a stationary atmosphere are summarized as a prelude to the more detailed steady flight analysis presented in the chapters that follow. In addition, the details of several aircraft are given that define the framework for case studies for steady flight and performance analysis that are studied in the subsequent chapters. These case studies are referred to as an executive jet aircraft and a propeller-driven general aviation aircraft.

The physics of steady aircraft flight is based on a balance between the forces and moments that act on an aircraft, taking into account any steady acceleration of the aircraft. This can be described by text and by diagrams, without appeal to any mathematics. This approach, taken in Barnard and Philpott (1995) and Wegener (1991), provides an intuitive insight into steady aircraft flight. On the other hand, such an approach does not provide a quantitative description of steady aircraft flight, which requires the use of mathematics.

The steady flight balance relationships can be expressed as mathematical equations that are derived by examining the forces and moments that act on the aircraft. Mathematical equations for steady aircraft flight are expressed as algebraic equations; this is in contrast to flight dynamics, where the mathematical equations are expressed as differential equations.

Suppose that the propulsive thrust is constant and the aerodynamic control surface deflections are maintained at constant values. Steady flight, by definition, occurs if the components of the translational velocity vector of the aircraft, when expressed in the aircraft-fixed frame, are constant in time, and if the components of the angular velocity vector, when expressed in the aircraft-fixed frame, are constant in time. In the case of turning flight, the components of the translational vector, when expressed in the Earth-fixed frame, are not necessarily constant in time.

Steady longitudinal flight occurs if the aircraft center of mass moves along a straight line in a fixed vertical plane. Steady longitudinal flight requires that the sum of all the force vectors on the aircraft is zero and the sum of all the moment vectors in the aircraft-fixed frame is zero. A special category of steady longitudinal flight occurs if the velocity vector is horizontal; this case is referred to as steady level longitudinal flight.

Steady level turning flight occurs if the aircraft center of mass moves along a circular path in a fixed horizontal plane. Consequently, the aircraft velocity vector has constant magnitude and is always tangent to the circular path. Steady level turning flight requires that the sum of all of the force vectors on the aircraft cause the aircraft to turn so that it moves along a circular path and the sum of all moment vectors in the aircraft-fixed frame is zero.

The most general type of steady turning flight occurs in three dimensions. The aircraft moves along a helical path whose axis is vertical; the projection of the path of the aircraft onto a horizontal plane is a circular path. This flight condition can be viewed as a combination of steady longitudinal flight and steady level turning flight; it is referred to as steady climbing or descending turning flight.

The basic approach in studying any of these steady flight conditions consists of construction of a free-body diagram of the aircraft that shows all of the forces acting on the aircraft. It is convenient to show the aircraft attitude and the velocity vector of the center of mass of the aircraft on the free-body diagram. Such diagrams are helpful in correctly describing and visualizing the direction of each force vector. These diagrams form the basis for developing the steady flight equations of motion based on Newton's laws.

In subsequent chapters, all types of steady flight are analyzed in detail. As a prelude to the subsequent development, this chapter summarizes the equations for steady flight. The equations for steady climbing or descending turning flight are presented; equations for steady longitudinal flight and for steady level turning flight are then presented as special cases of these more general equations.

It should also be mentioned that steady aircraft flight can be viewed as an equilibrium of differential equations that describe the translational dynamics of an aircraft. This perspective is developed in chapter 13, where the differential equations for translational flight dynamics are derived. In this chapter, which introduces the concept of steady flight, the results for steady flight are developed in a more direct and elementary way.

5.1 Aircraft Forces and Moments

The forces and moments that act on an aircraft are summarized in this section. There are three types of aircraft forces: aerodynamic force, a propulsion force, and a gravitational force.

Assuming steady coordinated flight, the aerodynamic forces are the lift force and the drag force. As described previously, the lift force vector has magnitude given by

$$L = \frac{1}{2}\rho V^2 S C_L, \tag{5.1}$$

which depends on the dynamic pressure, the wing surface area, and the lift coefficient. The lift force vector acts perpendicular to the velocity vector in the aircraft plane of mass symmetry. Recall also that the drag force vector has magnitude given by

$$D = \frac{1}{2}\rho V^2 S C_D, \tag{5.2}$$

which depends on the dynamic pressure, the wing surface area, and the drag coefficient. The drag force vector acts opposite to the velocity vector.

A propulsion system in the aircraft generates a propulsive force T that acts on the aircraft. As described in chapter 4, the propulsive force depends on the type of engine, the throttle setting, the flight altitude, and the airspeed. The propulsive force vector acts along the aircraft-fixed x_A-axis, in the plane of mass symmetry of the aircraft, and it is fixed with respect to the aircraft-fixed frame.

Finally, a gravitational force vector acts on the aircraft. It is a convenient approximation to assume that the Earth is flat and not rotating. The magnitude of the gravity force is the aircraft weight given by W. The direction of the gravity force is along the z_E-axis of the ground-fixed frame. It is also convenient to make the approximation that the weight of the aircraft is constant. This is a good approximation for time periods of the order of minutes, but it is not a good approximation for longer time periods where fuel consumption must be taken into account.

The aerodynamic roll, pitch, and yaw moments are described as follows. The aerodynamic roll moment has magnitude given by

$$\mathcal{L} = \frac{1}{2}\rho V^2 SbC_{\mathcal{L}}, \tag{5.3}$$

which depends on the dynamic pressure, the wing surface area, the wing span, and the roll moment coefficient. The roll moment vector acts along the roll or x_A-axis of the aircraft. The aerodynamic pitch moment has magnitude given by

$$M = \frac{1}{2}\rho V^2 ScC_M, \tag{5.4}$$

which depends on the dynamic pressure, the wing surface area, the mean wing chord, and the pitch moment coefficient. The pitch moment vector acts along the pitch or y_A-axis of the aircraft. The aerodynamic yaw moment has magnitude given by

$$N = \frac{1}{2}\rho V^2 SbC_N, \tag{5.5}$$

which depends on the dynamic pressure, the wing surface area, the wing span, and the yaw moment coefficient. The yaw moment vector acts along the yaw or z_A-axis of the aircraft.

5.2 Steady Flight Equations

As will be shown in chapter 10, the most general form of steady flight holds if the aircraft flies through a stationary atmosphere along a helical path with a vertical axis. For such steady flight, the algebraic equations that arise from Newton's laws are given by

$$-W \sin\gamma - D + T \cos\alpha = 0, \tag{5.6}$$

$$L \sin\mu + T \sin\alpha \sin\mu = \frac{W}{g}\frac{V^2 \cos^2\gamma}{r}, \tag{5.7}$$

$$W \cos\gamma - L \cos\mu - T \sin\alpha \cos\mu = 0. \tag{5.8}$$

Equation (5.6) states that the sum of the components of the force vector in the direction of the velocity vector is zero, since the aircraft is not accelerating in this direction. Equation (5.7) states that the sum of the components of the force vectors in the (horizontal) radial direction equals the mass of the aircraft multiplied by the (horizontal) radial, or centrifugal, acceleration of the aircraft. Equation (5.8) states that the sum of the components of the force vectors projected onto the vertical plane and normal to the velocity vector is zero, since the aircraft is not accelerating in this direction. This physical explanation of the equations for steady turning and climbing flight is illustrated by the free-body diagrams presented in chapter 10.

In equations (5.6)–(5.8), α denotes the angle of attack, γ denotes the flight path angle, μ denotes the bank angle about the velocity vector, and r denotes the turn radius, that is, the radius of the projected helical path onto a horizontal plane. As before, L is the aircraft lift, D is the aircraft drag, T is the thrust from the propulsion system, and V is the airspeed of the aircraft. Finally, W denotes the weight of the aircraft and g denotes the constant acceleration of gravity so that $\frac{W}{g}$ is the mass of the aircraft.

In addition to these three equations obtained from Newton's laws, expressions for aerodynamics relationships and constraints from chapters 2 and 3 and for propulsion relationships and constraints from chapter 4 are required.

Steady flight also requires that the sum of the moment vectors be zero. The aerodynamic moments have been described in equations (5.3), (5.4), and (5.5) and in chapter 3. Throughout the subsequent development, the propulsion force is assumed to act along the aircraft-fixed x_A-axis; hence there is no net moment on the aircraft due to the thrust. This assumption is made for simplicity; if the assumption is not valid it is easy to modify the subsequent development to incorporate a pitch moment due to the thrust. Finally, the weight of the aircraft (the force due to gravity) does not give a moment since the weight acts through the center of mass. In summary, steady flight requires that the sum of the moment vectors is zero. This gives the conditions that each of the components of the aerodynamic moments is zero, that is,

$$\mathcal{L} = M = N = 0, \tag{5.9}$$

or, equivalently, that the roll moment coefficient, the pitch moment coefficient, and the yaw moment coefficient are zero, that is,

$$C_{\mathcal{L}} = C_M = C_N = 0. \tag{5.10}$$

Since the aircraft is assumed to be in steady coordinated flight, the side-slip angle $\beta = 0$. This implies that the velocity vector of the aircraft lies in the plane of mass symmetry of the aircraft. Thus the yaw moment coefficient and the roll moment coefficient are zero if the differential ailerons deflection and the rudder deflection are zero. This is assumed throughout the subsequent steady flight analysis. Since the pitch moment in steady flight typically depends on the angle of attack and the elevator deflection, the elevator deflection needs to be selected, depending on the angle of attack, to guarantee that the pitch moment coefficient is zero. Trimmed flight occurs if the aerodynamic control surfaces are adjusted to achieve zero roll, pitch, and yaw moments.

In chapter 10, these fundamental equations are studied in detail to determine the properties of steady turning and climbing or descending flight.

5.3 Steady Longitudinal Flight

Steady longitudinal flight is steady flight in a fixed vertical flight; the aircraft flies along a straight line path. In this case, the aircraft bank angle is zero and the turn radius is infinite. Hence the aircraft velocity vector is constant in the Earth-fixed frame. Using these assumptions in equations (5.6), (5.8), and (5.7), the third equation is identically satisfied and the first two equations become

$$-W \sin \gamma - D + T \cos \alpha = 0, \tag{5.11}$$

$$W \cos \gamma - L - T \sin \alpha = 0, \tag{5.12}$$

which, together with the moment equations, describe steady longitudinal flight. In chapter 6, these equations are studied assuming that the thrust $T = 0$, which is referred to as steady gliding flight. In chapter 8, these equations for steady climbing or descending longitudinal flight are studied.

Steady level longitudinal flight occurs if the flight path angle $\gamma = 0$. The resulting equations are

$$-D + T \cos \alpha = 0, \tag{5.13}$$

$$W - L - T \sin \alpha = 0, \tag{5.14}$$

which, together with the moment equations, describe steady level longitudinal flight. In chapter 7, these equations for steady level longitudinal flight are studied.

5.4 Steady Level Turning Flight

Steady level turning flight is steady flight in a fixed horizontal plane; the aircraft flies along a circular path. In this case, the flight path angle is zero. Hence the aircraft velocity vector rotates at a constant rate about the vertical axis. Using these assumptions in equations (5.6), (5.8), and (5.7), we obtain

$$-D + T \cos \alpha = 0, \tag{5.15}$$

$$L \sin \mu + T \sin \alpha \sin \mu = \frac{W}{g} \frac{V^2}{r}, \tag{5.16}$$

$$W - L \cos \mu - T \sin \alpha \cos \mu = 0, \tag{5.17}$$

which, together with the moment equations, describe steady level turning flight. In chapter 9, these equations are studied to characterize steady level turning flight.

5.5 Flight Constraints

The algebraic equations provided in sections 5.2, 5.3, and 5.4 describe steady aircraft flight. As already indicated, the variables that describe a steady flight condition must also satisfy certain physical constraints. There are three physical constraints that are most important for steady flight: a stall constraint, propulsion constraints, and a wing loading constraint.

The stall constraint, described by expression (3.50), is expressed as an inequality involving the lift coefficient as

$$C_L \leq C_{L_{\max}}. \tag{5.18}$$

The propulsion constraints can be expressed either as inequalities on the thrust available for flight as in expression (4.4)

$$0 \leq T \leq \left(\frac{\rho}{\rho^s} \right)^m T_{\max}^s. \tag{5.19}$$

Or they can be expressed as inequalities involving the power available for flight, as in expression (4.10),

$$0 \leq P \leq \eta \left(\frac{\rho}{\rho^s} \right)^m P_{\max}^s, \tag{5.20}$$

depending on the type of propulsion system on the aircraft.

The wing loading constraint, of importance primarily for turning flight, limits the lift force on the aircraft; this constraint is primarily due to structural limits on the wings. The most common form for this constraint is as an inequality of the form

$$L \leq n_{max} W. \tag{5.21}$$

Here n_{max} denotes the maximum load factor that reflects the maximum possible lift force that the wing is allowed to provide. The wing loading constraint is effectively a structural safety constraint.

In some cases, it is natural to impose an explicit constraint on the maximum value of the dynamic pressure, the aircraft Mach number, or the aircraft airspeed. For example, this type of constraint might be used to restrict the aircraft to subsonic flight where the aerodynamics assumptions introduced previously are valid.

These flight constraints are important parts of the mathematical models for steady flight analysis. All physically meaningful steady flight conditions, a set of flight conditions referred to as the steady flight envelope, must satisfy not only the steady flight equations but also all of the flight constraints.

5.6 Aircraft Case Studies

In the remainder of the book, reference to a jet aircraft implies that the propulsion system consists of an ideal jet engine or several ideal jet engines. Reference to a propeller-driven general aviation aircraft (or general aviation aircraft) implies that the propulsion system consists of one or more propellers, each driven by an ideal internal combustion engine. These define the two most important categories of aircraft with idealized propulsion systems. Results for aircraft that require use of propulsion systems that are not ideal in the sense described can often be obtained by making suitable modifications of the results that are subsequently presented.

Figure 5.1. Executive jet aircraft.

In order to provide concrete illustrations of many of the results, two case studies are now introduced. An example of an executive jet aircraft is first introduced, followed by an example of a general aviation aircraft. A case study of an uninhabited aerial vehicle (UAV) is also introduced; this example is referred to in many of the end-of-chapter problems.

5.7 Characteristics of an Executive Jet Aircraft

This section introduces details that describe an executive jet aircraft, illustrated in Figure 5.1, that is capable of carrying up to eight passengers plus crew and cargo. The aircraft is fictitious in the sense that it does not represent any specific existing aircraft. The specific aircraft data are consistent with typical aircraft of this type. The flight analysis in the subsequent chapters is provided to illustrate the steady flight and flight performance concepts for a typical executive jet aircraft. These executive jet aircraft characteristics are also referred to in subsequent end-of-chapter problems.

The weight of the aircraft, with a full fuel tank, is 73,000 lbs. The wing surface area is 950 ft^2; the aspect ratio is 5.9. The aerodynamic drag polar is given by

$$C_D = 0.015 + 0.05C_L^2.$$

The lift coefficient, in terms of the angle of attack, is

$$C_L = 0.02 + 0.12\alpha,$$

where the angle of attack is measured in degrees. The maximum lift coefficient at stall is 2.8, which occurs at an angle of attack of 23.2 degrees. The pitch moment coefficient, expressed in terms of the angle of attack and the elevator deflection, is given by

$$C_M = 0.24 - 0.18\alpha + 0.28\delta_e,$$

where the angle of attack α and the elevator deflection δ_e are measured in degrees.

The jet aircraft is powered by two jet engines, each of which can provide a maximum sea level thrust of 6,250 lbs. The engines are configured in the aircraft so that they do not generate any pitch moment on the aircraft. The thrust produced by the propulsion system depends on the flight altitude and the throttle setting according to

$$T = \delta_t \left(\frac{\rho}{2.3769 \times 10^{-3}} \right)^{0.6} 12,500\,\text{lbs},$$

where δ_t is the throttle setting and ρ is the air density in slug/ft^3 at the flight altitude. The maximum fuel that can be carried in the aircraft is 28,000 lbs. The thrust specific fuel consumption rate for the jet engines is 0.69 lb fuel/hr/lb.

The maximum load factor for the executive jet aircraft is 2; this corresponds to a maximum lift force on the wing of twice the weight of the aircraft.

The data for the jet aircraft are used to analyze steady flight properties of the aircraft and to determine various performance measures of the aircraft.

5.8 Characteristics of a Single Engine Propeller-Driven General Aviation Aircraft

This section introduces details that describe a single engine propeller-driven aircraft, powered by an internal combustion engine, which is illustrated in Figure 5.2. This aircraft is capable of carrying two persons and minimal cargo. This general aviation aircraft is fictitious in the sense that it does not represent any specific existing aircraft. The specific aircraft data are consistent with typical aircraft of this type. The flight analysis in the subsequent chapters is provided to illustrate steady flight and flight performance concepts for a typical small propeller-driven aircraft. These single engine propeller-driven general aviation aircraft characteristics are also referred to in subsequent end-of-chapter problems.

The weight of the aircraft, with a full fuel tank, is 2,900 lbs. The wing surface area is 175 ft^2; the aspect ratio is 7.4. The aerodynamic drag polar is given by

$$C_D = 0.026 + 0.054C_L^2.$$

The lift coefficient, in terms of the angle of attack, is

$$C_L = 0.02 + 0.12\alpha,$$

where the angle of attack is measured in degrees. The maximum lift coefficient at stall is 2.4, which occurs at an angle of attack of 19.8 degrees. The pitch moment coefficient, in

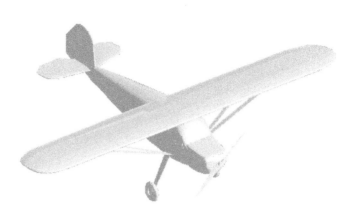

Figure 5.2. Propeller-driven general aviation aircraft.

terms of the angle of attack and the elevator deflection, expressed in terms of the angle of attack and the elevator deflection, is given by

$$C_M = 0.12 - 0.08\alpha + 0.075\delta_e,$$

where the angle of attack α and the elevator deflection δ_e are measured in degrees.

The single propeller is driven by an internal combustion engine that produces a maximum of 290 hp at sea level. The propulsion system is configured in the aircraft so that the thrust does not generate any pitch moment on the aircraft. The power produced by the propulsion system and available for flight, taking into account the propeller losses, depends on the flight altitude and the throttle setting according to

$$P = \delta_t \eta \left(\frac{\rho}{2.3769 \times 10^{-3}} \right)^{0.6} 290 \, \text{hp},$$

where δ_t is the throttle setting, $\eta = 0.8$ is the propeller efficiency, and ρ is the air density in slug/ft^3 at the flight altitude. The maximum fuel that can be carried in the aircraft is 370 lbs. The power specific fuel consumption rate for the internal combustion engine is 0.45 lb fuel/hr/hp.

The maximum load factor constraint for the general aviation aircraft is 2; this corresponds to a maximum lift force on the wing of twice the weight of the aircraft.

The data for a propeller-driven aircraft are used to analyze steady flight properties of the aircraft and to determine various performance measures of the aircraft.

5.9 Characteristics of an Uninhabited Aerial Vehicle (UAV)

This section introduces details of a small uninhabited aerial vehicle that might compete in a student aircraft design competition or might carry out surveillance and monitoring missions. This UAV is fictitious in the sense that it does not represent any specific existing aircraft. The steady flight analysis and flight performance of this aircraft are representative of typical UAVs. This case study is not developed in the subsequent text, but it does appear as a basis for some problems at the end of the subsequent chapters.

The weight of the UAV, with a full fuel tank, is 45 lbs. The wing surface area is 10.2 ft^2; the wing aspect ratio is 10. The aerodynamic drag polar is given by

$$C_D = 0.025 + 0.05\,(C_L - 0.7)^2\,.$$

Note that this drag polar differs slightly from the previously assumed form. The lift coefficient, expressed in terms of the angle of attack, is

$$C_L = 0.18 + 0.06\alpha,$$

where the angle of attack is measured in degrees. The maximum lift coefficient at stall is 1.7, which occurs at an angle of attack of 25.3 degrees. The pitch moment coefficient, expressed in terms of the angle of attack and the elevator deflection, is given by

$$C_M = 0.22 - 0.15\alpha + 0.23\delta_e,$$

where the angle of attack α and the elevator deflection δ_e are measured in degrees.

The UAV is powered by a propeller and an engine that are configured in the aircraft so that they do not generate any pitch moment on the aircraft. The thrust produced by the propulsion system, accounting for the propeller efficiency factor, depends on the airspeed, flight altitude, and throttle setting according to

$$T = \delta_t \left(\frac{\rho}{2.3769 \times 10^{-3}}\right)^{0.6} (22 - 0.126V)\ \text{lbs},$$

where δ_t is the throttle setting, V is the airspeed in ft/s, and ρ is the air density in slug/ft^3 at the flight altitude. This expression for the thrust is valid for airspeeds from zero up to a maximum of 170 ft/s. Note that this expression for the thrust provided by the propulsion system depends explicitly on the airspeed of the aircraft; this expression differs from the previously assumed form. The maximum weight of fuel that can be carried in the aircraft is 18 lbs. The thrust specific fuel consumption rate for the engine is 0.6 lb fuel/hr/lb.

The maximum load factor for the UAV is 2.37; this corresponds to a maximum lift force on the wing of 2.37 times the weight of the aircraft.

These UAV characteristics involve a drag polar expression and an engine thrust expression that differ from the standard assumptions that form the basis for the analytical developments in this book. This implies that some of the analytical formulas presented elsewhere in this book cannot be utilized in steady flight and performance analysis for the UAV. Rather, appropriate analytical formulas for this UAV must be developed using the basic concepts of steady flight and performance analysis that are introduced. Consequently, the subsequent end-of-chapter problems that involve the UAV are more advanced in that they require suitable modifications of the various analytical formulas.

5.10 Problems

5.1. This chapter introduces the basic concepts for steady flight based on certain clearly stated assumptions. Answer the following questions by describing how the development in this chapter would need to be modified based on the indicated change in the assumptions. Which equations would need to be modified? Give a brief description of the required modifications of the equations.

(a) Suppose that the aircraft is in steady flight and the atmosphere is not stationary. What modifications would need to be made to the development?

(b) Suppose that the aircraft is in steady flight and the Earth is assumed to be spherical. What modifications would need to be made to the development?

(c) Suppose that the aircraft is in steady flight and the Earth is assumed to be spherical and rotating. What modifications would need to be made to the development?

(d) Suppose the aircraft is in steady flight and the decrease in the weight of the aircraft as fuel is burned is taken into account in the analysis. What modifications would need to be made to the development?

5.2. Consider the executive jet aircraft.

(a) What are the required angle of attack and the required elevator deflection so that the aircraft flies with a constant lift to drag ratio of 12 in steady longitudinal flight?

(b) If the elevator deflection is 3 degrees, what is the required angle of attack for steady longitudinal flight of the aircraft? What are the resulting values of the lift and drag longitudinal coefficients?

(c) What are the values of the lift and drag coefficients so that the aircraft is in steady longitudinal flight with an angle of attack of 5 degrees? What is the required elevator deflection to maintain this steady flight condition?

(d) Suppose that the aircraft is to be in steady longitudinal flight at a stall flight condition. What are the values of the lift and drag coefficients? What is the elevator deflection required to keep the aircraft in steady flight at this stall flight condition?

5.3. Consider the general aviation aircraft.

(a) What are the required angle of attack and the required elevator deflection so that the aircraft flies with a constant lift to drag ratio of 10 in steady longitudinal flight?

(b) If the elevator deflection is 4 degrees, what is the required angle of attack for steady longitudinal flight of the aircraft? What are the resulting values of the lift and drag coefficients?

(c) What are the values of the lift and drag coefficients so that the aircraft is in steady longitudinal flight with an angle of attack of 4 degrees? What is the required elevator deflection to maintain this steady flight condition?

(d) Suppose that the aircraft is to be in steady longitudinal flight at a stall flight condition. What are the values of the lift and drag coefficients? What is the elevator deflection required to keep the aircraft in steady flight at this stall flight condition?

5.4. Consider the UAV.

(a) What are the required angle of attack and the required elevator deflection so that the aircraft flies with a constant lift to drag ratio of 14 in steady longitudinal flight?

(b) If the elevator deflection is 6 degrees, what is the required angle of attack for steady longitudinal flight of the aircraft? What are the resulting values of the lift and drag coefficients?

(c) What are the values of the lift and drag coefficients so that the aircraft is in steady longitudinal flight with an angle of attack of 6 degrees? What is the required elevator deflection to maintain this steady flight condition?

(d) Suppose that the aircraft is to be in steady longitudinal flight at a stall flight condition. What are the values of the lift and drag coefficients? What is the elevator deflection required to keep the aircraft in steady flight at this stall flight condition?

5.5. Suppose the weight of the executive jet aircraft is 73,000 lbs and the aircraft has an airspeed of 380 ft/s in steady longitudinal flight at an altitude of 15,000 ft.

(a) What are the required angle of attack and the required elevator deflection so that the aircraft wing generates a lift force of 73,000 lbs in steady flight?

(b) If the elevator deflection is 2 degrees, what is the required angle of attack for steady flight of the aircraft? What are the resulting values of the lift and drag forces?

(c) What are the values of the lift and drag forces so that the aircraft is in steady, trimmed flight with an angle of attack of 4 degrees? What is the required elevator deflection to maintain this steady flight condition?

(d) Suppose that the aircraft is to be in steady flight at a stall flight condition. What are the lift and drag forces? What is the elevator deflection required to keep the aircraft in steady flight at the stall flight condition?

5.6. Suppose the weight of the general aviation aircraft is 2,900 lbs and the aircraft has an airspeed of 240 ft/s in steady longitudinal flight at an altitude of 10,000 ft.

(a) What are the required angle of attack and the required elevator deflection so that the aircraft wing generates a lift force of 2,900 lbs in steady flight?

(b) If the elevator deflection is 2 degrees, what is the required angle of attack for steady flight of the aircraft? What are the resulting values of the lift and drag forces?

(c) What are the values of the lift and drag forces so that the aircraft is in steady, trimmed flight with an angle of attack of 4 degrees? What is the required elevator deflection to maintain this steady flight condition?

(d) Suppose that the aircraft is to be in steady flight at a stall flight condition. What are the lift and drag forces? What is the elevator deflection required to keep the aircraft in steady flight at this stall flight condition?

5.7. Suppose the weight of the UAV is 45 lbs and the aircraft has an airspeed of 90 ft/s in steady longitudinal flight at an altitude of 1,000 ft.

(a) What are the required angle of attack and the required elevator deflection so that the aircraft wing generates a lift force of 45 lbs in steady flight?

(b) If the elevator deflection is 4 degrees, what is the required angle of attack for steady flight of the aircraft? What are the resulting values of the lift and drag forces?

(c) What are the values of the lift and drag forces so that the aircraft is in steady, trimmed flight with an angle of attack of 4 degrees? What is the required elevator deflection to maintain this steady flight condition?

(d) Suppose that the aircraft is to be in steady flight at a stall flight condition. What are the lift and drag forces? What is the elevator deflection required to keep the aircraft in steady flight at this stall flight condition?

6

Aircraft Steady Gliding Longitudinal Flight

In this chapter, steady gliding longitudinal flight is studied in detail. Here gliding flight refers to the fact that there is no propulsive force acting on the aircraft, and longitudinal gliding flight refers to the fact that gliding flight occurs in a fixed vertical plane. This topic is of interest for aircraft gliders that do not contain an engine and for aircraft that contain an engine that is not operating. Conditions for steady gliding flight are derived; they are expressed in several different mathematical forms, and gliding flight performance is assessed.

6.1 Steady Gliding Longitudinal Flight

Mathematical descriptions of steady gliding longitudinal flight are developed. These descriptions are expressed in terms of algebraic equations that describe the fundamental flight variables that arise from the free-body diagram for steady gliding flight, equations that describe the aerodynamics, and the stall constraint for steady gliding flight.

Free-Body Diagram

An aircraft is assumed to be in steady gliding flight in a fixed vertical plane. Figure 6.1 gives the free-body diagram that illustrates all of the forces that act on the aircraft. The aircraft is assumed to have a velocity vector with magnitude V that makes an angle γ_{glide}, referred to as the glide angle, with the horizontal. Note that the glide angle is the negative of the flight path angle. As shown, in this case the only forces that act on the aircraft are the lift force, the drag force, and the weight of the aircraft.

Since the aircraft is in steady gliding longitudinal flight in a fixed vertical plane, it is not accelerating. That is, the velocity vector of the aircraft is constant in direction and magnitude. Based on Figure 6.1 the vector sum of the forces acting on the aircraft is zero. This leads to the algebraic equations (6.1) and (6.2) which show the sum of the components of the forces along the velocity vector is zero and the sum of the components of the forces normal to the velocity vector is zero:

$$W \sin \gamma_{glide} - D = 0, \tag{6.1}$$

$$W \cos \gamma_{glide} - L = 0. \tag{6.2}$$

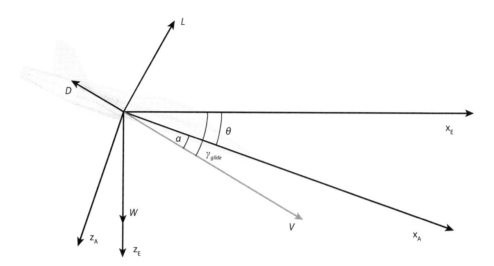

Figure 6.1. Free-body diagram of an aircraft in steady gliding longitudinal flight.

Steady gliding flight also requires that the aerodynamic moments are also zero; in particular, the pitch moment is zero:

$$M = 0. \tag{6.3}$$

Assuming the glide angle, in radian measure, is small so that the sin of the angle can be approximated by the angle and the cosine of the angle can be approximated as 1, we obtain the expressions

$$\gamma_{glide} = \frac{D}{W}, \tag{6.4}$$

$$L = W. \tag{6.5}$$

Aerodynamics

The equations that describe the lift force, the drag force, and the pitch moment from chapter 3 are

$$L = \frac{1}{2}\rho V^2 S C_L, \tag{6.6}$$

$$D = \frac{1}{2}\rho V^2 S C_D, \tag{6.7}$$

$$M = \frac{1}{2}\rho V^2 S c C_M. \tag{6.8}$$

The quadratic drag polar expression that relates the drag coefficient to the lift coefficient is given by

$$C_D = C_{D_0} + K\, C_L^2. \tag{6.9}$$

The lift coefficient is expressed as a linear function of the angle of attack as

$$C_L = C_{L_0} + C_{L_\alpha}\alpha. \tag{6.10}$$

The pitch moment coefficient is expressed as a linear function of the angle of attack and the elevator deflection as

$$C_M = C_{M_0} + C_{M_\alpha}\alpha + C_{M_{\delta_e}}\delta_e. \tag{6.11}$$

Flight Constraints

The only important constraint for steady gliding flight arises from the aerodynamics assumption that the aircraft avoids stalling. Based on the discussion in chapter 2, the stall constraint can be expressed as an inequality on the lift coefficient given by

$$C_L \le C_{L_{\max}}. \tag{6.12}$$

6.2 Steady Gliding Longitudinal Flight Analysis

In this section, flight conditions for steady gliding longitudinal flight are analyzed.

Equations for Steady Gliding Longitudinal Flight

Equations (6.4), (6.5), (6.6), (6.7), and (6.9) are now analyzed to ascertain the basic properties of an aircraft in steady gliding flight. These are five algebraic equations. The following perspective provides insight into these five equations. They involve the following six flight variables that define a steady gliding flight condition: the lift force L, the drag force D, the glide angle γ_{glide}, the airspeed V, the lift coefficient C_L, and the drag coefficient C_D.

Our objective is to determine the relations between these six variables using the conditions for steady gliding flight. In mathematical terms, we seek to determine the relations between these six flight variables that satisfy the five algebraic equations (6.4), (6.5), (6.6), (6.7), and (6.9). These algebraic equations also involve important flight parameters: the aircraft weight W, the aircraft wing surface S, the aircraft aerodynamic parameters C_{D_0} and $K = \frac{1}{\pi e AR}$, and the air density ρ. The air density parameter depends on the flight altitude according to the standard atmospheric model.

Based on equations (6.5) and (6.10), the airspeed can be expressed as

$$V = \sqrt{\frac{2W}{\rho S C_L}}. \tag{6.13}$$

During steady gliding flight, the weight is constant. According to the standard atmospheric model, the air density ρ depends on the altitude, which decreases slowly in steady gliding flight. For small changes in the altitude, the air density can be assumed constant.

Based on equations (6.4) and (6.5), the glide angle can be expressed as

$$\gamma_{glide} = \frac{D}{L} = \frac{C_D}{C_L} = \frac{C_{D_0} + KC_L^2}{C_L}. \tag{6.14}$$

Substituting the lift expression (6.6) into equation (6.5), it is easy to solve for the lift coefficient as

$$C_L = \frac{2W}{\rho V^2 S}. \tag{6.15}$$

The quadratic drag polar expression (6.9) gives the expression for the drag coefficient as

$$C_D = C_{D_0} + \frac{4KW^2}{\rho^2 V^4 S^2}. \tag{6.16}$$

The drag for steady gliding flight at airspeed V is obtained from equation (6.7) as

$$D = \frac{1}{2}\rho V^2 S C_{D_0} + \frac{2KW^2}{\rho V^2 S}. \tag{6.17}$$

The expression for the drag in steady gliding flight is important. The drag consists of two terms: the first term in equation (6.17) is the friction drag that is independent of the lift, and the second term in equation (6.17) is the induced drag that is dependent on the lift. The total drag is the sum of these two terms. Note that the drag is large at low airspeeds, large at high airspeeds, and has a minimum value at some intermediate airspeed.

The glide angle for steady gliding flight, in radians, can be expressed in terms of the airspeed as

$$\gamma_{glide} = \frac{1}{2}\frac{\rho V^2 S C_{D_0}}{W} + \frac{2KW}{\rho V^2 S}. \tag{6.18}$$

Since the glide angle is the drag to lift ratio, it is clear that the glide angle is large at low airspeeds, large at high airspeeds, and has a minimum value at some intermediate speed. The glide angle is always positive or, equivalently, the flight path angle is always negative. Thus, steady gliding flight always corresponds to steady descending flight.

Since the glide angle is a small angle in radians, the descent rate of an aircraft in steady gliding flight is approximated by

$$V_{descent} = V\gamma_{glide} = \frac{1}{2}\frac{\rho V^3 S C_{D_0}}{W} + \frac{2KW}{\rho V S}. \tag{6.19}$$

The descent rate is large at low airspeeds, large at high airspeeds, and has a minimum value at some intermediate speed. The descent rate is always positive.

Equation (6.18) can be expressed in an alternative form. If the glide angle γ_{glide} is assumed known, then equation (6.18) can be written as

$$\frac{1}{2}\frac{\rho V^4 S C_{D_0}}{W} - \gamma_{glide} V^2 + \frac{2KW}{\rho S} = 0, \qquad (6.20)$$

which can be viewed as an equation whose solution is the airspeed for steady gliding flight at the given value of the glide angle. Solving this equation requires finding the roots of a fourth degree polynomial; in fact, the fourth degree polynomial can be viewed as a quadratic polynomial in the square of the airspeed. This implies that the roots of the fourth degree polynomial can be easily determined. In general, equation (6.20) has two real and positive solutions corresponding to a low-speed steady gliding flight solution and a high-speed steady gliding flight solution.

If the descent rate $V_{descent}$ is assumed known, then equation (6.19) can be written as

$$\frac{1}{2}\frac{\rho V^4 S C_{D_0}}{W} - V_{descent} V + \frac{2KW}{\rho S} = 0. \qquad (6.21)$$

This equation can be viewed as an equation whose solution is the airspeed for steady gliding flight at the given value of the descent rate. Solving this equation requires finding the roots of a fourth degree polynomial. The roots of the fourth degree polynomial can be determined. In general, equation (6.21) has two real and positive solutions corresponding to a low-speed steady gliding flight solution and a high-speed steady gliding flight solution.

Each of the above low-speed solutions must be checked to guarantee that it satisfies the stall constraint. Using equations (6.15) and (6.12), the stall constraint is satisfied if the low-speed solution satisfies

$$\sqrt{\frac{2W}{\rho S C_{L_{max}}}} \leq V. \qquad (6.22)$$

If this stall inequality is not satisfied, then there may be only one airspeed, the high-speed solution, for which steady gliding flight is possible at the given glide angle or at the given descent rate. Consequently, for steady gliding flight the stall constraint imposes a lower bound on the aircraft airspeed, given by the inequality (6.22).

For relatively large glide angles, it is important to check that the Mach number corresponding to a steady gliding flight condition is subsonic so that the assumed quadratic drag polar is valid.

For any set of flight variables corresponding to steady gliding flight, the pitch moment must be zero and the pitch moment coefficient in equation (6.8) must be zero, that is,

$$0 = C_{M_0} + C_{M_\alpha} \alpha + C_{M_{\delta_e}} \delta_e. \qquad (6.23)$$

If the value of the lift coefficient is known as from equation (6.15), then equations (6.23) and (6.10) can be used to determine the values of the angle of attack α and the elevator deflection δ_e, corresponding to steady gliding flight. This analysis requires solving two linear algebraic equations.

6.3 Minimum Glide Angle

An important measure of steady gliding flight performance is the minimum possible glide angle. The minimum glide angle can be determined using calculus by setting the derivative of the glide angle with respect to the airspeed to zero:

$$\frac{d\gamma_{glide}}{dV} = \frac{\rho V S C_{D_0}}{W} - \frac{4KW}{\rho V^3 S} = 0.$$

This equation can be solved to obtain the airspeed at which the glide angle is minimum:

$$V = \sqrt{\frac{2W}{\rho S}\sqrt{\frac{K}{C_{D_0}}}}. \tag{6.24}$$

The resulting minimum glide angle in steady gliding flight is given by

$$\left(\gamma_{glide}\right)_{min} = 2\sqrt{KC_{D_0}}. \tag{6.25}$$

The airspeed for a minimum glide angle is a function of the altitude. In particular, the required airspeed is inversely related to the square root of the air density, which depends on the altitude according to the standard atmospheric model. The minimum glide angle depends only on the aircraft aerodynamics parameters, not on the altitude.

　　It is important to check that the lift coefficient corresponding to the minimum glide angle does not exceed the maximum lift coefficient at stall which, in terms of the airspeed, is given by the inequality (6.22).

6.4 Minimum Descent Rate

A different measure of steady gliding flight performance is the minimum descent rate. The flight condition for minimum steady descent rate can be determined using calculus by setting the derivative of the descent rate with respect to the airspeed to zero:

$$\frac{dV_{descent}}{dV} = \frac{3}{2}\frac{\rho V^2 S C_{D_0}}{W} - \frac{2KW}{\rho V^2 S} = 0.$$

This equation can be solved to obtain the airspeed at which the descent rate is minimum:

$$V = \sqrt{\frac{2W}{\rho S}\sqrt{\frac{K}{3C_{D_0}}}}. \tag{6.26}$$

The resulting minimum descent rate in steady gliding flight is given by

$$(V_{descent})_{min} = \frac{4\sqrt{3}}{3}\sqrt{\frac{2W}{\rho S}K\sqrt{\frac{KC_{D_0}}{3}}}. \tag{6.27}$$

The minimum descent rate depends on the aircraft aerodynamics parameters and on the altitude. Note that the flight conditions for minimum glide angle and for minimum descent rate are different.

It is important to check that the lift coefficient corresponding to the minimum descent rate does not exceed the maximum lift coefficient at stall which, in terms of the airspeed, is given by the inequality (6.22).

6.5 Maximum Glide Angle

The maximum glide angle, or, equivalently, the minimum flight path angle, is now determined, assuming steady gliding longitudinal flight. According to equation (6.18), the glide angle in steady gliding flight is large at low airspeeds. The smallest steady airspeed is determined by the stall constraint; this provides a locally maximum glide angle. If the stall constraint is active then the airspeed is

$$ V = \sqrt{\frac{2W}{\rho S C_{L_{\max}}}}. \tag{6.28} $$

Equation (6.18) gives the expression for the local maximum glide angle as

$$ \left(\gamma_{glide} \right)_{max} = \frac{C_{D0}}{C_{L_{\max}}} + K C_{L_{\max}}. \tag{6.29} $$

This local maximum glide angle depends only on the aerodynamics of the aircraft; it does not depend on the flight altitude.

6.6 Maximum Descent Rate

The maximum descent rate, or, equivalently, the minimum climb rate, is now determined, assuming steady gliding longitudinal flight. According to equation (6.19), the descent rate in steady gliding flight is large at low airspeeds. The smallest possible airspeed occurs at stall; this provides a locally maximum descent rate. If the stall constraint is active then the airspeed is

$$ V = \sqrt{\frac{2W}{\rho S C_{L_{\max}}}}. \tag{6.30} $$

Equation (6.19) gives the expression for the local maximum descent rate as

$$ (V_{descent})_{max} = \sqrt{\frac{2W}{\rho S C_{L_{\max}}}} \left(\frac{C_{D0}}{C_{L_{\max}}} + K C_{L_{\max}} \right). \tag{6.31} $$

This local maximum descent rate depends on the aerodynamics of the aircraft and on the flight altitude.

6.7 Steady Gliding Longitudinal Flight Envelopes

The geometry of several steady gliding flight envelopes is described. A flight condition described by a triple consisting of an altitude, an airspeed, and a glide angle is said to be physically feasible, or to lie within the steady gliding flight envelope, if the physical equations and constraints for steady gliding flight are satisfied.

It is possible to provide a graphical representation of the steady gliding flight envelope using the analysis developed in this chapter, in particular the steady gliding flight equation (6.18) and the stall inequality (6.22). If the flight envelope is viewed as a set in three dimensions consisting of altitude, airspeed, and glide angle variables, then the flight envelope is a two-dimensional surface in the three variables that satisfies the stall inequality. One part of the surface corresponds to low-speed gliding flight; the other part of the surface corresponds to high-speed gliding flight. Only points in the steady gliding flight envelope that satisfy the stall constraint are feasible flight conditions.

It is possible to provide a simplified graphical representation of the steady gliding flight envelope in two dimensions. It is most convenient to represent the steady gliding flight envelope, for a fixed value of the glide angle, as a set in two dimensions defined by the altitude and airspeed. Each such set, for a fixed value of the glide angle, is referred to as a cross section of the steady gliding flight envelope. A cross section typically consists of two one-dimensional curves in the two dimensions defined by altitude and airspeed. One of the curves represents low-speed gliding flight; the other curve represents high-speed gliding flight.

6.8 Steady Gliding Longitudinal Flight: Executive Jet Aircraft

In this section, steady gliding longitudinal flight of the executive jet aircraft is analyzed. The thrust provided by the jet engine is assumed zero throughout this analysis. The aircraft weight is 73,000 lbs.

Steady Gliding Flight

Suppose that the jet aircraft has a constant glide angle of $\gamma_{glide} = 9$ degrees. The steady gliding flight conditions are analyzed for this glide angle.

The glide angle γ_{glide} is related to the lift coefficient C_L according to

$$\gamma_{glide} = \frac{D}{L} = \frac{C_D}{C_L} = \frac{C_{D_0} + K C_L^2}{C_L}.$$

Multiplying the above equation by C_L, the second degree polynomial equation is obtained:

$$K C_L^2 - \gamma_{glide} C_L + C_{D_0} = 0.$$

This quadratic can be factored to obtain two solutions, $C_L = 0.099$ and $C_L = 3.043$. The second solution exceeds the stall limit and hence is not feasible. The physically meaningful solution for the lift coefficient is $C_L = 0.099$. Since the lift equals the weight in steady gliding flight, the airspeed for this steady gliding flight condition depends on the altitude according to

$$V = \sqrt{\frac{2W}{\rho C_L S}}.$$

Since $C_M = 0$, the aircraft angle of attack α and the elevator deflection δ_e satisfy the equations

$$0.02 + 0.12\alpha = 0.099,$$

$$0.24 - 0.18\alpha + 0.28\delta_e = 0,$$

which has the solution of 0.655 degrees for the angle of attack and -0.436 degrees for the elevator deflection.

In summary, the conditions for steady gliding flight of the executive jet aircraft with a glide angle of 9 degrees requires the following pilot inputs: elevator deflection of -0.436 degrees, zero thrust, and zero bank angle. These results do not depend on the flight altitude. The airspeed for such steady gliding flight does depend on the altitude.

Minimum Glide Angle

The minimum glide angle for the executive jet aircraft occurs when the drag to lift ratio is a minimum. The minimum glide angle is

$$\left(\gamma_{glide}\right)_{min} = 2\sqrt{K C_{D_0}} = 0.055 \, \text{rad} = 3.14 \, \text{degrees}.$$

The corresponding airspeed depends on the altitude according to

$$V = \sqrt{\frac{2W}{\rho S}\sqrt{\frac{K}{C_{D_0}}}}.$$

At the minimum drag to lift ratio, the lift coefficient is $C_L = \sqrt{\frac{C_{D_0}}{K}} = 0.548$, and $C_M = 0$. The aircraft angle of attack and the elevator deflection satisfy

$$0.02 + 0.12\alpha = 0.548,$$

$$0.24 - 0.18\alpha + 0.28\delta_e = 0,$$

which has the solution of 4.4 degrees for the angle of attack and 1.97 degrees for the elevator deflection.

In summary, the minimum glide angle of the executive jet aircraft is 3.14 degrees, and it occurs with the following pilot inputs: elevator deflection of 1.97 degrees, zero throttle setting, and zero bank angle. These results do not depend on the flight altitude. The airspeed for gliding flight at the minimum glide angle does depend on the altitude.

Maximum Glide Angle

A local maximum glide angle for the executive jet aircraft occurs at stall when the airspeed is

$$V = \sqrt{\frac{2W}{\rho S C_{L_{\max}}}},$$

which depends on the flight altitude. The maximum glide angle is

$$\left(\gamma_{glide}\right)_{max} = \frac{C_{D0}}{C_{L_{\max}}} + K C_{L_{\max}} = 0.145 \,\text{rad} = 8.33 \,\text{degrees}.$$

Since $C_L = C_{L_{\max}} = 2.8$ and $C_M = 0$, the aircraft angle of attack and the elevator deflection satisfy

$$0.02 + 0.12\alpha = 2.8,$$
$$0.24 - 0.18\alpha + 0.28\delta_e = 0,$$

which has the solution $\alpha = 23.2$ degrees and $\delta_e = 14.06$ degrees.

In summary, the local maximum glide angle of the executive jet aircraft is 8.33 degrees, and it occurs with the following pilot inputs: elevator deflection of 14.06 degrees, zero throttle setting, and zero bank angle. These results do not depend on the flight altitude. The airspeed for gliding flight at the maximum glide angle does depend on the altitude.

Steady Gliding Longitudinal Flight Envelope

The steady gliding longitudinal flight envelope of the executive jet aircraft can now be computed. The steady gliding flight envelope is represented in three dimensions defined by the flight altitude, the airspeed, and the glide angle. Values of the altitude, the airspeed, and the glide angle that satisfy the steady gliding flight equation

$$\frac{1}{2} \frac{\rho V^4 S C_{D_0}}{W} - \gamma_{glide} V^2 + \frac{2KW}{\rho S} = 0$$

and the stall constraint

$$V \geq \sqrt{\frac{2W}{\rho S C_{Lmax}}}$$

lie in the steady gliding flight envelope. In geometric terms, the steady gliding flight envelope is a two-dimensional surface in three dimensions. Any values of flight altitude, airspeed, and glide angle that lie on this two-dimensional surface in three dimensions correspond to a steady gliding flight condition.

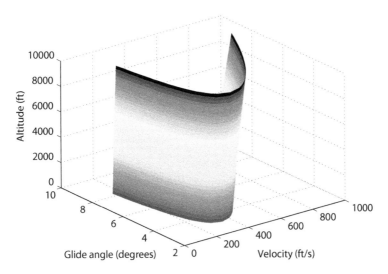

Figure 6.2. *Steady gliding longitudinal flight envelope: executive jet aircraft.*

Define the Matlab function SGFJ that computes the gliding airspeed and the stall speed for the executive jet aircraft gliding at a specified altitude and glide angle.

Matlab function 6.1 SGFJ.m
```
function [V Vstall]=SGFJ(h,gangle);
%Input: altitude h (ft), glide angle gangle (rad)
%Output: gliding airspeed V (ft/s), stall speed Vstall (ft/s)
W=73000; S=950; CD0=0.015; K=0.05; Tsmax=12500; m=0.6; CLmax=2.8;
[T p rho]=StdAtpUS(h);
tmp=sort(roots([1/2*rho*S*CD0/W 0 -gangle 0 2*K*W/rho/S]));
V=tmp(3:4); Vstall=sqrt(2*W/rho/S/CLmax);
```

The steady gliding flight envelope is developed by evaluating the SGFJ function for various altitudes and glide angles. Figure 6.2 shows this steady gliding flight envelope as a surface in three dimensions; it is generated by the following Matlab m-file.

Matlab function 6.2 FigSGFJ.m
```
CD0=0.015; K=0.05; ganglemin=2*sqrt(K*CD0);
gangle=linspace(ganglemin+1e-10,10*pi/180,40);
for i=1:size(gangle,2)
    h(i,:)=linspace(0,10000,40);
    for j=1:size(h,2)
        [V Vstall(i,j)]=SGFJ(h(i,j),gangle(i));
        Vlow(i,j)=V(1);
    Vhig(i,j)=V(2);
```

(continued)

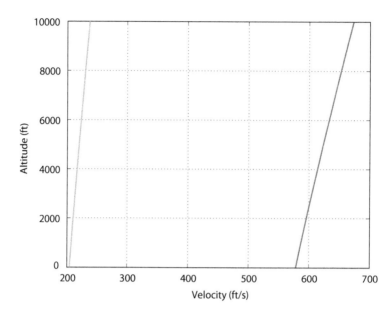

Figure 6.3. *Steady gliding longitudinal flight envelope: executive jet aircraft ($\gamma_{glide} = 5$ degrees).*

(continued)

```
      %Check stall limit
      if Vlow(i,j) < Vstall(i,j)
          h(i,j)=NaN;
      end
   end
end
surf(Vlow,gangle*180/pi,h,'LineStyle','none');hold on;
surf(Vhig,gangle*180/pi,h,'LineStyle','none');
ylabel('Glide angle (deg)');
zlabel('Altitude (ft)');
xlabel('Velocity (ft/s)');
```

A two-dimensional cross section of the steady gliding flight envelope is developed corresponding to the fixed glide angle of 5 degrees. This cross section is described in terms of the flight altitude and the airspeed for the fixed glide angle. A graphical representation of this two-dimensional cross section of the steady gliding flight envelope is obtained by evaluating the SGFJ function at various altitudes for the fixed glide angle of 5 degrees. This cross section, shown in Figure 6.3, is generated by the following Matlab m-file.

Matlab function 6.3 FigSGFEJ.m
```
gangle=5*pi/180; h=linspace(0,10000,100);
for
k=1:size(h,2)
    [V(:,k) Vstall(k)]=SGFJ(h(k),gangle);
```

(continued)

(continued)

```
end
plot(V(2,:),h,'r-',V(1,:),h,'b-');
grid on;
xlabel('Velocity (ft/s)');
ylabel('Altitude (ft)');
```

The cross section of the gliding flight envelope shown in Figure 6.3 consists of a high-speed curve and a low-speed curve. Both sets of solutions satisfy the stall constraint so that the cross section of the steady gliding flight envelope is the union of the points on the two curves. Different glide angles give different cross sections of the steady gliding flight envelope.

6.9 Steady Gliding Longitudinal Flight: General Aviation Aircraft

In this section, steady gliding longitudinal flight of the general aviation aircraft is analyzed. The aircraft weight is 2,900 lbs. The power provided by the engine is assumed zero throughout this analysis.

Steady Gliding Flight

Suppose that the aircraft has a constant glide angle of 12 degrees in longitudinal flight. The steady gliding flight conditions are analyzed for this glide angle.

The glide angle γ_{glide} is related to the lift coefficient C_L according to

$$\gamma_{glide} = \frac{D}{L} = \frac{C_D}{C_L} = \frac{C_{D_0} + KC_L^2}{C_L}.$$

Multiplying the above equation by C_L, the second degree polynomial equation is obtained,

$$KC_L^2 - \gamma_{glide}C_L + C_{D_0} = 0,$$

which has the two solutions $C_L = 0.128$ and $C_L = 3.75$. The second solution exceeds the stall limit and hence is not feasible. The physically meaningful solution for the lift coefficient is $C_L = 0.128$. Since the lift equals the weight in steady gliding flight, the airspeed for this steady gliding flight condition depends on the altitude according to

$$V = \sqrt{\frac{2W}{\rho C_L S}}.$$

Since $C_M = 0$, the aircraft angle of attack and the elevator deflection satisfy the equations

$$0.02 + 0.12\alpha = 0.128,$$
$$0.12 - 0.08\alpha + 0.075\delta_e = 0,$$

which has the solution for the angle of attack of 0.9 degrees and the elevator deflection of -0.64 degrees.

In summary, the conditions for steady gliding flight of the general aviation aircraft with a glide angle of 12 degrees requires the following pilot inputs: elevator deflection of -0.64 degrees, zero thrust, and zero bank angle. Note that these results do not depend on the flight altitude. The airspeed for such steady gliding flight does depend on the altitude.

Minimum Glide Angle

The minimum glide angle for the aircraft in longitudinal flight occurs when the drag to lift ratio is a minimum. The minimum glide angle is

$$\left(\gamma_{glide}\right)_{min} = 2\sqrt{K C_{D_0}} = 0.075 \text{ rad} = 4.29 \text{ degrees.}$$

The corresponding airspeed depends on the altitude according to

$$V = \sqrt{\frac{2W}{\rho S}}\sqrt{\frac{K}{C_{D_0}}}.$$

At the minimum drag to lift ratio, the lift coefficient is $C_L = \sqrt{\frac{C_{D_0}}{K}} = 0.694$, and $C_M = 0$. The aircraft angle of attack α and the elevator deflection δ_e satisfy

$$0.02 + 0.12\alpha = 0.694,$$

$$0.12 - 0.08\alpha + 0.075\delta_e = 0,$$

which has the solution for the angle of attack of 5.62 degrees and the elevator deflection of 4.39 degrees.

In summary, the minimum glide angle of the general aviation aircraft is 4.29 degrees, and it occurs with the following pilot inputs: elevator deflection of 4.39 degrees, zero throttle setting, and zero bank angle. These results do not depend on the flight altitude. The airspeed for gliding flight at the minimum glide angle does depend on the altitude; the airspeed increases as the flight altitude increases.

Maximum Glide Angle

A local maximum glide angle for the general aviation aircraft in longitudinal flight occurs at stall when the airspeed is

$$V = \sqrt{\frac{2W}{\rho S C_{L_{max}}}},$$

which depends on the flight altitude. The maximum glide angle is

$$\left(\gamma_{glide}\right)_{max} = \frac{C_{D0}}{C_{L_{max}}} + K C_{L_{max}} = 0.14 \text{ rad} = 8.05 \text{ degrees.}$$

At stall, $C_L = C_{L_{max}} = 2.4$ and $C_M = 0$, the aircraft angle of attack and the elevator deflection satisfy

$$0.02 + 0.12\alpha = 2.4,$$
$$0.12 - 0.08\alpha + 0.075\delta_e = 0,$$

which has the solution for the angle of attack of 19.8 degrees and the elevator deflection of 19.52 degrees.

In summary, the local maximum glide angle of the general aviation aircraft is 8.05 degrees, and it occurs with the following pilot inputs: elevator deflection of 19.52 degrees, zero throttle setting, and zero bank angle. These results do not depend on the flight altitude. The airspeed for gliding flight at the maximum glide angle does depend on the altitude.

Steady Gliding Longitudinal Flight Envelope

The steady gliding longitudinal flight envelope of the general aviation aircraft can now be computed. The steady gliding flight envelope is represented in three dimensions defined by the flight altitude, the airspeed, and the glide angle. Any set of values of the altitude, the airspeed, and the glide angle that satisfy the steady gliding flight equation

$$\frac{1}{2} \frac{\rho V^4 S C_{D_0}}{W} - \gamma_{glide} V^2 + \frac{2KW}{\rho S} = 0$$

and the stall constraint

$$V \geq \sqrt{\frac{2W}{\rho S C_{Lmax}}}$$

lies in the steady gliding flight envelope. In geometric terms, the steady gliding flight envelope is a two-dimensional surface in three dimensions. Any values of flight altitude, airspeed, and glide angle that lie on this surface correspond to a steady gliding flight condition.

Define the Matlab function SGFP that computes the gliding airspeed and the stall speed for a specified altitude and glide angle.

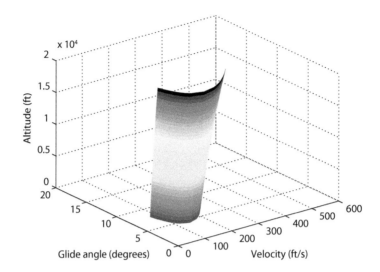

Figure 6.4. *Steady gliding longitudinal flight envelope: general aviation aircraft.*

Matlab function 6.4 SGFP.m
```
function [V Vstall]=SGFP(h,gangle);
%Input: altitude h (ft) glide angle gangle (rad)
%Output: gliding airspeed V (ft/s), stall speed Vstall (ft/s)
W=2900; S=175; CD0=0.026; K=0.054; CLmax=2.4;
[T p rho]=StdAtpUS(h);
tmp=sort(roots([1/2*rho*S*CD0/W 0 -gangle 0 2*K*W/rho/S]));
V=tmp(3:4); Vstall=sqrt(2*W/rho/S/CLmax);
```

The steady gliding flight envelope is developed by evaluating the SGFP function for various altitudes and glide angles. Figure 6.4 shows this steady gliding flight envelope as a surface in three dimensions; it is generated by the following Matlab m-file.

Matlab function 6.5 FigSGFP.m
```
CD0=0.026; K=0.054; ganglemin=2*sqrt(K*CD0);
gangle=linspace(ganglemin+1e-10,20*pi/180,40);
for
i=1:size(gangle,2)
    h(i,:)=linspace(0,20000,40);
    for j=1:size(h,2)
        [V Vstall(i,j)]=SGFP(h(i,j),gangle(i));
        Vlow(i,j)=V(1);
```

(continued)

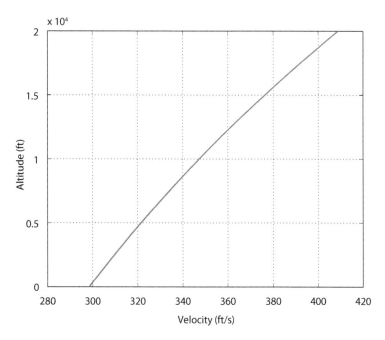

Figure 6.5. Steady gliding longitudinal flight envelope: general aviation aircraft (γ_{glide} = 10 degrees).

(continued)

```
        Vhig(i,j)=V(2);
        %Check stall limit
        if Vlow(i,j) < Vstall(i,j)
            h(i,j)=NaN;
        end
    end
end
surf(Vlow,gamma*180/pi,h,'LineStyle','none');hold on;
surf(Vhig,gamma*180/pi,h,'LineStyle','none');
ylabel('Glide angle (deg)');
zlabel('Altitude (ft)');
xlabel('Velocity (ft/s)');
```

A two-dimensional cross section of the steady gliding flight envelope is developed corresponding to the fixed glide angle of 10 degrees. This set is described in terms of flight altitude and the airspeed for the fixed glide angle. A graphical representation of this two-dimensional cross section of the steady gliding flight envelope is obtained by evaluating the SGFP function at various altitudes and the fixed glide angle of 10 degrees. This cross section of the steady gliding flight envelope, shown in Figure 6.5, is generated by the following Matlab m-file.

Matlab function 6.6 FigSGFEP.m
```
gangle=10*pi/180; h=linspace(0,20000,100);
for
k=1:size(h,2)
    [V(:,k) Vstall(k)]=SGFP(h(k),gangle);
end
plot(V(2,:),h,'r-',V(1,:),h,'b-');
grid on;
xlabel('Velocity (ft/s)');
ylabel('Altitude (ft)');
```

The points that lie on the curve in Figure 6.5 constitute the high-speed part of the steady gliding flight envelope corresponding to a glide angle of 10 degrees. There are no low-speed solutions at this glide angle since such solutions do not satisfy the stall constraint; consequently, they are not feasible. In this case the cross section of the steady gliding flight envelope is a single curve in two dimensions. Different glide angles give different cross sections of the steady gliding flight envelope.

6.10 Conclusions

This chapter has presented steady longitudinal flight results for an aircraft with zero thrust, that is, the aircraft is in steady gliding flight. These results are important in characterizing steady level flight for gliders and for aircraft in an emergency glide. As seen in the development, the steady gliding flight properties are strongly dependent on the aerodynamic properties of the aircraft, primarily the lift coefficient and the drag coefficient, and their relation through the drag polar.

This chapter also sets the stage for the subsequent analysis of steady powered flight, the subject of the remaining chapters.

6.11 Problems

6.1. Consider the executive jet aircraft, with a weight of 73,000 lbs, in steady gliding longitudinal flight with the engine off.
   (a) What is the minimum possible glide angle at an altitude of 20,000 ft. What is the airspeed in this flight condition? What is the descent rate? What are the angle of attack and the elevator deflection to maintain this steady gliding flight condition?
   (b) Estimate the maximum time required for the aircraft to glide from an altitude of 20,000 ft to an altitude of 15,000 ft.
   (c) Estimate the maximum gliding range for the aircraft to glide from an altitude of 20,000 ft to an altitude of 15,000 ft.
   (d) Plot the steady gliding flight envelope when the glide angle has its minimum possible value.
   (e) What is the maximum possible glide angle at an altitude of 20,000 ft? What is the airspeed in this flight condition? What is the descent rate?

What are the angle of attack and the elevator deflection to maintain this steady gliding flight condition?

(f) Estimate the minimum time required for the aircraft to glide from an altitude of 20,000 ft to an altitude of 15,000 ft.

(g) Estimate the minimum gliding range for the aircraft to glide from an altitude of 20,000 ft to an altitude of 15,000 ft.

(h) Plot the steady gliding flight envelope when the glide angle has its maximum possible value.

6.2. Consider the executive jet aircraft, with a weight of 73,000 lbs, in steady gliding longitudinal flight with the engine off.

(a) What is the airspeed for a gliding flight condition with glide angle of 11 degrees at an altitude of 20,000 ft? What is the descent rate? What are the angle of attack and the elevator deflection to maintain this steady gliding flight condition?

(b) Estimate the time required for the aircraft to glide from an altitude of 20,000 ft to an altitude of 15,000 ft assuming the glide angle is maintained at 11 degrees.

(c) Estimate the gliding range for the aircraft to glide from an altitude of 20,000 ft to an altitude of 15,000 ft assuming the glide angle is maintained at 11 degrees.

(d) Plot the steady gliding flight envelope when the glide angle is 11 degrees.

6.3. Consider the propeller-driven general aviation aircraft, with a weight of 2,900 lbs, in steady gliding longitudinal flight with the engine off.

(a) What is the minimum possible glide angle at an altitude of 4,000 ft? What is the airspeed in this flight condition? What is the descent rate? What are the angle of attack and the elevator deflection to maintain this steady gliding flight condition?

(b) Estimate the maximum time required for the aircraft to glide from an altitude of 4,000 ft to sea level.

(c) Estimate the maximum gliding range for the aircraft to glide from an altitude of 4,000 ft to sea level.

(d) Plot the steady gliding flight envelope when the glide angle has its minimum possible value.

(e) What is the maximum possible glide angle at an altitude of 4,000 ft? What is the airspeed in this flight condition? What is the descent rate? What are the angle of attack and the elevator deflection to maintain this steady gliding flight condition?

(f) Estimate the minimum time required for the aircraft to glide from an altitude of 4,000 ft to sea level.

(g) Estimate the minimum gliding range for the aircraft to glide from an altitude of 4,000 ft to sea level.

(h) Plot the steady gliding flight envelope when the glide angle has its maximum possible value.

6.4. Consider the propeller-driven general aviation aircraft, with a weight of 2,900 lbs, in steady gliding longitudinal flight with the engine off.

(a) What is the airspeed for a gliding flight condition with glide angle of 8 degrees at an altitude of 4,000 ft? What is the descent rate? What are the angle of attack and the elevator deflection to maintain this steady gliding flight condition?

(b) Estimate the time required for the aircraft to glide from an altitude of 4,000 ft to sea level assuming the glide angle is maintained at 8 degrees.

(c) Estimate the gliding range for the aircraft to glide from an altitude of 4,000 ft to sea level assuming the glide angle is maintained at 8 degrees.

(d) Plot the steady gliding flight envelope for a glide angle of 8 degrees.

6.5. Consider the UAV, with a weight of 45 lbs, in steady gliding longitudinal flight with the engine off.

(a) What is the minimum possible glide angle at an altitude of 2,000 ft? What is the airspeed in this flight condition? What is the descent rate? What are the angle of attack and the elevator deflection to maintain this steady gliding flight condition?

(b) Estimate the maximum time required for the aircraft to glide from an altitude of 2,000 ft to sea level.

(c) Estimate the maximum gliding range for the aircraft to glide from an altitude of 2,000 ft to sea level.

(d) Plot the steady gliding flight envelope when the glide angle has its minimum possible value.

(e) What is the maximum possible glide angle at an altitude of 2,000 ft? What is the airspeed in this flight condition? What is the descent rate? What are the angle of attack and the elevator deflection to maintain this steady gliding flight condition?

(f) Estimate the minimum time required for the aircraft to glide from an altitude of 2,000 ft to sea level.

(g) Estimate the minimum gliding range for the aircraft to glide from an altitude of 2,000 ft to sea level.

(h) Plot the steady gliding flight envelope when the glide angle has its maximum possible value.

6.6. Consider the UAV, with a weight of 45 lbs, in steady gliding longitudinal flight with the engine off.

(a) What is the airspeed for a gliding flight condition with a glide angle of 7 degrees at an altitude of 2,000 ft? What is the descent rate? What are the angle of attack and the elevator deflection to maintain this steady gliding flight condition?

(b) Estimate the time required for the aircraft to glide from an altitude of 2,000 ft to sea level assuming the glide angle is maintained at 7 degrees.

(c) Estimate the gliding range for the aircraft to glide from an altitude of 2,000 ft to sea level assuming the glide angle is maintained at 7 degrees.

(d) Plot the steady gliding flight envelope for a glide angle of 7 degrees.

7

Aircraft Cruise in Steady Level Longitudinal Flight

In this chapter, steady level longitudinal aircraft flight through a stationary atmosphere is studied assuming there is a propulsive force that acts on the aircraft. The equations for steady level longitudinal flight are derived and studied in detail. Several steady level flight performance measures are also studied.

7.1 Steady Level Longitudinal Flight

Steady level flight is the most common type of flight, sometimes referred to as cruising flight. It occurs when the aircraft velocity vector is a constant horizontal vector so that the path along which the aircraft moves is a straight line in a horizontal plane. Coordinated flight of the aircraft is assumed; that is, the side-slip angle is zero.

Mathematical descriptions for steady level flight are developed. These descriptions, expressed in terms of algebraic equations that describe the fundamental flight variables, arise from the free-body diagram for steady level flight. Equations that describe the aerodynamics and important flight constraints for both a jet aircraft and a propeller-driven general aviation aircraft are presented.

Free-Body Diagram

Consider the free-body diagram in Figure 7.1 showing the forces acting on an aircraft in steady level flight. The aircraft has a velocity vector that is horizontal; the magnitude of the aircraft velocity vector is the airspeed V. The lift force on the aircraft acts normal to the velocity vector, and the drag force on the aircraft acts opposite to the velocity vector. The thrust vector, with magnitude T, acts along the aircraft-fixed x_A-axis; that is, the angle between the thrust vector and the velocity vector is the aircraft angle of attack α.

The algebraic equations for steady level flight are now obtained. Since steady level flight implies zero acceleration of the aircraft, the vector sum of the forces on the aircraft is zero; that is, the sum of the forces along the velocity vector is zero, and the sum of the forces

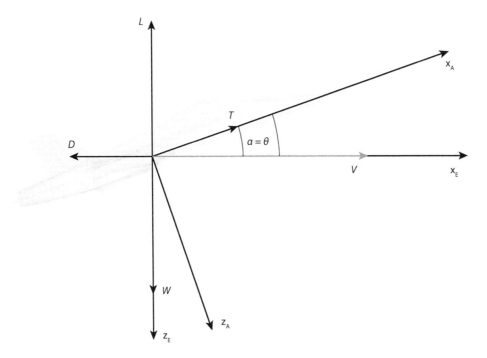

Figure 7.1. Free-body diagram of an aircraft in steady level longitudinal flight.

normal to the velocity vector is zero:

$$-D + T \cos \alpha = 0, \tag{7.1}$$

$$W - L - T \sin \alpha = 0. \tag{7.2}$$

The pitch moment is also zero:

$$M = 0. \tag{7.3}$$

The roll moment and the yaw moment are also zero in steady level flight. Since the side-slip angle is zero, the roll moment and the yaw moment are zero if there are no ailerons deflection and no rudder deflection.

Make the approximation that the thrust is much smaller than the lift and the angle of attack is a small angle in radian measure. The first approximation is justified for most conventional fixed wing aircraft; typical aircraft parameter data illustrate the validity of this approximation. The assumption that the angle of attack is small, when measured in radians, is consistent with the use of the linear aerodynamics approximations introduced in chapter 3. Thus $\cos \alpha$ is approximated by 1 and $T \sin \alpha$ is ignored as a higher-order term. These approximations imply that equations (7.1)–(7.3) can be written in the form

$$-D + T = 0, \tag{7.4}$$

$$W - L = 0, \tag{7.5}$$

and

$$M = 0. \tag{7.6}$$

These three equations provide the basis for studying steady level flight. They are conceptually simple equations for steady level flight:

- The lift must equal the weight of the aircraft.

- The thrust must equal the drag.

- The aerodynamic pitch moment must be zero.

It is important to keep in mind, however, that these equations are only approximations to the exact, but more complicated, steady level fight equations given in equations (7.1)–(7.3). Equations (7.5) and (7.6) are used in our subsequent analysis.

Aerodynamics

Recall the equations that describe the lift force, the drag force, and the pitch moment from chapter 3:

$$L = \frac{1}{2}\rho V^2 S C_L, \tag{7.7}$$

$$D = \frac{1}{2}\rho V^2 S C_D, \tag{7.8}$$

$$M = \frac{1}{2}\rho V^2 S c C_M. \tag{7.9}$$

The quadratic drag polar expression that relates the drag coefficient to the lift coefficient is given by

$$C_D = C_{D_0} + K\,C_L^2, \tag{7.10}$$

where $K = \frac{1}{\pi e AR}$.
    The lift coefficient is expressed as a linear function of the angle of attack as

$$C_L = C_{L_0} + C_{L_\alpha}\alpha. \tag{7.11}$$

The pitch moment coefficient is expressed as a linear function of the angle of attack and the elevator deflection as

$$C_M = C_{M_0} + C_{M_\alpha}\alpha + C_{M_{\delta_e}}\delta_e. \tag{7.12}$$

Flight Constraints

There are two types of important flight constraints that apply to steady level flight. These constraints play a central role in analyzing the performance properties of an aircraft in steady level flight.

The first constraint is that the aircraft must avoid stalling. As in chapter 5, the stall constraint can be expressed as an inequality on the lift coefficient given by

$$C_L \leq C_{L_{\max}}. \tag{7.13}$$

The second set of constraints arises from limits on the operation of the propulsion system. As discussed in chapter 4, a jet engine can provide a limited thrust for steady level flight, and an internal combustion engine driving a propeller can provide a limited power for steady level flight.

A jet engine can provide a maximum rated thrust, and the thrust required for steady level flight must not exceed this maximum thrust. The maximum rated thrust corresponds to its sea level value. The maximum thrust that the engine can provide decreases with altitude according to equation (4.3), where the air density ρ depends on the altitude according to the standard atmospheric model. In mathematical form, the thrust T required for steady level flight must satisfy

$$0 \leq T \leq T_{\max}^s \left(\frac{\rho}{\rho^s} \right)^m. \tag{7.14}$$

This condition is subsequently referred to as the propulsion constraints or the jet engine thrust constraints. The jet engine thrust constraints (7.14) can also be expressed as constraints on the throttle setting. The jet engine thrust constraint is equivalent to the throttle inequalities

$$0 \leq \delta_t \leq 1.0, \tag{7.15}$$

where

$$\delta_t = \frac{T}{T_{\max}^s} \left(\frac{\rho^s}{\rho} \right)^m. \tag{7.16}$$

An internal combustion engine can provide a maximum rated power; the power required for steady level flight must not exceed this maximum power. The maximum rated power is usually given as a sea level value. The maximum power that the engine can provide decreases with altitude according to expression (4.10), where the air density ρ depends on the altitude according to the standard atmospheric model. In mathematical form, the power P required for steady level flight must satisfy the inequalities

$$0 \leq P \leq \eta P_{\max}^s \left(\frac{\rho}{\rho^s} \right)^m, \tag{7.17}$$

where η is the efficiency of the propeller. This condition is subsequently referred to as the propulsion constraints or as the power constraints for a propeller-driven general aviation aircraft. The engine power constraints (7.17) can also be expressed as constraints on the throttle setting

$$0 \leq \delta_t \leq 1.0, \tag{7.18}$$

where

$$\delta_t = \frac{P}{\eta P_{\max}^s} \left(\frac{\rho^s}{\rho} \right)^m. \tag{7.19}$$

7.2 Steady Level Longitudinal Flight Analysis

In this section, flight conditions for steady level longitudinal flight are analyzed.

Equations for Steady Level Longitudinal Flight

Equations (7.4), (7.5), (7.7), (7.8), and (7.10) are now analyzed to ascertain the basic properties of an aircraft in steady level flight. The following perspective provides insight into these five algebraic equations. They involve the following six flight variables that define a steady level flight condition: the lift force L, the drag force D, the thrust T, the airspeed V, the lift coefficient C_L, and the drag coefficient C_D.

Our objective is to determine the relations between these six variables using the conditions for steady level flight. In mathematical terms, we seek to determine the relations between these six flight variables that satisfy the five algebraic equations (7.4), (7.5), (7.7), (7.8), and (7.10). These algebraic equations also involve important flight parameters: the aircraft weight W, the aircraft wing surface S, the aircraft aerodynamic parameters C_{D_0} and $K = \frac{1}{\pi e AR}$, and the air density ρ. Recall that the air density parameter depends on the flight altitude according to the standard atmospheric model. It should be emphasized that these algebraic equations are not linear algebraic equations. This implies that a priori statements cannot be made about existence or uniqueness of solutions of these equations; the nonlinear algebraic equations must be analyzed carefully.

It is important to make a distinction between the flight parameters and the flight variables. This perspective is consistent with the viewpoint that the flight parameters are given constants that characterize a given aircraft in flight at a given altitude, while the flight variables are constants that satisfy the five algebraic equations for steady level flight and hence define the specific steady level flight conditions.

Motivated by the fact that there are five equations and six variables, it is reasonable to select a fixed value for one of the flight variables and then to solve the five equations for the other five flight variables. In this way, it is possible to express five flight variables in terms of the prior selected flight variable. This is the procedure that is followed.

The previous equations that describe conditions for steady level flight are solved in two different forms. One expression provides the thrust required for steady level flight in terms of the airspeed of the aircraft. The other expression provides the power required for steady level flight in terms of the airspeed of the aircraft.

Thrust Required

Consider the conditions for steady level flight supposing that V is a given airspeed. Substituting the lift expression (7.7) into equation (7.5) it is easy to solve for the lift coefficient as

$$C_L = \frac{2W}{\rho V^2 S}. \tag{7.20}$$

The quadratic drag polar expression (7.10) gives the expression for the drag coefficient as

$$C_D = C_{D_0} + \frac{4KW^2}{\rho^2 V^4 S^2}.$$

(7.21)

Consequently, the drag for steady level flight at airspeed V is obtained from equation (7.8) as

$$D = \frac{1}{2}\rho V^2 S C_{D_0} + \frac{2KW^2}{\rho V^2 S}.$$

(7.22)

The thrust required to maintain steady level flight is obtained using equation (7.4) as

$$T = \frac{1}{2}\rho V^2 S C_{D_0} + \frac{2KW^2}{\rho V^2 S}.$$

(7.23)

This equation gives the thrust T required for steady level flight as it depends on the airspeed V and the flight parameters. Since steady level flight corresponds to flight at a constant altitude, it is clear that the required air density ρ in the thrust expression corresponds to that altitude according to the standard atmospheric model.

There are two terms in the expression (7.23) for the required thrust. The first term on the right side of equation (7.23) describes the part of the drag that is due to aerodynamic friction and is independent of the lift; this drag term increases as the airspeed increases. The second term on the right side of equation (7.23) is the drag that is induced by the lift; this drag term decreases as the airspeed increases. Thus the total thrust required to maintain steady level flight is the sum of both drag terms.

Equation (7.23) can be expressed in an alternative form. If the thrust T is assumed to be known, then equation (7.23) can be written as

$$\frac{1}{2}\rho V^4 S C_{D_0} - TV^2 + \frac{2KW^2}{\rho S} = 0,$$

(7.24)

which can be viewed as an equation whose solution is the airspeed for steady level flight at the given value of the thrust. Solving this equation requires finding the roots of a fourth degree polynomial; in fact, the fourth degree polynomial can be viewed as a quadratic polynomial in the square of the airspeed. In general, equation (7.24) has two real and positive solutions corresponding to a low-speed steady level flight solution and a high-speed steady level flight solution.

The low-speed solution must be checked to guarantee that it satisfies the stall constraint. Using equation (7.20) and inequality (7.13), the stall constraint can be expressed as an inequality in the aircraft airspeed given by

$$\sqrt{\frac{2W}{\rho S C_{L_{max}}}} \leq V.$$

(7.25)

The angle of attack and the elevator deflection for an aircraft in a steady level flight can be determined based on knowledge of the value of the lift coefficient computed from equation (7.20),

$$C_{L_0} + C_{L_\alpha}\alpha = \frac{2W}{\rho V^2 S}, \tag{7.26}$$

and knowledge that the pitch moment coefficient must be zero for steady level flight:

$$0 = C_{M_0} + C_{M_\alpha}\alpha + C_{M_{\delta_e}}\delta_e. \tag{7.27}$$

The stall constraint can also be checked using the computed value of the angle of attack. Equation (7.11) can be used to express the stall constraint as an inequality in the aircraft angle of attack given by

$$\alpha \le \frac{C_{L_{max}} - C_{L0}}{C_{L\alpha}}. \tag{7.28}$$

The throttle setting is easily determined from the above analysis. Since steady level flight is to be maintained, there should be no deflection of the ailerons and no deflection of the rudder.

The thrust required for steady level flight is equal to the drag in steady level flight. The thrust versus airspeed curve has the characteristic feature that it has a minimum value. This minimum value can be obtained using the methods of calculus: at the minimum, the thrust versus airspeed curve has zero slope; that is, the derivative of the thrust with respect to the airspeed is zero. In mathematical terms

$$\frac{dT}{dV} = \rho V S C_{D_0} - \frac{4KW^2}{\rho V^3 S} = 0. \tag{7.29}$$

The solution of this equation is given by

$$V = \sqrt{\frac{2W}{\rho S}\sqrt{\frac{K}{C_{D_0}}}}, \tag{7.30}$$

which is the airspeed for which the thrust, or, equivalently, the drag, required for steady level flight is a minimum. This airspeed value depends on the air density and hence on the flight altitude; this airspeed increases with an increase in the flight altitude.

The minimum value of the thrust, or, equivalently, the minimum drag, required for steady level flight is obtained by substituting the expression (7.30) into equation (7.23) to obtain

$$T_{\min} = 2W\sqrt{KC_{D_0}}. \tag{7.31}$$

The minimum value of the thrust, or, equivalently, the minimum value of the drag, required for steady level flight does not depend on the air density and hence does not depend on the flight altitude.

Power Required

The prior development characterizes the thrust required for steady level flight. It is also possible to characterize the power required for steady level flight using the basic relationship between thrust and power. In steady level flight, the power required is the product of the thrust required and the airspeed of the aircraft. Hence, equation (7.23) can be multiplied by the airspeed to obtain the expression for the power required for steady level flight:

$$P = \frac{1}{2}\rho V^3 SC_{D_0} + \frac{2KW^2}{\rho VS}. \tag{7.32}$$

There are two terms in expression (7.32) for the required power, leading to the following interpretation. The first term on the right of equation (7.32) describes the power required to overcome the drag that is due to aerodynamic friction and is independent of the lift; this term increases as the airspeed increases. The second term on the right of equation (7.32) is the power required to overcome the drag that is induced by the lift; this term decreases as the airspeed increases. Thus the total power required to maintain steady level flight is the sum of both terms.

Equation (7.32) can be expressed in an alternative form. If the power P is assumed known, then equation (7.32) can be written as

$$\frac{1}{2}\rho V^4 SC_{D_0} - PV + \frac{2KW^2}{\rho S} = 0, \tag{7.33}$$

which can be viewed as an equation whose solution is the airspeed for steady level flight at the given value of the power. Solving this equation requires finding the roots of a fourth degree polynomial. In general, equation (7.33) has two real and positive solutions corresponding to a low-speed steady level flight solution and a high-speed steady level flight solution. The low-speed solution should be checked to see that the stall constraint is satisfied. The stall constraint can be expressed as an inequality in the aircraft airspeed given by

$$\sqrt{\frac{2W}{\rho SC_{L_{\max}}}} \leq V. \tag{7.34}$$

The angle of attack and the elevator deflection for an aircraft in steady level flight can be determined based on knowledge of the value of the lift coefficient computed from equation (7.20) so that

$$C_{L_0} + C_{L_\alpha}\alpha = \frac{2W}{\rho V^2 S}, \tag{7.35}$$

and knowledge that the pitch moment coefficient must be zero for steady level flight:

$$0 = C_{M_0} + C_{M_\alpha}\alpha + C_{M_{\delta_e}}\delta_e. \tag{7.36}$$

The stall constraint can also be checked using the computed value of the angle of attack. Equation (7.11) can be used to express the stall constraint as an inequality in the aircraft angle of attack given by

$$\alpha \leq \frac{C_{L_{max}} - C_{L0}}{C_{L\alpha}}. \tag{7.37}$$

The throttle setting is easily determined from the above analysis. Since steady level flight is to be maintained, there should be no deflection of the ailerons and no deflection of the rudder.

The power required versus airspeed curve has the characteristic feature that it has a minimum value. This minimum value can be obtained using the methods of calculus: at the minimum the power versus airspeed curve has zero slope; that is, the derivative of the power with respect to the airspeed is zero. In mathematical terms,

$$\frac{dP}{dV} = \frac{3}{2}\rho V^2 SC_{D_0} - \frac{2KW^2}{\rho V^2 S} = 0. \tag{7.38}$$

The solution of this equation is given by

$$V = \sqrt{\frac{2W}{\rho S}\sqrt{\frac{K}{3C_{D_0}}}}, \tag{7.39}$$

which is the airspeed for which the power required for steady level flight is a minimum. The airspeed value at which the minimum power for steady level flight occurs depends on the air density and hence on the flight altitude; this airspeed increases with an increase in the flight altitude.

The minimum value of the power for steady level flight is obtained by substituting expression (7.39) into equation (7.32) to obtain

$$P_{min} = \frac{4}{3}\sqrt{\frac{2W^3}{\rho S}\sqrt{3K^3 C_{D_0}}}. \tag{7.40}$$

The minimum value of the power required for steady level flight depends on the air density and hence on the flight altitude; this power level increases with an increase in the flight altitude.

Equations (7.23) and (7.32) are simply different ways of expressing what the propulsion system needs to provide to maintain steady level flight at a given airspeed. Since jet engines are rated in terms of the maximum thrust that they provide, it is natural to use equation (7.23) in the analysis of steady flight conditions for aircraft with jet engines. Since internal combustion engines driving propellers are rated in terms of the maximum power that they provide, it is natural to use equation (7.32) in the analysis of steady flight conditions for aircraft with internal combustion engines and propellers.

7.3 Jet Aircraft Steady Level Longitudinal Flight Performance

In this section, steady level flight performance measures are studied, taking into account that the propulsion system is a jet engine. The following flight performance measures are analyzed: maximum flight speed, minimum flight speed, and flight ceiling. The steady level flight envelope is also introduced.

Maximum and Minimum Flight Speeds

Flight conditions for minimum and maximum airspeeds of a jet aircraft in steady level flight are analyzed. The maximum thrust that the jet engine can provide limits the minimum and the maximum possible airspeeds for which the aircraft can maintain steady level flight. In particular, the minimum airspeed and the maximum airspeed are characterized by the fact that the thrust required for steady level flight is equal to the maximum thrust that the engine can provide at full throttle; this leads to the equation

$$\frac{1}{2}\rho V^2 S C_{D_0} + \frac{2KW^2}{\rho V^2 S} = T_{\max}^s \left(\frac{\rho}{\rho^s}\right)^m . \tag{7.41}$$

This equation typically has two positive solutions. The high-speed solution denotes the maximum possible airspeed of the aircraft in steady level flight; the low-speed solution denotes the minimum possible airspeed of the aircraft in steady level flight. Equation (7.41) can be solved to obtain these two solutions. Steady level flight at an airspeed higher than this high-speed solution is not possible since the required thrust would exceed the maximum thrust that the jet engine can deliver. Steady level flight at an airspeed slower than this low-speed solution is not possible since the required thrust would exceed the maximum thrust that the jet engine can deliver.

Two important qualifications must be made about the maximum airspeed and the minimum airspeed obtained from this analysis. This computed minimum airspeed is the minimum airspeed if it satisfies the stall constraint given by the inequality (7.25); otherwise the stall speed is the minimum airspeed for steady level flight at the given thrust value. The computed maximum airspeed may be near to or exceed the speed of sound; if this case occurs, the computed maximum airspeed is inaccurate since supersonic aerodynamics factors, not taken into account in our analysis, become important.

Flight Ceiling

The above analysis is based on steady flight analysis at a fixed altitude. It is possible to assess the dependence of the steady level flight conditions on the altitude. As the flight altitude increases the maximum thrust provided by the jet engine decreases; this implies that the maximum airspeed due to the jet engine constraints tends to decrease and the minimum airspeed due to the jet engine constraints tends to increase. The airspeed at the stall limit also tends to increase. These facts suggest that there is a maximum altitude, referred to as the steady level flight ceiling, at which steady level flight can be maintained. The flight condition at the flight ceiling is characterized by the fact that the minimum thrust required for steady level flight, given in equation (7.31), is equal to the maximum thrust that the engine can produce at this altitude. This leads to the mathematical equation

$$2W\sqrt{KC_{D_0}} = T_{\max}^s \left(\frac{\rho}{\rho^s}\right)^m. \tag{7.42}$$

Since the air density depends on the altitude according to the standard atmospheric model, equation (7.42) can be viewed as an implicit equation in the altitude; the flight ceiling solves this equation.

Flight Envelope

The steady level flight envelope for a jet aircraft consists of all flight conditions for which steady level flight can be maintained.

It is possible to provide a graphical representation of the steady level flight envelope. It is most convenient to represent the steady level flight envelope as a set in two dimensions defined by the altitude and airspeed variables. In particular, specific values of flight altitude and airspeed lie in the steady level flight envelope if both the stall constraint and the engine thrust constraints are satisfied at this flight condition. Since the prior analysis provides expressions for the thrust required for steady level flight and the lift coefficient for steady level flight, these constraints can easily be checked.

The flight ceiling is the maximum altitude for all flight conditions that lie within the steady level flight envelope. The steady level flight envelope provides a convenient graphical summary of all possible steady level flight conditions.

7.4 General Aviation Aircraft Steady Level Longitudinal Flight Performance

In this section, steady level flight is analyzed, taking into account that the propulsion system consists of a propeller driven by an internal combustion engine. The following flight performance measures are analyzed: maximum flight speed, minimum flight speed, and flight ceiling. The steady level flight envelope is also introduced.

Maximum and Minimum Flight Speeds

Flight conditions for maximum and minimum airspeeds of a propeller-driven general aviation aircraft in steady level flight are analyzed. The maximum power that the propeller and internal combustion engine can provide limits the minimum and maximum possible airspeeds for which the aircraft can maintain steady level flight. In particular, the maximum and minimum airspeeds are characterized by the fact that the power required for steady level flight is equal to the maximum power that the engine can provide at full throttle. In mathematical terms, this corresponds to the satisfaction of the equation

$$\frac{1}{2}\rho V^3 S C_{D_0} + \frac{2KW^2}{\rho V S} = \eta P_{\text{max}}^s \left(\frac{\rho}{\rho^s}\right)^m. \tag{7.43}$$

This equation typically has two positive solutions. The high-speed solution denotes the maximum possible airspeed of the aircraft in steady level flight; the low-speed solution denotes the minimum possible airspeed of the aircraft in steady level flight. Equation (7.43) can be solved to obtain these two solutions. Steady level flight at an airspeed higher than this high-speed solution is not possible since the required power would exceed the maximum power that the internal combustion engine can deliver. Steady level flight at an airspeed slower than this low speed solution is not possible since the required power would also exceed the maximum power that the internal combustion engine can deliver.

Two important qualifications must be made about the maximum airspeed and the minimum airspeed obtained from this analysis. This computed minimum airspeed is the minimum airspeed if it satisfies the stall constraint given by the inequality (8.40); otherwise the stall speed is the minimum airspeed for steady level flight. The computed maximum airspeed may be near to or exceed the speed of sound; if this case occurs, the computed maximum airspeed is inaccurate since supersonic aerodynamics factors, not taken into account in our analysis, become important.

Flight Ceiling

The above analysis is based on steady flight analysis at a fixed altitude. It is possible to assess the dependence of the steady level flight conditions on the altitude. Note that as the altitude increases the maximum power provided by the internal combustion engine decreases; this implies that the maximum airspeed due to the internal combustion engine constraints tends to decrease and the minimum airspeed due to the internal combustion engine constraints tends to increase. The airspeed at the stall limit also tends to increase. These facts suggest that there is a maximum altitude, referred to as the flight ceiling, at which steady level flight can be maintained. The flight condition at the flight ceiling is characterized by the fact that the minimum power required for steady level flight, given in equation (7.40), is exactly equal to the maximum power that the engine can produce at this altitude. This leads to the mathematical equation

$$\frac{4}{3}\sqrt{\frac{2W^3}{\rho S}}\sqrt{3K^3 C_{D_0}} = \eta P_{\text{max}}^s \left(\frac{\rho}{\rho^s}\right)^m. \tag{7.44}$$

Since the air density depends on the altitude according to the standard atmospheric model, equation (7.44) can be viewed as an implicit equation in the altitude; the flight ceiling satisfies this equation.

Flight Envelope

The steady level flight envelope for a propeller-driven general aviation aircraft consists of all flight conditions for which steady level flight can be maintained.

It is possible to provide a graphical representation of the steady level flight envelope. It is most convenient to represent the steady level flight envelope as a set in two dimensions defined by the altitude and airspeed variables. In particular, specific values of flight altitude and airspeed lie in the steady level flight envelope if both the stall constraint and the engine power constraints are satisfied at this flight condition. Since the prior analysis provides expressions for the power required for steady level flight and the lift coefficient for steady level flight, these constraints can easily be checked.

The flight ceiling is the maximum altitude for all flight conditions that lie within the steady level flight envelope. The steady level flight envelope provides a convenient graphical summary of all possible steady level flight conditions.

7.5 Steady Level Longitudinal Flight: Executive Jet Aircraft

Suppose the executive jet aircraft is in steady level longitudinal flight at an altitude of 10,000 ft. The aircraft weight is 73,000 lbs. From the standard atmospheric table, the air density at this altitude is $\rho = 1.7553 \times 10^{-3}$ slug/ft^3.

Thrust Required

First, suppose that the airspeed $V = 500$ ft/s. Then the thrust required for steady level flight at this altitude is given by

$$T = \frac{1}{2}\rho V^2 S C_{D_0} + \frac{2KW^2}{\rho V^2 S} = 4,404.9 \text{ lbs.}$$

The lift coefficient and the pitch moment coefficient are

$$C_L = \frac{2W}{\rho V^2 S} = 0.35,$$
$$C_M = 0.$$

These expressions allow determination of the aircraft angle of attack and the elevator deflection according to

$$0.02 + 0.12\alpha = 0.35,$$
$$0.24 - 0.18\alpha + 0.28\delta_e = 0,$$

to give $\alpha = 2.75$ degrees and $\delta_e = 0.91$ degrees. The corresponding throttle setting is

$$\delta_t = \frac{T}{T_{\max}^s}\left(\frac{\rho^s}{\rho}\right)^m = 0.42.$$

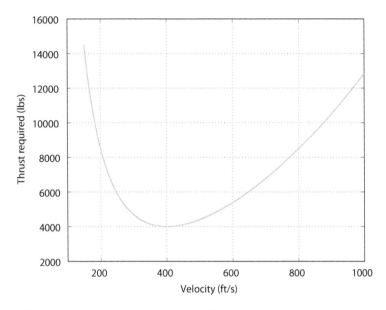

Figure 7.2. Thrust required for steady level longitudinal flight: executive jet aircraft.

In summary, if the pilot inputs are elevator deflection of 0.91 degrees and throttle setting of 0.42, with zero bank angle, then the aircraft is in steady level flight at an altitude of 10,000 ft with an airspeed of 500 ft/s.

Next, the required thrust is determined for various airspeeds. The thrust required curve, shown in Figure 7.2, is generated by the following Matlab m-file.

Matlab function 7.1 TRSLFJ.m
```
V=linspace (150, 1000, 500);
W=73000; S=950; CD0=0.015; K=0.05; rho=1.7553e-3;
T=1/2*rho*V.^2*S*CD0+2*K*W^2/rho./V.^2/S
plot (V,T);
xlabel ('Velocity (ft/s)');
ylabel ('Thrust required (lb)');
grid on;
```

This thrust required curve for steady level flight exhibits the standard features; high thrust is required for steady level flight at low airspeeds and high thrust is required at high airspeeds.

Minimum Required Thrust

As shown graphically in Figure 7.2, there is a minimum required thrust for steady level flight at an altitude of 10,000 ft. The airspeed to achieve minimum required thrust for steady level flight is given by:

$$V = \sqrt{\frac{2W}{\rho S} \sqrt{\frac{K}{C_{D_0}}}} = 399.81 \text{ ft/s}.$$

The lift coefficient and pitch moment coefficient at this airspeed are

$$C_L = \frac{2W}{\rho V^2 S} = 0.548,$$
$$C_M = 0.$$

These expressions allow determination of the aircraft angle of attack and the elevator deflection according to

$$0.02 + 0.12\alpha = 0.548,$$
$$0.24 - 0.18\alpha + 0.28\delta_e = 0.$$

The results are $\alpha = 4.4$ degrees and $\delta_e = 1.97$ degrees.

The minimum value of the thrust required for steady level flight at this altitude is given by

$$T_{\min} = 2W\sqrt{KC_{D_0}} = 3{,}998.4 \text{ lbs}.$$

The corresponding throttle setting is

$$\delta_t = \frac{T}{T_{\max}^s}\left(\frac{\rho^s}{\rho}\right)^m = 0.384.$$

In summary, if the pilot inputs are elevator deflection of 1.97 degrees and throttle setting of 0.384, with zero bank angle, then the aircraft is in steady level flight at an altitude 10,000 ft with an airspeed of 399.81 ft/s; this flight condition uses the smallest possible thrust of 3,998.4 lbs to maintain steady level flight.

Maximum Aircraft Airspeed

The maximum airspeed of the jet aircraft in steady level flight at an altitude of 10,000 ft occurs at full throttle. This maximum airspeed can be obtained graphically from the intersection of the thrust required curve for steady level flight and the maximum engine thrust curve. The intersection of these two curves is shown in Figure 7.3.

The maximum air speed can also be obtained by solving the fourth degree polynomial equation

$$\frac{1}{2}\rho S C_{D_0} V^4 - T_{\max}^s\left(\frac{\rho}{\rho^s}\right)^m V^2 + \frac{2KW^2}{\rho S} = 0.$$

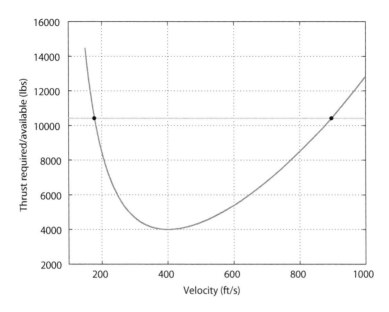

Figure 7.3. Maximum airspeed for steady level longitudinal flight: executive jet aircraft.

This equation can be solved numerically using the Matlab function MMSSLFJ as follows:

Matlab function 7.2 MMSSLFJ.m
```
function [Vmin,Vmax]=MMSSLFJ
%Output: Minimum air speed and maximum air speed
W=73000; S=950; CD0=0.015; K=0.05; Tsmax=12500; m=0.6;
[Ts ps rhos]=StdAtpUS (0);
[Th ph rhoh]=StdAtpUS (h);
[Vmin,Vmax]=roots ([1/2*rho*S*CD0 0-Tsmax* (rho/rhos)^m 0
2*K*W^2/rho/S]);
```

Matlab computes two positive solutions with values 895.19 ft/s and 178.57 ft/s. The high-speed solution is the maximum airspeed of the aircraft. At the given altitude, the speed of sound is 1,077.39 ft/s; the corresponding Mach number at this steady level flight condition is 0.83. At the maximum airspeed, the lift coefficient and the pitch moment coefficient are

$$C_L = \frac{2W}{\rho V^2 S} = 0.109,$$

$$C_M = 0.$$

This allows determination of the aircraft angle of attack and the elevator deflection as $\alpha = 0.74$ degrees and $\delta_e = -0.38$ degrees.

In summary, the maximum airspeed for steady level flight of the executive jet aircraft at an altitude of 10,000 ft is 895.19 ft/s and it occurs with pilot inputs given by elevator deflection of -0.38 degrees, full throttle, and zero bank angle.

Minimum Airspeed

The minimum airspeed of the executive jet aircraft at an altitude of 10,000 ft must satisfy both the stall constraint and the jet engine thrust constraints. At stall, the maximum value of the lift coefficient is $C_{L_{\max}} = 2.8$. The airspeed at stall is

$$V_{\text{stall}} = \sqrt{\frac{2W}{\rho S C_{L_{\max}}}} = 176.83 \text{ ft/s.}$$

The minimum airspeed due to the engine thrust constraint was calculated as 178.57 ft/s. In this case, the minimum airspeed due to the thrust constraint is larger than the stall speed; consequently, the maximum thrust constraint must be active. Hence, the minimum airspeed is equal to the minimum airspeed due to the thrust constraint, namely 178.57 ft/s. At this minimum airspeed, the lift coefficient and the pitch moment coefficient are

$$C_L = \frac{2W}{\rho V^2 S} = 2.746,$$

$$C_M = 0.$$

This allows determination of the aircraft angle of attack and the elevator deflection; the results are $\alpha = 22.72$ degrees and $\delta_e = 13.75$ degrees.

In summary, the minimum airspeed for steady level flight of the jet aircraft at an altitude of 10,000 ft is 178.57 ft/s and it occurs with pilot inputs given by elevator deflection of 13.75 degrees, full throttle, and zero bank angle.

7.6 Steady Level Longitudinal Flight Envelopes: Executive Jet Aircraft

We characterize the set of all possible steady level longitudinal flight conditions for the executive jet aircraft. The steady level longitudinal flight envelope is developed for the executive jet aircraft.

Steady Level Longitudinal Flight Ceiling

The flight ceiling for steady level flight can be determined by solving the equation

$$2W\sqrt{KC_{D_0}} - T^s_{\max}\left(\frac{\rho}{\rho^s}\right)^m = 0,$$

where ρ is a function of altitude according to the standard atmospheric model. The Matlab function `fsolve` is employed to solve the nonlinear equation defined by the Matlab function, eqnFCLJ as follows:

Matlab function 7.3 eqnFCLJ.m
```
function error=eqnFCLJ(h)
%Input: altitude (ft)
%Output: Error
W=73000; CD0=0.015; K=0.05; Tsmax=12500; m=0.6;
[Ts ps rhos]=StdAtpUS (0);
[Th ph rhoh]=StdAtpUS (h);
error=2*W*sqrt (K*CD0)-Tsmax*(rhoh/rhos)^m;
```

The following Matlab command computes the steady level flight ceiling where the value of eqnFCLJ is zero.

```
hmax=fsolve (@eqnFCLJ,10000)
```

where 10,000 is an initial guess for the value of the flight ceiling. After a few iterations, Matlab provides the solution of this equation, namely the value of the flight ceiling is 50,361.36 ft.

Steady Level Longitudinal Flight Envelope

Any pair of values for the altitude and aircraft airspeed that lies within the steady level longitudinal flight envelope corresponds to a feasible steady level flight condition for which the engine constraints and the stall constraint are satisfied.

A graphical representation of the steady level flight envelope is developed using the previous analysis. By following the same procedure described above, the maximum airspeed and the minimum airspeed due to the thrust constraints and the stall speed can be determined for various altitudes. Define a Matlab function SLFJ that computes the maximum airspeed and the minimum airspeed due to the thrust constraints, and the stall speed for a specified altitude.

Matlab function 7.4 SLFJ.m
```
function [Vmax VminTC Vstall]=SLFJ (h);
%Input: altitude h (ft)
%Output: max air speed Vmax (ft/s), min air speed due to thrust
constraint VminTC (ft/s),
stall speed Vstall (ft/s)
W=73000; S=950; CD0=0.015; K=0.05; Tsmax=12500; m=0.6; CLmax=2.8;
[Ts ps rhos]=StdAtpUS (0);
[T p rho]=StdAtpUS (h);
tmp=sort (roots ([1/2*rho*S*CD0 0 $-$Tsmax* (rho/rhos)^m 0
2*K*W^2/rho/S]));
Vmax=tmp (4);
VminTC=tmp (3);
Vstall=sqrt (2*W/rho/S/CLmax);
```

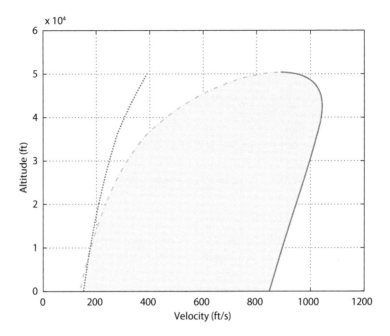

Figure 7.4. *Steady level longitudinal flight envelope: executive jet aircraft.*

The steady level flight envelope is generated by using the SLFJ function to compute the maximum and minimum airspeeds of the aircraft for various altitudes. The steady level flight envelope, shown in Figure 7.4, is generated by the following Matlab m-file.

Matlab function 7.5 FigSLFJ.m
```
h=linspace (0,50361,500);
for k=1:size (h,2)
  [Vmax (k) VminTC (k) Vstall (k)]=SLFJ (h(k));
end
Vmin=max (VminTC,Vstall);
area ([Vmin Vmax (end:-1:1)],[h h (end:-1:1)],...
  'FaceColor',[0.8 1 1],'LineStyle','none');
hold on;
plot (Vmax,h,VminTC,h,Vstall,h);
grid on;
xlim ([0 1200]);
xlabel ('Velocity (ft/s)');
ylabel ('Altitude (ft)');
```

For each such feasible pair of altitude and airspeed values that lie within the steady level flight envelope of the executive jet aircraft, methodology has been provided to compute the pilot inputs, namely the elevator deflection, the throttle setting, and zero bank angle, that give this steady level flight condition. In addition, the methodology also allows computation of the associated angle of attack, values of the lift and drag coefficients, and values of the lift and drag at this steady level flight condition.

7.7 Steady Level Longitudinal Flight: General Aviation Aircraft

Suppose the single engine propeller-driven general aviation aircraft is in steady level longitudinal flight at an altitude of 10,000 ft. The aircraft weight is 2,900 lbs. From the standard atmospheric table, the air density is 1.7553×10^{-3} slug/ft^3.

Power Required

Suppose that the airspeed $V = 200$ ft/s. Then the power required for steady level flight at this altitude is

$$P = \frac{1}{2}\rho V^3 S C_{D_0} + \frac{2K W^2}{\rho V S} = 46{,}731 \text{ ft lbs/s} = 84.97 \text{ hp.}$$

The lift coefficient and the pitch moment coefficient are

$$C_L = \frac{2W}{\rho V^2 S} = 0.472,$$

$$C_M = 0.$$

These expressions allow determination of the aircraft angle of attack and the elevator deflection according to

$$0.02 + 0.12\alpha = 0.472,$$

$$0.12 - 0.08\alpha + 0.075\delta_e = 0,$$

to give the angle of attack of 3.77 degrees and the elevator deflection of 2.42 degrees. The corresponding throttle setting is

$$\delta_t = \frac{P}{P^s_{\max}} \left(\frac{\rho^s}{\rho}\right)^m = 0.439.$$

In summary, if the pilot inputs are elevator deflection of 2.42 degrees, throttle setting of 0.439, and zero bank angle, then the aircraft is in steady level flight at an altitude of 10,000 ft, with an airspeed of 200 ft/s.

Next, the required power is determined for various airspeeds. The power required curve, shown in Figure 7.5, is generated by the following Matlab m-file.

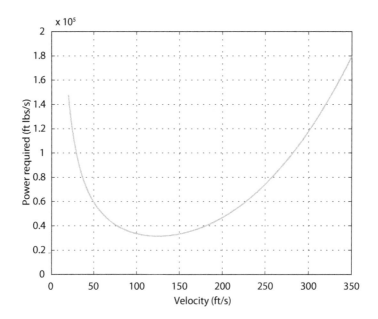

Figure 7.5. Power required for steady level longitudinal flight: general aviation aircraft.

Matlab function 7.6 PRSLFP.m
```
V=linspace (20,350,500);
W=2900; S=175; CD0=0.026; K=0.054; rho=1.7553e-3;
P=1/2*rho*V.^3*S*CD0+2*K*W^2/rho./V/S
plot (V,P);
xlabel ('Velocity (ft/s)');
ylabel ('Power required (ft lb/s)');
grid on;
```

This power required curve for steady level flight exhibits the standard features; high power is required for steady level flight at low airspeeds and high airspeeds.

Minimum Required Power

As shown graphically in Figure 7.5, there is a minimum required power for steady level flight at an altitude of 10,000 ft. The airspeed to achieve minimum required power is

$$V = \sqrt{\frac{2W}{\rho S} \sqrt{\frac{K}{3C_{D_0}}}} = 125.34 \, \text{ft/s}.$$

The lift coefficient and pitch moment coefficient at this air speed are

$$C_L = \frac{2W}{\rho V^2 S} = 1.202,$$

$$C_M = 0.$$

These expressions allow determination of the aircraft angle of attack α and the elevator deflection δ_e according to

$$0.02 + 0.12\alpha = 1.202,$$

$$0.12 - 0.08\alpha + 0.075\delta_e = 0.$$

The results are an angle of attack of 9.85 degrees and an elevator deflection of 8.91 degrees. The minimum value of the power required for steady level flight at this altitude is

$$P_{\min} = \frac{4}{3}\sqrt{\frac{2W^3}{\rho S}}\sqrt{3K^3 C_{D_0}} = 31{,}454 \,\text{ft lbs/s} = 57.19\,\text{hp}.$$

The corresponding throttle setting is

$$\delta_t = \frac{P}{\eta P_{\max}^s}\left(\frac{\rho^s}{\rho}\right)^m = 0.296.$$

    In summary, if the pilot inputs are elevator deflection of 8.91 degrees, throttle setting of 0.296, and zero bank angle, then the aircraft is in steady level flight at an altitude of 10,000 ft with an airspeed of 125.34 ft/s; this flight condition uses the smallest possible power, namely 57.19 hp, to maintain steady level flight.

Maximum Airspeed

The maximum airspeed of the general aviation aircraft in steady level flight at an altitude of 10,000 ft occurs at full throttle. This maximum airspeed can be obtained graphically from the intersection of the power required curve for steady level flight and the curve defined by the maximum power that the internal combustion engine and propeller can deliver at this altitude. The intersection of these two curves is shown in Figure 7.6.

    The maximum airspeed can also be obtained by solving the following fourth degree polynomial equation:

$$\frac{1}{2}\rho S C_{D_0} V^4 - \eta P_{\max}^s \left(\frac{\rho}{\rho^s}\right)^m V + \frac{2K W^2}{\rho S} = 0.$$

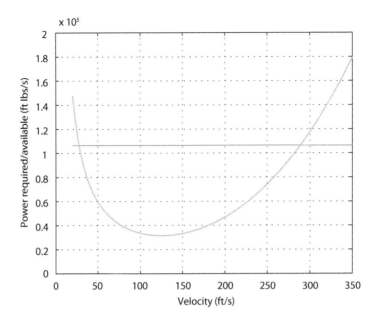

Figure 7.6. Maximum airspeed for steady level longitudinal flight: general aviation aircraft.

This equation can be solved numerically using the Matlab function MMSSLFP as follows:

Matlab function 7.7 MMSSLFP.m
```
function [Vmin,Vmax]=MMSSLFP
%Output: Minimum air speed and maximum air speed
W=2900; S=175; CD0=0.026; K=0.054; Psmax=290*550; m=0.6;
[Ts ps rhos]=StdAtpUS (0);
[Th ph rhoh]=StdAtpUS (h);
[Vmin,Vmax]=sort (roots ([1/2*rho*S*CD0 0 0 -eta*Psmax*
(rho/rhos)^m 2*K*W^2/rho/S]));
```

Matlab computes two positive solutions with values of 288.75 ft/s and 27.82 ft/s. The high-speed solution is the maximum airspeed of the aircraft. At the given altitude, the speed of sound is 1,077.39 ft/s; the corresponding Mach number at this steady level flight condition is 0.27. At the maximum airspeed, the lift coefficient and the pitch moment coefficient are

$$C_L = \frac{2W}{\rho V^2 S} = 0.265,$$

$$C_M = 0.$$

This allows determination of an aircraft angle of attack of 1.72 degrees and an elevator deflection of 0.24 degrees.

In summary, the maximum airspeed for steady level flight of the general aviation aircraft at an altitude of 10,000 ft occurs with pilot inputs given by an elevator deflection of 0.24 degrees, full throttle, and with zero bank angle. The maximum airspeed is 288.75 ft/s.

Minimum Airspeed

The minimum airspeed of the general aviation aircraft must satisfy both the stall constraint and the engine power constraints. At stall, the maximum value of the lift coefficient is $C_{L_{max}} = 2.4$. The airspeed at stall is

$$V_{stall} = \sqrt{\frac{2W}{\rho S C_{L_{max}}}} = 88.7 \text{ ft/s.}$$

The minimum airspeed due to the engine power constraints was calculated as 27.82 ft/s. In this case, the stall speed is larger than the minimum airspeed due to the power constraints; consequently the stall constraint is active. Hence, the minimum airspeed is equal to the stall speed, namely 88.7 ft/s. At the minimum airspeed, the lift coefficient and the pitch moment coefficient are

$$C_L = 2.4,$$
$$C_M = 0.$$

This allows determination of an aircraft angle of attack of 19.83 degrees and an elevator deflection of 19.55 degrees.

Thus the minimum airspeed for steady level flight of the general aviation aircraft at an altitude of 10,000 ft is 88.7 ft/s and occurs with pilot inputs given by an elevator deflection of 19.55 degrees, full throttle, and with zero bank angle.

7.8 Steady Level Longitudinal Flight Envelopes: General Aviation Aircraft

We characterize the set of all possible steady level longitudinal flight conditions for the general aviation aircraft. The steady level longitudinal flight envelope is developed for the general aviation aircraft.

Steady Level Flight Ceiling

The flight ceiling for steady level flight can be determined by solving the equation

$$\frac{4}{3}\sqrt{\frac{2W^3}{\rho S}}\sqrt{3K^3 C_{D_0}} = \eta P_{max}^s \left(\frac{\rho}{\rho^s}\right)^m,$$

where ρ is a function of altitude according to the standard atmospheric model. Define the Matlab function eqnFCLP as follows.

Matlab function 7.8 eqnFCLP.m
```
function error=eqnFCLP (h)
%Input: altitude (ft)
%Output: error
W=2900; S=175; CD0=0.026; K=0.054; Psmax=290*550; m=0.6; eta=0.8;
[Ts ps rhos]=StdAtpUS (0);
[Th ph rhoh]=StdAtpUS (h);
error=4/3*sqrt (2*W^3/rhoh/S*sqrt (3*K^3*CD0))-eta*Psmax*
(rhoh/rhos)^m;
```

The following Matlab command computes the altitude, the steady level flight ceiling, such that the value of eqnFCLP is zero.

```
    hmax=fsolve(@eqnFCLP,10000)
```

where 10,000 is an initial guess of h_{max}. After a few iterations, Matlab provides the solution of this equation, namely the value of the steady level flight ceiling is 40,190.05 ft.

Steady Level Longitudinal Flight Envelope

Any pair of values for the altitude and aircraft airspeed that lies within the steady level longitudinal flight envelope corresponds to a steady level flight condition for which the engine power constraints and the stall constraint are satisfied.

A graphical representation of the steady level flight envelope is now developed using the previous analysis. The maximum airspeed and the minimum airspeed due to the power constraint and the stall speed can be determined for various altitudes. Define a Matlab function SLFP that computes the maximum airspeed and the minimum airspeed due to the power constraints, and the stall speed for a specified altitude.

Matlab function 7.9 SLFP.m
```
function [Vmax VminPC Vstall]=SLFP(h);
%Input: altitude h (ft)
%Output: max air speed Vmax (ft/s), min air speed due to thrust
constraint VminPC (ft/s),
% stall speed Vstall (ft/s)
W=2900; S=175; CD0=0.026; K=0.054; Psmax=290*550; m=0.6;
CLmax=2.4; eta=0.8;
[Ts ps rhos]=StdAtpUS (0);
[T p rho]=StdAtpUS (h);
tmp=sort (roots ([1/2*rho*S*CD0 0 0 -eta*Psmax* (rho/rhos)^m
2*K*W^2/rho/S]));
Vmax=tmp (2);
VminPC=tmp (1);
Vstall=sqrt (2*W/rho/S/CLmax);
```

The steady level flight envelope is generated by using the SLFP function to compute the maximum and minimum airspeeds of the aircraft for various altitudes. The steady level flight envelope, shown in Figure 7.7, is generated by the following Matlab m-file.

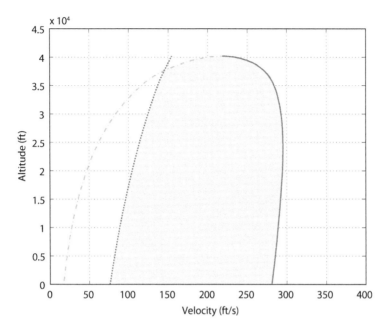

Figure 7.7. *Steady level longitudinal flight envelope: general aviation aircraft.*

Matlab function 7.10 FigSLFP.m
```
h=linspace (0,40190,500);
for k=1:size(h,2)
 [Vmax(k) VminTC(k) Vstall(k)]=SLFP(h(k));
end
Vmin=max (VminTC,Vstall);
area ([Vmin Vmax(end:-1:1)],[h h(end:-1:1)],...
 'FaceColor', [0.8 1 1],'LineStyle','none');
hold on;
plot (Vmax,h,VminTC,h,Vstall,h);
grid on;
xlim ([0 400]);
xlabel ('Velocity (ft/s)');
ylabel ('Altitude (ft)');
```

For each such feasible pair of altitude and airspeed values that lie within the steady level flight envelope of the general aviation aircraft, methodology has been provided to compute the pilot inputs, namely the elevator deflection, the throttle setting, and zero bank angle, that characterize this steady level flight condition. In addition, the methodology also allows computation of the associated angle of attack, values of the lift and drag coefficients, and values of the lift and drag at this steady level flight condition.

7.9 Conclusions

This chapter has treated the principles of steady level longitudinal flight. Since most aircraft spend the majority of the time in steady level cruise, the steady flight analysis presented here provides the baseline comparison for all steady flight analyses. The various mathematical relationships for steady level flight have been derived from the basic assumptions and the physics of flight. These mathematical relationships are quantitative, but they also provide important qualitative insight into steady level flight. These relationships have been illustrated in detail by studying steady level flight properties of concrete aircraft examples.

7.10 Problems

7.1. Consider the executive jet aircraft, with a weight of 73,000 lbs, in steady level flight.

(a) What is the expression for the thrust required to maintain steady level flight at an altitude of 28,000 ft? Plot the thrust required to maintain steady level flight as a function of the airspeed. What is the airspeed at which this required thrust is minimum? What is the minumum value of the thrust required to maintain steady level flight?

(b) If the elevator deflection is 2 degrees and the throttle setting is 75%, what are the possible flight conditions for which steady level flight can be maintained at an altitude of 28,000 ft? For each such flight condition, what are the airspeed and the angle of attack?

(c) For steady level flight at an altitude of 28,000 ft, what is the maximum airspeed for which there is sufficient thrust produced by the jet engine to maintain steady level flight?

(d) For steady level flight at an altitude of 28,000 ft, what is the minimum airspeed for which there is sufficient thrust produced by the jet engine to maintain steady level flight?

(e) What is the flight ceiling of the executive jet aircraft?

(f) Plot the flight envelope for steady level flight of the executive jet aircraft.

7.2. Consider the general aviation propeller-driven aircraft, with a weight of 2,900 lbs, in steady level flight.

(a) What is the expression for the power required to maintain steady level flight at an altitude of 14,000 ft? Plot the power required to maintain steady level flight as a function of airspeed. What is the airspeed at which this required power is a minimum? What is the minimum value of the power required to maintain steady level flight?

(b) If the elevator deflection is 3 degrees and the throttle setting is 80%, what are the possible flight conditions for which steady level flight can be maintained at an altitude of 14,000 ft? For each such flight condition, what are the airspeed and the angle of attack?

(c) For steady level flight at an altitude of 14,000 ft, what is the maximum airspeed for which there is sufficient power produced by the engine to maintain steady level flight?

(d) For steady level flight at an altitude of 14,000 ft, what is the minimum airspeed for which there is sufficient power produced by the engine to maintain steady level flight?

(e) What is the flight ceiling of the general aviation propeller-driven aircraft?

(f) Plot the flight envelope for steady level flight of the general aviation propeller-driven aircraft.

7.3. Consider the UAV, with a weight of 45 lbs, in steady level flight.

(a) What is the expression for the thrust required to maintain steady level flight at an altitude of 5,000 ft? Plot the thrust required to maintain steady level flight as a function of airspeed. What is the airspeed at which this required thrust is a minimum? What is the minimum value of the thrust required to maintain steady level flight?

(b) If the elevator deflection is 4 degrees and the throttle setting is 90%, what are the possible flight conditions for which steady level flight can be maintained at an altitude of 5,000 ft? For each such flight condition, what are the airspeed and the angle of attack?

(c) For steady level flight at an altitude of 5,000 ft, what is the expression for the angle of attack required to maintain steady level flight as a function of the airspeed?

(d) For steady level flight at an altitude of 5,000 ft, what is the maximum airspeed for which the engine can maintain steady level flight?

(e) For steady level flight at an altitude of 5,000 ft, what is the minimum airspeed for which the engine can maintain steady level flight?

(f) What is the flight ceiling of the UAV?

(g) Plot the flight envelope for steady level flight of the UAV.

7.4. Consider the executive jet aircraft in steady level longitudinal flight. Is it possible that the aircraft can remain in steady level longitudinal flight at an airspeed of 300 ft/s and an altitude of 30,000 ft? In other words, is this flight condition in the steady level longitudinal flight envelope?

If this flight condition is in the steady level longitudinal flight envelope, determine the throttle setting and the elevator deflection required to maintain this flight condition.

If this flight condition is not in the steady level longitudinal flight envelope, which constraints are violated?

7.5. Consider the executive jet aircraft in steady level longitudinal flight. Is it possible that the aircraft can remain in steady level longitudinal flight at an airspeed of 750 ft/s and an altitude of 40,000 ft? In other words, is this flight condition in the steady level longitudinal flight envelope?

If this flight condition is in the steady level longitudinal flight envelope, determine the throttle setting and the elevator deflection required to maintain this flight condition.

If this flight condition is not in the steady level longitudinal flight envelope, which constraints are violated?

7.6. Consider the propeller-driven general aviation aircraft in steady level longitudinal flight. Is it possible that the aircraft can remain in steady level longitudinal flight at an airspeed of 350 ft/s and an altitude of 25,000 ft? In other words, is this flight condition in the steady level longitudinal flight envelope?

If this flight condition is in the steady level longitudinal flight envelope, determine the throttle setting and the elevator deflection required to maintain this flight condition.

If this flight condition is not in the steady level longitudinal flight envelope, which constraints are violated?

7.7. Consider the propeller-driven general aviation aircraft in steady level longitudinal flight. Is it possible that the aircraft can remain in steady level longitudinal flight at an airspeed of 700 ft/s and an altitude of 30,000 ft? In other words, is this flight condition in the steady level longitudinal flight envelope?

If this flight condition is in the steady level longitudinal flight envelope, determine the throttle setting and the elevator deflection required to maintain this flight condition.

If this flight condition is not in the steady level longitudinal flight envelope, which constraints are violated?

7.8. Consider the executive jet aircraft, with a weight of 73,000 lbs, in steady level flight through a constant horizontal headwind of 50 ft/s.

(a) What is the relation between the steady level airspeed and the steady level ground speed of the aircraft?

(b) What is the expression for the thrust required to maintain steady level flight at an altitude of 28,000 ft? Plot the thrust required to maintain steady level flight as a function of the ground speed. What is the ground speed at which this required thrust is minimum? What is the minumum value of the thrust required to maintain steady level flight?

(c) If the elevator deflection is 2 degrees and the throttle setting is 75%, what are the possible flight conditions for which steady level flight can be maintained at an altitude of 28,000 ft? For each such flight condition, what are the ground speed and the angle of attack?

(d) For steady level flight at an altitude of 28,000 ft, what is the maximum ground speed for which there is sufficient thrust produced by the jet engine to maintain steady level flight?

(e) For steady level flight at an altitude of 28,000 ft, what is the minimum ground speed for which there is sufficient thrust produced by the jet engine to maintain steady level flight?

7.9. Consider the executive jet aircraft, with a weight of 73,000 lbs, in steady level flight through a constant horizontal tailwind of 50 ft/s.

    (a) What is the relation between the steady level airspeed and the steady level ground speed of the aircraft?

    (b) What is the expression for the thrust required to maintain steady level flight at an altitude of 28,000 ft? Plot the thrust required to maintain steady level flight as a function of the ground speed. What is the ground speed at which this required thrust is minimum? What is the minimum value of the thrust required to maintain steady level flight?

    (c) If the elevator deflection is 2 degrees and the throttle setting is 75%, what are the possible flight conditions for which steady level flight can be maintained at an altitude of 28,000 ft? For each such flight condition, what are the ground speed and the angle of attack?

    (d) For steady level flight at an altitude of 28,000 ft, what is the maximum ground speed for which there is sufficient thrust produced by the jet engine to maintain steady level flight?

    (e) For steady level flight at an altitude of 28,000 ft, what is the minimum ground speed for which there is sufficient thrust produced by the jet engine to maintain steady level flight?

7.10. Consider the general aviation propeller-driven aircraft, with a weight of 2,900 lbs, in steady level flight through a constant horizontal headwind of 25 ft/s.

    (a) What is the relation between the steady level airspeed and the steady level ground speed of the aircraft?

    (b) What is the expression for the thrust required to maintain steady level flight at an altitude of 14,000 ft? Plot the thrust required to maintain steady level flight as a function of the ground speed. What is the ground speed at which this required thrust is minimum? What is the minumum value of the thrust required to maintain steady level flight?

    (c) If the elevator deflection is 3 degrees and the throttle setting is 80%, what are the possible flight conditions for which steady level flight can be maintained at an altitude of 14,000 ft? For each such flight condition, what are the ground speed and the angle of attack?

    (d) For steady level flight at an altitude of 14,000 ft, what is the maximum ground speed for which there is sufficient thrust produced by the jet engine to maintain steady level flight?

    (e) For steady level flight at an altitude of 14,000 ft, what is the minimum ground speed for which there is sufficient thrust produced by the jet engine to maintain steady level flight?

7.11. Consider the general aviation propeller-driven aircraft, with a weight of 2,900 lbs, in steady level flight through a constant horizontal tailwind of 25 ft/s.

    (a) What is the relation between the steady level airspeed and the steady level ground speed of the aircraft?

(b) What is the expression for the thrust required to maintain steady level flight at an altitude of 14,000 ft? Plot the thrust required to maintain steady level flight as a function of the ground speed. What is the ground speed at which this required thrust is minimum? What is the minumum value of the thrust required to maintain steady level flight?

(c) If the elevator deflection is 3 degrees and the throttle setting is 80%, what are the possible flight conditions for which steady level flight can be maintained at an altitude of 14,000 ft? For each such flight condition, what are the ground speed and the angle of attack?

(d) For steady level flight at an altitude of 14,000 ft, what is the maximum ground speed for which there is sufficient thrust produced by the jet engine to maintain steady level flight?

(e) For steady level flight at an altitude of 14,000 ft, what is the minimum ground speed for which there is sufficient thrust produced by the jet engine to maintain steady level flight?

8

Aircraft Steady Longitudinal Flight

This chapter treats steady longitudinal flight through a stationary atmosphere with the aircraft in either a steady climb or a steady descent. The equations for steady longitudinal flight are derived and studied in detail. Several steady longitudinal flight performance measures are studied. Steady climbing flight and descending flight are studied in a unified way, since steady climbing flight corresponds to a positive flight path angle (or positive rate of climb) while steady descending flight corresponds to a negative flight path angle (or negative rate of climb). Steady climbing or descending flight occurs when the velocity vector is a constant vector so that the path along which the aircraft moves is a straight line. Steady level flight is a special case of steady climbing or descending flight for which the flight path angle is zero or, equivalently, the climb rate is zero.

The main development on steady climbing or steady descending flight in this chapter is based on the assumption that the flight time of interest is sufficiently short such that the air density is approximately constant.

8.1 Steady Longitudinal Flight

Descriptions of steady climbing or descending longitudinal flight of an aircraft are expressed in terms of equations that describe the fundamental flight variables that arise from the free-body diagram for steady climbing or descending flight. Equations that describe the aerodynamics, propulsion characteristics, and important flight constraints for both a jet aircraft and a propeller-driven general aviation aircraft are developed.

Free-Body Diagrams

Consider the free-body diagram in Figure 8.1 that shows the forces acting on an aircraft in steady climbing flight. The center of mass of the aircraft moves along a straight line with constant airspeed, constant flight path angle or climb rate, and constant thrust. Coordinated flight of the aircraft is assumed, that is, the side-slip angle is zero.

The aircraft has a velocity vector that makes an angle γ, the flight path angle, with the horizontal; the vertical component of the aircraft velocity vector is the rate of climb V_{climb}. The magnitude of the aircraft velocity vector is the airspeed V. The lift force on the aircraft acts normal to the velocity vector in the plane of mass symmetry, and the drag force on the aircraft acts opposite to the velocity vector. The thrust vector, with magnitude T, acts

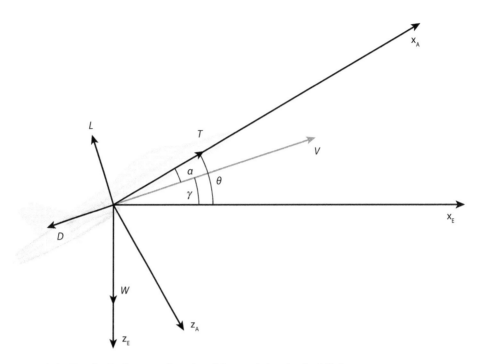

Figure 8.1. Free-body diagram of an aircraft in steady longitudinal flight.

along the aircraft-fixed x_A-axis of the aircraft; the angle between the thrust vector and the velocity vector is the aircraft angle of attack α. The algebraic equations for steady climbing or steady descending flight are now obtained. Since steady climbing or descending flight implies zero acceleration of the aircraft, the vector sum of the forces on the aircraft is zero. The sum of the force components along the velocity vector is zero:

$$-W \sin \gamma - D + T \cos \alpha = 0. \tag{8.1}$$

The sum of the force components normal to the velocity vector is also zero:

$$W \cos \gamma - L - T \sin \alpha = 0. \tag{8.2}$$

Steady climbing or descending flight also requires that the pitch moment be zero:

$$M = 0. \tag{8.3}$$

The roll moment and the yaw moment are also zero in steady climbing or descending flight. Since the side-slip angle is zero, the roll moment and the yaw moment are zero if there is no deflection of the ailerons and no rudder deflection.

Make the approximation that the thrust is much smaller than the lift and the angle of attack is a small angle in radian measure. The first approximation is justified for most conventional fixed wing aircraft. The assumption that the angle of attack is small, when measured in radians, is consistent with the use of the linear aerodynamics introduced in chapter 3. Thus $\sin \alpha$ is approximated by α and $\cos \alpha$ is approximated by 1. These approximations imply that the force balance equations (8.1) and (8.2) can be approximated as

$$T = D + W\gamma, \tag{8.4}$$

$$L = W, \tag{8.5}$$

and the pitch moment balance equation is

$$M = 0. \tag{8.6}$$

These three equations provide the basis for studying steady climbing or descending flight in a fixed vertical plane. They provide the conceptually simple relationships for steady longitudinal flight:

- The lift must equal the weight of the aircraft.

- The thrust must equal the drag plus the component of the aircraft weight along the velocity vector.

- The aerodynamic pitch moment must be zero.

These equations are approximations to the exact, but more complicated, steady climbing flight equations (8.1), (8.2), and (8.3). Equations (8.4)–(8.6) are used in the subsequent flight analysis.

Aerodynamics

Recall the equations that describe the lift force, the drag force, and the pitch moment from chapter 3 are

$$L = \frac{1}{2}\rho V^2 S C_L, \tag{8.7}$$

$$D = \frac{1}{2}\rho V^2 S C_D, \tag{8.8}$$

$$M = \frac{1}{2}\rho V^2 S c C_M. \tag{8.9}$$

The quadratic drag polar expression that relates the drag coefficient to the lift coefficient is given by

$$C_D = C_{D_0} + K C_L^2, \tag{8.10}$$

where $K = \frac{1}{\pi e AR}$. The lift coefficient is expressed as a linear function of the angle of attack as

$$C_L = C_{L_0} + C_{L_\alpha}\alpha. \tag{8.11}$$

The pitch moment coefficient is expressed as a linear function of the angle of attack and the elevator deflection as

$$C_M = C_{M_0} + C_{M_\alpha}\alpha + C_{M_{\delta_e}}\delta_e. \tag{8.12}$$

Flight Constraints

There are two types of important flight constraints that apply to steady climbing and descending flight. These constraints play a central role in determing the performance properties of an aircraft in steady climbing or descending flight.

The first constraint arises from the aerodynamics assumption that the flight of the aircraft must avoid stalling. Based on the discussion in chapter 5, the stall constraint can be expressed as an inequality on the lift coefficient given by

$$C_L \leq C_{L_{\max}}. \tag{8.13}$$

The second set of constraints arise from limits on the operation of the propulsion system. For a jet aircraft, the thrust T required for steady climbing or descending flight must satisfy

$$0 \leq T \leq T_{\max}^s \left(\frac{\rho}{\rho^s}\right)^m. \tag{8.14}$$

The jet engine thrust constraints (8.14) can also be expressed as constraints on the throttle setting

$$0 \leq \delta_t \leq 1, \tag{8.15}$$

where

$$\delta_t = \frac{T}{T_{\max}^s}\left(\frac{\rho^s}{\rho}\right)^m. \tag{8.16}$$

For internal combustion engine and propeller-driven aircraft, the power P required for steady climbing or descending flight must satisfy

$$0 \leq P \leq \eta P_{\max}^s \left(\frac{\rho}{\rho^s}\right)^m, \tag{8.17}$$

where η is the propeller efficiency. This is subsequently referred to as the propulsion constraints or as the power constraints for a propeller-driven general aviation aircraft. The engine power constraints (8.17) can also be expressed as constraints on the throttle setting

$$0 \leq \delta_t \leq 1, \tag{8.18}$$

where

$$\delta_t = \frac{P}{\eta P_{\max}^s} \left(\frac{\rho^s}{\rho} \right)^m . \tag{8.19}$$

8.2 Steady Longitudinal Flight Analysis

Flight conditions for steady climbing or descending longitudinal flight are analyzed. This analysis is based on the equations of motion and constraints described in section 8.1.

Equations for Steady Climbing or Descending Flight

Equations (8.4)–(8.12) form the fundamental equations for steady climbing or descending flight; these equations are now analyzed to ascertain the basic properties of an aircraft in steady climbing or descending flight. It is convenient to study equations (8.4), (8.5), (8.7), (8.8), and (8.10). These are five algebraic equations that are expressed in terms of the following seven flight variables that define a steady climbing or descending flight condition: the lift force L, the drag force D, the thrust T, the airspeed V, the flight path angle γ, the lift coefficient C_L, and the drag coefficient C_D.

Our objective is to determine the relation between these seven variables using the conditions for steady climbing or descending flight. These algebraic equations also involve important flight parameters: the aircraft weight W, the aircraft wing surface area S, the aircraft aerodynamic parameters C_{D_0} and $K = \frac{1}{\pi e AR}$, and the air density ρ. The air density parameter depends on the flight altitude according to the standard atmospheric model. Since the aircraft is in a steady climb, or a steady descent, the altitude is not exactly constant. However, for time periods on the order of several minutes or less the change in altitude results in only a small change in the air density. Hence, in the subsequent analysis of steady climbing or steady descending flight the air density parameter is approximated as a constant.

The five algebraic flight equations for steady climbing or descending flight, namely equations (8.4), (8.5), (8.7), (8.8), and (8.10), are expressed in terms of seven flight variables. It is reasonable to select fixed values for two of the flight variables and then to solve the five equations for the other five flight variables. In this way, it is possible to express the five flight variables in terms of the prior selected two flight variables. This is the procedure that is followed.

The previous equations that describe conditions for steady climbing or descending flight are solved in several different forms. One expression provides the thrust required for steady climbing flight in terms of the airspeed of the aircraft and in terms of the flight path angle (or climb rate). The other expression provides the power required for steady climbing or descending flight in terms of the airspeed of the aircraft and in terms of the climb rate (or flight path angle).

We first make use of equations (8.7) and (8.5) to obtain an expression for the airspeed of the aircraft:

$$V = \sqrt{\frac{2W}{\rho S C_L}}. \tag{8.20}$$

This expression for the airspeed holds for any steady climbing or descending flight condition.

Thrust Required

Consider the conditions for steady climbing or descending flight supposing that γ is the flight path angle. Using equations (8.5) and (8.4), the required thrust can be expressed as

$$T = \left(\gamma + \frac{C_D}{C_L}\right) W. \tag{8.21}$$

This equation gives the thrust required for steady climbing or descending flight as it depends on the flight path angle, and the aerodynamic parameters.

Another expression for the thrust required for steady climbing or descending flight can be obtained. Substituting the lift expression (8.7) into equation (8.5) it is easy to solve for the lift coefficient as

$$C_L = \frac{2W}{\rho V^2 S}. \tag{8.22}$$

The quadratic drag polar expression (8.10) gives the expression for the drag coefficient as

$$C_D = C_{D_0} + \frac{4K W^2}{\rho^2 V^4 S^2}. \tag{8.23}$$

Consequently, the drag for steady climbing or descending flight at airspeed V is obtained from equation (8.8) as

$$D = \frac{1}{2}\rho V^2 S C_{D_0} + \frac{2K W^2}{\rho V^2 S}. \tag{8.24}$$

The thrust required to maintain steady climbing or descending flight is obtained using equation (8.4) as

$$T = W\gamma + \frac{1}{2}\rho V^2 S C_{D_0} + \frac{2K W^2}{\rho V^2 S}. \tag{8.25}$$

This equation gives the thrust required for steady climbing or descending flight as it depends on the airspeed, the flight path angle, and the flight parameters.

Assuming a small flight path angle as measured in radians, the flight path angle and the climb rate are related by

$$V_{climb} = V \sin \gamma = V \gamma. \tag{8.26}$$

Thus the thrust required for steady climbing or descending flight can also be expressed in terms of the airspeed and climb rate as

$$T = W \left(\frac{V_{climb}}{V}\right) + \frac{1}{2}\rho V^2 S C_{D_0} + \frac{2K W^2}{\rho V^2 S}. \tag{8.27}$$

There are three terms in the expressions (8.25) and (8.27) for the required thrust. The first term on the right of the equations describes the part of the thrust that is required to compensate for the fact that the aircraft is climbing or descending; this thrust component is positive for climbing flight and negative for descending flight. The second term represents the drag component that is due to aerodynamic friction and is independent of the lift; this drag term increases as the airspeed increases. The third term represents the drag component that is induced by the lift; this drag term decreases as the airspeed increases. Thus the total thrust required to maintain steady climbing or descending flight is the sum of all three terms.

Equation (8.25) can be rewritten to express the flight path angle in terms of the airspeed and thrust as

$$\gamma = \frac{T}{W} - \frac{1}{2} \frac{\rho V^2 S C_{D_0}}{W} - \frac{2KW}{\rho V^2 S}. \tag{8.28}$$

Equation (8.27) can be rewritten to express the climb rate in terms of the airspeed and thrust as

$$V_{climb} = \frac{TV}{W} - \frac{1}{2} \frac{\rho V^3 S C_{D_0}}{W} - \frac{2KW}{\rho V S}. \tag{8.29}$$

Equation (8.25) can be expressed in an alternative form. If the thrust T and the flight path angle γ are assumed known, then equation (8.25) can be written as

$$\frac{1}{2} \rho V^4 S C_{D_0} - (T - W\gamma)V^2 + \frac{2KW^2}{\rho S} = 0, \tag{8.30}$$

which can be viewed as an equation whose solution is the airspeed for steady climbing or descending flight at the given value of thrust and flight path angle. Equation (8.30) has two real and positive solutions corresponding to a low-speed steady climbing or descending flight solution and a high-speed steady climbing or descending flight solution. Solving this equation requires finding the roots of a fourth degree polynomial. The low-speed solution should be checked to see that the stall constraint is satisfied. The stall constraint can be expressed as an inequality in the aircraft airspeed given by

$$\sqrt{\frac{2W}{\rho S C_{L_{max}}}} \leq V. \tag{8.31}$$

The throttle setting is easily determined from the above analysis. Since steady longitudinal flight is to be maintained, there should be no deflection of the ailerons and no deflection of the rudder.

For a specified value of the flight path angle, the thrust required for steady climbing or descending flight has a minimum value. This minimum value can be obtained using the methods of calculus: at the minimum the thrust required curve has zero slope; that is, the derivative of the thrust with respect to the airspeed is zero. In particular, the minimum thrust occurs when the drag is minimized. This occurs if the airspeed is

$$V = \sqrt{\frac{2W}{\rho S} \sqrt{\frac{K}{C_{D_0}}}}.$$
(8.32)

The airspeed value at which the minimum thrust for steady climbing or descending flight occurs depends on the air density and hence on the flight altitude; this airspeed increases with an increase in the flight altitude.

The minimum value of the thrust required for steady climbing or descending flight is obtained by substituting the expression (8.32) into equation (8.25) to obtain

$$T_{\min} = W\gamma + 2W\sqrt{KC_{D_0}}.$$
(8.33)

The minimum value of the thrust required for steady climbing or descending flight for a fixed flight path angle does not depend on the air density and hence does not depend on the flight altitude. The minimum value of the thrust required for steady climbing or descending flight can also be expressed in terms of the climb rate by substituting the expression (8.32) into equation (8.27) to obtain

$$T_{\min} = V_{climb}\sqrt{\frac{\rho SW}{2}\sqrt{\frac{C_{D0}}{K}}} + 2W\sqrt{KC_{D_0}}.$$
(8.34)

The minimum value of the thrust required for steady climbing or descending flight for a fixed value of the climb rate depends on the air density and hence on the flight altitude; the minimum thrust value decreases with an increase in the flight altitude.

For a jet aircraft, the thrust expressions given above must satisfy the thrust constraints given in (8.14). In particular, for a positive flight path angle the most important thrust constraint is typically the upper thrust limit, while for a negative flight path angle the lower limit of zero thrust may be an important constraint.

Power Required

The prior development characterizes the thrust required for steady climbing or descending flight. It is also possible to characterize the power required for steady climbing or descending flight using the basic relationship between thrust and power. In steady climbing or descending flight, the power required is the product of the thrust required and the airspeed of the aircraft.

The power required for steady climbing or descending flight, expressed in terms of the airspeed and the flight path angle, is obtained by multiplying equation (8.25) by the airspeed to obtain

$$P = WV\gamma + \frac{1}{2}\rho V^3 SC_{D_0} + \frac{2KW^2}{\rho VS}.$$
(8.35)

The power required for steady climbing or descending flight can also be expressed in terms of the airspeed and the climb rate as

$$P = W V_{climb} + \frac{1}{2}\rho V^3 S C_{D_0} + \frac{2 K W^2}{\rho V S}. \tag{8.36}$$

There are three terms in expressions (8.35) and (8.36) for the required power, leading to the following interpretation. The first term on the right of the equations describes the part of the power that is required to compensate for the fact that the aircraft is climbing or descending. The second term represents the power required to overcome the drag that is due to aerodynamic friction and is independent of the lift. The third term represents the power required to overcome the drag that is induced by the lift. Thus the total power required to maintain steady climbing or descending flight is the sum of all three terms.

Equation (8.35) can be rewritten to express the flight path angle in terms of the power and the airspeed as

$$\gamma = \frac{P}{W V} - \frac{1}{2}\frac{\rho V^2 S C_{D_0}}{W} - \frac{2 K W}{\rho V^2 S}. \tag{8.37}$$

Equation (8.36) can be rewritten to express the climb rate in terms of the power and the airspeed as

$$V_{climb} = \frac{P}{W} - \frac{1}{2}\frac{\rho V^3 S C_{D_0}}{W} - \frac{2 K W}{\rho V S}. \tag{8.38}$$

Equation (8.36) can be expressed in an alternative form. If the power and the climb rate are assumed known, then equation (8.36) can be written as

$$\frac{1}{2}\rho V^4 S C_{D_0} - (P - W V_{climb}) V + \frac{2 K W^2}{\rho S} = 0, \tag{8.39}$$

which can be viewed as a fourth degree polynomial equation whose solution is the airspeed for steady climbing or descending flight at the given value of power and climb rate. Equation (8.39) typically has two real and positive solutions corresponding to a low-speed steady climbing or descending flight solution and a high-speed steady climbing or descending flight solution. Solving this equation requires finding the roots of a fourth degree polynomial. The low-speed solution should be checked to see that the stall constraint is satisfied:

$$\sqrt{\frac{2 W}{\rho S C_{L_{max}}}} \leq V. \tag{8.40}$$

The throttle setting is easily determined from the above analysis. Since steady longitudinal flight is to be maintained, there should be no differential deflection of the ailerons and no rudder deflection.

For a specified value of the climb rate, the power required for steady climbing or descending flight has a minimum value. This minimum value can be obtained using the methods of calculus: at the minimum the power required curve has zero slope; that is, the derivative of the power with respect to the airspeed is zero. This occurs if the airspeed is

$$V = \sqrt{\frac{2W}{\rho S} \sqrt{\frac{K}{3C_{D_0}}}}. \tag{8.41}$$

The airspeed value, corresponding to the minimum power for steady climbing or descending flight, depends on the air density and hence on the flight altitude; this airspeed increases with an increase in the flight altitude. The minimum value of the power required for steady climbing or descending flight, expressed in terms of the climb rate, is obtained by substituting expression (8.41) into equation (8.36) to obtain

$$P_{\min} = W V_{climb} + \frac{4}{3} \sqrt{\frac{2W^3}{\rho S} \sqrt{3K^3 C_{D_0}}}. \tag{8.42}$$

The minimum value of the power required for steady climbing or descending flight can also be expressed in terms of the flight path angle; this expression is obtained by substituting expression (8.41) into equation (8.35) to obtain

$$P_{\min} = \gamma \sqrt{\frac{2W^3}{\rho S} \sqrt{\frac{K}{3C_{D_0}}}} + \frac{4}{3} \sqrt{\frac{2W^3}{\rho S} \sqrt{3K^3 C_{D_0}}}. \tag{8.43}$$

For a fixed value of the flight path angle, the minimum value of the power required for steady climbing or descending flight depends on the air density and hence on the flight altitude; the minimum value of the power increases with an increase in the flight altitude. For a fixed value of the climb rate, the minimum value of the power required for steady climbing or descending flight depends on the air density and hence on the flight altitude; the minimum value of the power increases with an increase in the flight altitude.

For an aircraft powered by an internal combustion engine and propeller, the power expressions given above must satisfy the power constraints given in expression (8.17). In particular, for a positive climb rate the most important power constraint is typically the upper power limit, while for a negative climb rate the lower limit of zero power may be an important constraint.

8.3 Jet Aircraft Steady Longitudinal Flight Performance

Steady climbing or descending flight performance measures are studied, taking into account that the propulsion system is a jet engine. The following flight performance measures are analyzed: maximum airspeed at a given flight path angle, minimum airspeed at a given flight path angle, maximum flight path angle, minimum flight path angle, and flight ceiling. The steady climbing or descending flight envelope is also introduced.

Maximum and Minimum Airspeeds

Flight conditions for maximum and minimum airspeeds of a jet aircraft are analyzed for steady climbing or descending flight with a fixed flight path angle γ. The maximum thrust that the jet engine can provide limits the maximum and minimum possible airspeeds for which the aircraft can maintain steady climbing or descending flight. In particular, the maximum and minimum airspeeds are characterized by the fact that the thrust required for steady climbing or descending flight is equal to the maximum thrust that the engine can provide at full throttle; this leads to the equation

$$W\gamma + \frac{1}{2}\rho V^2 SC_{D_0} + \frac{2KW^2}{\rho V^2 S} = T^s_{max}\left(\frac{\rho}{\rho^s}\right)^m. \tag{8.44}$$

This equation typically has two positive solutions. The high-speed solution denotes the maximum possible airspeed of the aircraft at the fixed flight path angle; the low-speed solution denotes the minimum possible airspeed of the aircraft at the fixed flight path angle.

Two important qualifications must be made about the maximum airspeed and the minimum airspeed obtained from this analysis. This computed minimum airspeed is the minimum airspeed if it satisfies the stall constraint given by the inequality (8.31); otherwise the stall speed is the minimum airspeed for steady climbing or descending flight at the given thrust and flight path angle. The computed maximum airspeed may be near to or exceed the speed of sound; if this case occurs, the computed maximum airspeed is inaccurate since supersonic aerodynamics factors, not taken into account in our analysis, are important.

Maximum and Minimum Flight Path Angles

At a given flight altitude, a jet aircraft has a maximum flight path angle. Since the flight path angle is the ratio of the excess thrust and the weight, maximization of the flight path angle requires maximization of the excess thrust; that is, the thrust provided by the jet engine at that altitude is a maximum and the drag at that altitude is a minimum. The airspeed for which the flight path is maximum is identical with the airspeed for which the drag is a minimum. According to equation (8.32), this airspeed is

$$V = \sqrt{\frac{2W}{\rho S}\sqrt{\frac{K}{C_{D_0}}}}. \tag{8.45}$$

Using the results in equation (8.14) for the maximum thrust and in equation (8.33) for the minimum value of the drag, the maximum flight path angle is

$$\gamma_{max} = \frac{T^s_{max}}{W}\left(\frac{\rho}{\rho^s}\right)^m - 2\sqrt{KC_{D_0}}. \tag{8.46}$$

This flight condition for an aircraft with maximum flight path angle should be checked to make sure that the stall constraint given by inequality (8.31) is satisfied. This expression shows that the maximum flight path angle decreases as the flight altitude increases.

At a given flight altitude, the aircraft also has a minimum possible flight path angle. Equation (8.4) can be rewritten as

$$\gamma = \frac{T - D}{W}, \tag{8.47}$$

so that the minimum flight path angle occurs when the thrust is zero and the drag is maximum. A local maximum of the drag occurs at the stall condition, where the airspeed is

$$V = \sqrt{\frac{2W}{\rho S C_{L_{\max}}}}. \tag{8.48}$$

Using equation (8.28), the minimum value of the flight path angle is

$$\gamma_{\min} = -\frac{C_{D0}}{C_{L_{\max}}} - K C_{L_{\max}}. \tag{8.49}$$

This is the negative value of the maximum glide angle obtained for steady gliding flight.

Flight Ceiling

The above analysis is based on steady flight analysis at a fixed altitude. It is possible to assess the dependence of the flight path angle on the altitude. The maximum flight altitude, referred to as the flight ceiling of a jet aircraft in steady longitudinal flight with a fixed flight path angle, is determined. The flight condition at this flight ceiling is characterized by the fact that the minimum thrust required for steady climbing or descending flight, given in equation (8.33), is exactly equal to the maximum thrust that the engine can produce at this altitude. This leads to the equation

$$W\gamma + 2W\sqrt{KC_{D_0}} = T_{\max}^s \left(\frac{\rho}{\rho^s}\right)^m. \tag{8.50}$$

Since the air density depends on the altitude according to the standard atmospheric model, equation (8.50) can be viewed as an implicit equation in the altitude; the flight ceiling solves this equation.

Flight Envelope

The steady climbing or descending longitudinal flight envelope for a jet aircraft consists of all flight conditions for which steady climbing or descending flight can be maintained. The steady climbing or descending flight envelope is a set in three dimensions defined by the altitude, airspeed, and flight path angle that satisfy the steady flight equations and the stall and engine thrust constraints.

Since the prior analysis in this chapter provides the basis for computation of the thrust required and the lift coefficient for steady climbing or descending flight, these constraints can easily be checked. The boundary of the steady climbing or descending longitudinal flight envelope is defined by surfaces that are obtained by making the maximum thrust constraint active or by making the stall constraint active. Each of these surfaces is defined in three dimensions; the steady climbing or descending flight envelope is the intersection of these two regions whose boundaries are defined by the two constraint surfaces. Graphical tools are available to visualize this steady climbing or descending flight envelope in three dimensions.

It is possible to provide a two-dimensional graphical representation of the steady climbing or descending flight envelope for a jet aircraft. It is most convenient to represent a cross section of the steady climbing or descending flight envelope, for a fixed value of the flight path angle, as a set in two dimensions defined by the altitude and airspeed variables. In particular, specific values of flight altitude and airspeed lie in the cross section of the steady climbing or descending flight envelope if both the stall constraint and the engine thrust constraints are satisfied at the flight condition corresponding to the fixed value of the flight path angle. The steady level flight envelope is such a cross section if the flight path angle is zero.

8.4 General Aviation Aircraft Steady Longitudinal Flight Performance

Steady climbing or descending flight performance measures are studied, taking into account that the propulsion system is an internal combustion engine and propeller. The following flight performance measures are analyzed: maximum airspeed at a given climb rate, minimum airspeed at a given climb rate, maximum rate of climb, maximum descent rate, and flight ceiling. The steady climbing or descending flight envelope is also introduced.

Maximum and Minimum Airspeeds

Flight conditions for maximum and minimum airspeeds of a general aviation aircraft are analyzed for steady climbing or descending flight with a fixed climb rate V_{climb}. The maximum power that the engine can provide limits the maximum and minimum possible airspeeds for which the aircraft can maintain steady climbing or descending flight. In particular, the maximum and minimum airspeeds are characterized by the fact that the power required for steady climbing or descending flight at a fixed climb rate is equal to the maximum power that the engine can provide at full throttle; this leads to the equation

$$W V_{climb} + \frac{1}{2}\rho V^3 S C_{D_0} + \frac{2KW^2}{\rho V S} = \eta P^s_{max} \left(\frac{\rho}{\rho^s} \right)^m . \tag{8.51}$$

This equation typically has two positive solutions. The high-speed solution denotes the maximum possible airspeed of the aircraft at the fixed climb rate; the low-speed solution denotes the minimum possible airspeed of the aircraft at the fixed climb rate.

Two important qualifications must be made about the maximum and minimum airspeeds obtained from this analysis. The computed minimum airspeed is the minimum airspeed if it satisfies the stall constraint given by the inequality (8.40); otherwise the stall speed is the minimum airspeed for steady climbing or descending flight at the given thrust and climb rate. The computed maximum airspeed may be near to or exceed the speed of sound; if this case occurs, the computed maximum airspeed is inaccurate since supersonic aerodynamics factors, not taken into account in our analysis, are important.

Maximum Rate of Climb and Maximum Descent Rate

At a given flight altitude, a general aviation aircraft has a maximum rate of climb. Since the climb rate is the ratio of the excess power and the weight, maximization of the climb rate requires maximization of the excess power; that is, the power provided by the engine is a maximum and the power to overcome the drag is a minimum. The airspeed for which the climb rate is maximum is identical to the airspeed for which the power to overcome the drag is a minimum. According to equation (8.41), this airspeed is

$$V = \sqrt{\frac{2W}{\rho S} \sqrt{\frac{K}{3C_{D_0}}}}. \tag{8.52}$$

Assuming full throttle to achieve maximum engine power and using equation (8.42), the maximum climb rate is

$$(V_{climb})_{max} = \frac{\eta P_{max}^s}{W} \left(\frac{\rho}{\rho^s}\right)^m - \frac{4}{3}\sqrt{\frac{2W}{\rho S}\sqrt{3K^3 C_{D_0}}}. \tag{8.53}$$

This flight condition for maximum climb rate should be checked to make sure that the stall constraint given by inequality (8.40) is satisfied. This expression shows that the maximum climb rate decreases as the flight altitude increases.

At a given flight altitude, the aircraft also has a maximum possible steady descent rate. Equations (8.4) and (8.26) can be combined to obtain

$$V_{climb} = \frac{P - DV}{W}, \tag{8.54}$$

where P is the power available for flight that is produced by the engine and propeller.

The minimum climb rate, or, equivalently, the maximum descent rate, occurs when the power produced by the engine is zero and the power due to drag is maximum. The power due to drag has a local maximum at the stall condition, where the airspeed is

$$V = \sqrt{\frac{2W}{\rho S C_{L_{max}}}}. \tag{8.55}$$

Using equation (8.28), the maximum value of the descent rate in steady longitudinal flight is

$$(V_{descent})_{max} = \sqrt{\frac{2W}{\rho S C_{L_{max}}} \left(\frac{C_{D0}}{C_{L_{max}}} + K C_{L_{max}} \right)}. \tag{8.56}$$

This is also the maximum descent rate for steady gliding flight.

Flight Ceiling

The above analysis is based on steady flight analysis at a fixed altitude. There is a maximum altitude, referred to as the steady climbing or descending flight ceiling, at which steady climbing or descending flight with a fixed rate of climb can be maintained. The flight condition at this flight ceiling is characterized by the fact that the minimum power required for steady climbing or descending flight, given in equation (8.41), is exactly equal to the maximum power that the engine can produce at this altitude. This leads to the equation

$$W V_{climb} + \frac{4}{3} \sqrt{\frac{2W^3}{\rho S}} \sqrt{3K^3 C_{D_0}} = \eta P^s_{max} \left(\frac{\rho}{\rho^s} \right)^m. \tag{8.57}$$

Since the air density depends on the altitude according to the standard atmospheric model, equation (8.57) can be viewed as an implicit equation in the altitude; the flight ceiling solves this equation.

Flight Envelope

The steady climbing or descending longitudinal flight envelope for a general aviation aircraft consists of all flight conditions for which steady climbing or descending flight can be maintained. The steady climbing or descending longitudinal flight envelope is a set in three dimensions defined by the altitude, airspeed, and climb rate that satisfy the steady flight equations and the stall and engine power constraints.

Since the prior analysis in this chapter provides the basis for computation of the power required and the lift coefficient for steady climbing or descending flight, these constraints can easily be checked. The boundary of the steady climbing or descending flight envelope is defined by surfaces that are obtained by making the maximum power constraint active or by making the stall constraint active. Each of these surfaces is defined in three dimensions; the steady climbing or descending flight envelope is the intersection of these two regions whose boundaries are defined by the two constraint surfaces. Graphical tools are available to visualize this steady climbing or descending flight envelope in three dimensions.

It is also possible to provide a two-dimensional graphical representation of the steady climbing or descending flight envelope for a general aviation aircraft. It is most convenient to represent a cross section of the steady climbing or descending flight envelope, for a fixed value of the climb rate, as a set in two dimensions defined by the altitude and airspeed variables. In particular, specific values of flight altitude and airspeed lie in the cross section of the steady climbing or descending flight envelope if both the stall constraint and the engine power constraints are satisfied at the flight condition corresponding to the fixed value of the climb rate. The steady level flight envelope is such a cross section if the climb rate is zero.

8.5 Steady Climbing Longitudinal Flight: Executive Jet Aircraft

In this section, steady climbing longitudinal flight of the executive jet aircraft with a weight of 73,000 lbs is analyzed. Suppose that the aircraft has a flight path angle of 3 degrees at an altitude of 10,000 ft; our analysis holds over a time period for which the change in air density is insignificant.

Thrust Required

Suppose that the airspeed in this steady climbing flight condition is $V = 500$ ft/s. The thrust required for this steady climbing flight condition is

$$T = W\gamma + \frac{1}{2}\rho V^2 S C_{D_0} + \frac{2KW^2}{\rho V^2 S} = 8,227.2 \text{ lbs.}$$

This computation verifies the obvious fact that the thrust required for a steady climb with a flight path angle of 3 degrees exceeds the thrust required for steady level flight at the same altitude.

More generally, a graphical plot of the thrust required for this steady climbing flight condition as it depends on the aircraft airspeed can be developed. The thrust required curve, given in Figure 8.2, is generated by the following Matlab m-file.

Matlab function 8.1 TRSCFJ.m
```
V=linspace(150,1000,500);
W=73000; S=950; CD0=0.015; K=0.05;
[T p rho]=StdAtpUS(10000);
gamma=3*pi/180;
T=W*gamma+1/2*rho*V.^2*S*CD0+2*K*W^2/rho./V.^2/S;
plot(V,T);
xlabel('Velocity (ft/s)');
ylabel('Thrust required (lbs)');
grid on;
```

Thus Figure 8.2 shows the required thrust to achieve a steady climb with a flight path angle of 3 degrees as a function of the airspeed. It has the typical form: large thrust is required at low airspeeds and at high airspeeds.

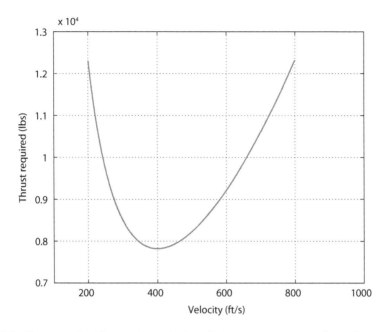

Figure 8.2. Thrust required for steady longitudinal flight: executive jet aircraft ($\gamma = 3$ degrees).

Minimum Required Thrust

A steady climb with a flight path angle of 3 degrees can be achieved by the executive jet aircraft for a range of different airspeeds. As seen in Figure 8.2, there is an airspeed for which the required thrust for a steady climb with flight path angle of 3 degrees is a minimum. This optimal flight condition is now analyzed.

The airspeed to achieve the minimum required thrust for this steady climb can be obtained graphically from Figure 8.2. This airspeed to achieve minimum required thrust for this steady climb occurs at the airspeed that minimizes the drag, namely

$$V = \sqrt{\frac{2W}{\rho S} \sqrt{\frac{K}{C_{D_0}}}} = 399.81 \text{ ft/s.}$$

At this airspeed, the lift coefficient and the pitch moment coefficient are given by

$$C_L = \frac{2W}{\rho V^2 S} = 0.548,$$
$$C_M = 0.$$

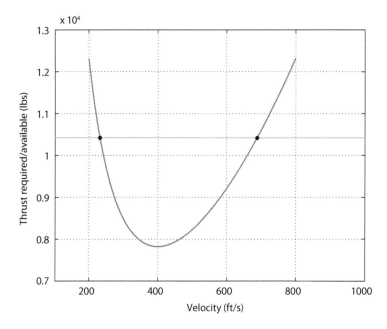

Figure 8.3. Maximum airspeed for steady longitudinal flight: executive jet aircraft ($\gamma = 3$ degrees).

The aircraft angle of attack and the elevator deflection are determined according to

$$0.02 + 0.12\alpha = 0.548,$$
$$0.24 - 0.18\alpha + 0.28\delta_e = 0.$$

Consequently, the angle of attack is 4.40 degrees and the elevator deflection is 1.97 degrees. The minimum thrust required for this steady climb is computed as

$$T_{\min} = W\gamma + 2W\sqrt{KC_{D_0}} = 7,820.6 \text{ lbs.}$$

The corresponding throttle setting for this steady climb is

$$\delta_t = \frac{T_{\min}}{T^s_{\max}}\left(\frac{\rho^s}{\rho}\right)^m = 0.751.$$

In summary, the executive jet aircraft can achieve a minimum thrust steady climb with a flight path angle of 3 degrees at an altitude of 10,000 ft using the following pilot inputs: elevator deflection of 1.97 degrees, throttle setting of 0.751, and zero bank angle. This steady climb condition occurs at an airspeed of 399.81 ft/s and the minimum thrust value is 7,820.6 lbs.

Maximum Airspeed

The maximum airspeed of the executive jet aircraft in a steady climb with a flight path angle of 3 degrees at an altitude of 10,000 ft is limited by the engine thrust. This maximum airspeed can be obtained graphically from the intersection of the thrust required curve and the curve for maximum thrust available from the jet engine at this altitude, as shown in Figure 8.3.

The maximum airspeed can also be obtained by solving the equation

$$W\gamma + \frac{1}{2}\rho V^2 SC_{D_0} + \frac{2KW^2}{\rho V^2 S} = T_{max}^s \left(\frac{\rho}{\rho^s}\right)^m.$$

Multiplying the above equation by V^2, the following fourth degree polynomial equation is obtained:

$$\frac{1}{2}\rho SC_{D_0} V^4 + \left[W\gamma - T_{max}^s \left(\frac{\rho}{\rho^s}\right)^m\right] V^2 + \frac{2KW^2}{\rho S} = 0.$$

This equation can be solved numerically using the Matlab function MMSSCFJ.

Matlab function 8.2 MMSSCFJ.m
```
function [Vmin,Vmax]=MMSSCFJ
%Output: Minimum air speed and maximum air speed in a climb
W=73000; S=950; CD0=0.015; K=0.05; Tsmax=12500; m=0.6;
[Ts ps rhos]=StdAtpUS(0);
[Th ph rhoh]=StdAtpUS(h);
gamma=3*pi/180;
[Vmin,Vmax]=sort(roots([1/2*rho*S*CD0 0 W*gamma-Tsmax*
(rho/rhos)^m 0 2*K*W^2/rho/S]));
```

Matlab computes two positive solution values of 688.25 ft/s and 232.26 ft/s. The high-speed solution is the maximum airspeed of the aircraft. At the given altitude, the speed of sound is 1,077.39 ft/s. Consequently, the corresponding Mach number of the aircraft in this steady climbing flight condition is 0.64. The maximum airspeed is not close to the speed of sound.

At this maximum airspeed, the lift coefficient and the pitch moment coefficient are

$$C_L = \frac{2W}{\rho V^2 S} = 0.185,$$

$$C_M = 0.$$

The aircraft angle of attack and the elevator deflection are determined to be $\alpha = 1.37$ degrees and $\delta_e = 0.026$ degrees.

In summary, the maximum airspeed for the executive jet aircraft in a steady climb with a flight path angle of 3 degrees at an altitude of 10,000 ft is achieved with the following pilot inputs: elevator deflection of 0.026 degrees, full throttle, and zero bank angle. The maximum airspeed in this steady climbing flight condition is 688.25 ft/s.

Minimum Airspeed

The minimum airspeed for the executive jet aircraft in a steady climb with a flight path angle of 3 degrees is limited by the jet engine thrust constraint and by the stall constraint.

Since the condition for stall occurs at the maximum value of the lift coefficient $C_{L_{max}} = 2.8$, the airspeed at stall with a flight path angle is of 3 degrees is:

$$V_{stall} = \sqrt{\frac{2W}{\rho S C_{L_{max}}}} = 176.83 \text{ ft/s.}$$

As seen in the prior section, the minimum airspeed due to the thrust constraint is 232.26 ft/s. In this case, the minimum airspeed due to the thrust constraint is larger than the stall speed, so the maximum thrust constraint is active. Consequently, the minimum airspeed in this climbing flight condition is 232.26 ft/s. At this minimum airspeed, the lift coefficient and the pitch moment coefficient are

$$C_L = \frac{2W}{\rho V^2 S} = 1.62,$$
$$C_M = 0.$$

This allows determination of the aircraft angle of attack and the elevator deflection as $\alpha = 13.36$ degrees and $\delta_e = 7.73$ degrees, respectively.

In summary, the minimum airspeed for the executive jet aircraft in a steady climb with a flight path angle of 3 degrees at an altitude of 10,000 ft is achieved with the following pilot inputs: elevator deflection of 7.73 degrees, full throttle, and zero bank angle. The minimum airspeed in this steady climbing flight condition is 232.26 ft/s.

Maximum Flight Path Angle

The executive jet aircraft can maintain steady longitudinal flight at many different flight path angles. Due to the flight constraints that must be satisfied, especially the limit on the maximum thrust that the jet engines can provide, there is a maximum flight path angle. Flight conditions for a maximum flight path angle in longitudinal flight are analyzed in this section.

The flight path angle is the ratio of the excess thrust and the weight, where the excess thrust is the difference between the engine thrust and the drag. To maximize the flight path angle, it is clear that the thrust provided by the engine should be a maximum; that is, full throttle should be used. In addition, the airspeed should be selected to minimize the drag.

The maximum flight path angle in steady longitudinal flight at an altitude of 10,000 ft is

$$\gamma_{max} = \frac{T^s_{max}}{W}\left(\frac{\rho}{\rho^s}\right)^m - 2\sqrt{KC_{D_0}} = 8.7982 \times 10^{-2} \text{ rad} = 5.04 \text{ degrees.}$$

The airspeed for which the flight path angle is maximum is given by

$$V = \sqrt{\frac{2W}{\rho S}} \sqrt{\frac{K}{C_{D_0}}} = 399.8 \text{ ft/s}.$$

At this flight condition for maximum flight path angle, the lift coefficient and the pitch moment coefficient are

$$C_L = \frac{2W}{\rho V^2 S} = 0.548,$$
$$C_M = 0.$$

The aircraft angle of attack $\alpha = 4.39$ degrees and the elevator deflection $\delta_e = 1.97$ degrees.

In summary, the maximum flight path angle of the executive jet aircraft at 10,000 ft is 5.04 degrees corresponding to the following pilot inputs: elevator deflection of 1.97 degrees, full throttle, and zero bank angle.

Steady Climbing Flight Ceiling

The flight ceiling for the executive jet aircraft in a steady climbing flight condition with a flight path angle of 3 degrees occurs when the engine thrust is maximum. This flight ceiling can be determined by solving

$$W\gamma + 2W\sqrt{KC_{D_0}} - T_{\text{max}}^s \left(\frac{\rho}{\rho^s}\right)^m = 0,$$

where ρ is a function of altitude according to the standard atmospheric model. The Matlab function fsolve is used to solve the nonlinear equation defined by a zero value of the Matlab function eqnFCCJ . The solution is the desired value of the flight ceiling.

Matlab function 8.3 eqnFCCJ.m
```
function error=eqnFCCJ(h,gamma)
W=73000; CD0=0.015; K=0.05; Tsmax=12500; m=0.6;
[Ts ps rhos]=StdAtpUS(0);
[Th ph rhoh]=StdAtpUS(h);
error=W*gamma+2*W*sqrt(K*CD0)-Tsmax*(rhoh/rhos)^m;
```

The following Matlab command computes the value of the flight ceiling.

```
hmax=fsolve(@(h) eqnFCCJ(h,3*pi/180),10000)
```

where 10,000 is an initial guess of the flight ceiling. After a few iterations, the value of the flight ceiling of 24,401.6 ft is obtained.

In summary, the flight ceiling for the executive jet aircraft in a steady climb with a flight path angle of 3 degrees is 24,401.6 ft.

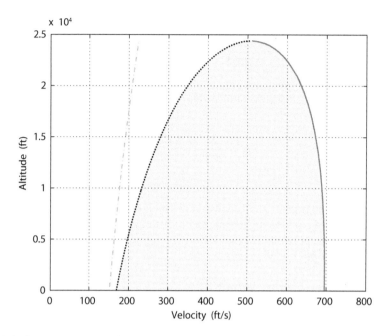

Figure 8.4. Steady longitudinal flight envelope: executive jet aircraft ($\gamma = 3$ degrees).

Steady Climbing Flight Envelope

The steady climbing longitudinal flight envelope with a flight path angle of 3 degrees consists of the set of all pairs of altitude and airspeed values, satisfying the jet engine constraints and the stall constraint, for which this steady climbing flight condition can be maintained.

A graphical representation of the steady climbing flight envelope using the previous analysis is now developed. By following the same procedure described previously, the maximum airspeed, the minimum airspeed due to the thrust constraint, and the stall speed are computed for various altitudes. Define the Matlab function SCFJ that computes the maximum airspeed, the minimum airspeed due to the thrust constraint, and the stall speed for a specified altitude.

Matlab function 8.4 SCFJ.m

```
function [Vmax VminTC Vstall]=SCFJ(h,gamma);
%Input: altitude h (ft) flight path angle gamma (rad)
%Output: max air speed Vmax (ft/s), min air speed due to the
thrust constraint VminTC (ft/s)
%stall speed Vstall (ft/s);
W=73000; S=950; CD0=0.015; K=0.05; Tsmax=12500; m=0.6;
CLmax=2.8;
[Ts ps rhos]=StdAtpUS(0);
```

(continued)

(continued)

```
[T p rho]=StdAtpUS(h);
tmp=sort(roots([1/2*rho*S*CD0 0 W*gamma-Tsmax*(rho/rhos)^m 0
2*K*W^2/rho/S]));
Vmax=tmp(4);
VminTC=tmp(3);
Vstall=sqrt(2*W/rho/S/CLmax);
```

The steady climbing flight envelope is determined by evaluating the SCFJ function at various altitudes. A graphical representation of this steady climbing flight envelope is obtained using the following Matlab m-file.

Matlab function 8.5 FigSCFJ.m
```
gamma=3*pi/180;
h=linspace(0,24400,500);
for k=1:size(h,2)
    [Vmax(k) VminTC(k) Vstall(k)]=SCFJ(h(k),gamma);
end
Vmin=max(VminTC,Vstall);
area([Vmin Vmax(end:-1:1)],[h h(end:-1:1)],...
    'FaceColor',[0.8 1 1],'LineStyle','none');
hold on;
plot(Vmax,h,VminTC,h,Vstall,h);
grid on;
xlim([0 1200]);
xlabel('Velocity (ft/s)');
ylabel('Altitude (ft)');
```

The steady climbing flight envelope is shown in Figure 8.4 assuming that the jet aircraft is in a steady climb with a flight path angle of 3 degrees.

8.6 Steady Descending Longitudinal Flight: Executive Jet Aircraft

In this section, an analysis is made of the executive jet aircraft in a steady descending flight condition with a flight path angle of -2 degrees at an altitude of 10,000 ft. The weight of the aircraft is 73,000 lbs.

Thrust Required

Suppose that the airspeed in this steady descending flight condition is 500 ft/s. The thrust required for this steady descending flight condition is

$$T = W\gamma + \frac{1}{2}\rho V^2 S C_{D_0} + \frac{2KW^2}{\rho V^2 S} = 1,856.7 \text{ lbs.}$$

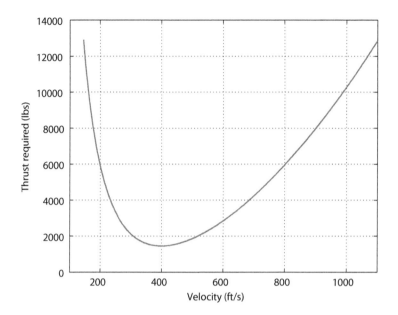

Figure 8.5. Thrust required for steady longitudinal flight: executive jet aircraft ($\gamma = -2$ degrees).

This confirms the obvious fact that the thrust required for steady descending flight is smaller than the thrust required for steady level flight.

Next, the thrust required for this steady descending flight condition is computed for varying airspeeds. The thrust required curve, shown in Figure 8.5, is developed by the following Matlab m-file.

Matlab function 8.6 TRSDFJ.m
```
V=linspace(145,1100,500);
W=73000; S=950; CD0=0.015; K=0.05;
[T p rho]=StdAtpUS(10000);
gamma=-2*pi/180;
T=W*gamma+1/2*rho*V.^2*S*CD0+2*K*W^2/rho./V.^2/S;
plot(V,T);
xlabel('Velocity (ft/s)');
ylabel('Thrust required (lbs)');
grid on;
```

The thrust required curve for steady descending flight shows the typical features: the required thrust for steady descending flight is large at low airspeeds and at high airspeeds.

Minimum Required Thrust

The executive jet aircraft can achieve steady descent with a flight path angle of -2 degrees at a range of airspeeds. As seen in Figure 8.5, there is an airspeed at which the thrust required for steady descending flight at the specified negative flight path angle is a minimum. This minimum could be determined graphically from Figure 8.5.

The minimum required thrust for steady descending flight occurs at the airspeed for which the drag is a minimum. The value of the airspeed for which the drag is a minimum can be computed according to

$$V = \sqrt{\frac{2W}{\rho S} \sqrt{\frac{K}{C_{D_0}}}} = 399.81 \text{ ft/s.}$$

At this airspeed, the lift coefficient and the pitch moment coefficient are

$$C_L = \frac{2W}{\rho V^2 S} = 0.548,$$

$$C_M = 0.$$

These expressions allow determination of the aircraft angle of attack and the elevator deflection according to

$$0.02 + 0.12\alpha = 0.548,$$

$$0.24 - 0.18\alpha + 0.28\delta_e = 0.$$

Thus, the angle of attack is 4.40 degrees and the elevator deflection is 1.97 degrees.

The minimum thrust for this steady descending flight condition is computed from

$$T_{\min} = W\gamma + 2W\sqrt{KC_{D_0}} = 1,450.2 \text{ lbs.}$$

The corresponding throttle setting is

$$\delta_t = \frac{T_{\min}}{T_{\max}^s} \left(\frac{\rho^s}{\rho}\right)^m = 0.14.$$

In summary, the minimum required thrust for steady descending flight with a flight path angle of -2 degrees at an altitude of 10,000 ft is achieved using the following pilot inputs: elevator deflection of 1.97 degrees, throttle setting of 0.14, and zero bank angle. The value of the minimum required thrust is 1,450.2 lbs.

Maximum Airspeed

As shown in Figure 8.6, the maximum airspeed of the executive jet aircraft for steady descending flight with a flight path angle of -2 degrees at an altitude of 10,000 ft occurs when the maximum engine thrust curve intersects the thrust required curve.

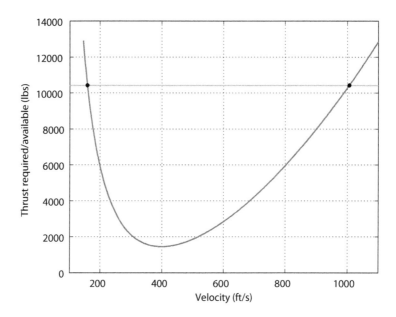

Figure 8.6. Maximum airspeed for steady longitudinal flight: executive jet aircraft ($\gamma = -2$ degrees).

The airspeed at which this steady descending flight condition holds is found by solving

$$W\gamma + \frac{1}{2}\rho V^2 SC_{D_0} + \frac{2KW^2}{\rho V^2 S} = T^s_{\max}\left(\frac{\rho}{\rho^s}\right)^m.$$

Multiplying the above equation by V^2, the following fourth degree polynomial equation is obtained:

$$\frac{1}{2}\rho SC_{D_0} V^4 + \left[W\gamma - T^s_{\max}\left(\frac{\rho}{\rho^s}\right)^m\right] V^2 + \frac{2KW^2}{\rho S} = 0.$$

This equation can be solved numerically using the Matlab function MMSSDFJ.

Matlab function 8.7 MMSSDFJ.m
```
function [Vmin,Vmax]=MMSSDFJ
%Output: Minimum air speed and maximum air speed in descent
W=73000; S=950; CD0=0.015; K=0.05; Tsmax=12500; m=0.6;
[Ts ps rhos]=StdAtpUS(0);
[Th ph rhoh]=StdAtpUS(h);
gamma=-2*pi/180;
[Vmin,Vmax]=sort(roots([1/2*rho*S*CD0 0 W*gamma-Tsmax*(rho/rhos)
^m 0 2*K*W^2/rho/S]));
```

Matlab provides two positive values that solve this polynomial equation: 1,005.86 ft/s and 158.92 ft/s. The high-speed solution is the maximum airspeed of the aircraft in this steady descending flight condition. At the given altitude, the speed of sound is 1,077.39 ft/s. The corresponding Mach number of the executive jet aircraft in this steady descending flight condition is 0.93. In this case, the maximum airspeed is close to the speed of sound. In this flight region, compressibility effects are likely to be important. The computed maximum airspeed for this steady descending flight condition may not be accurate.

At this maximum airspeed in this steady descending flight condition, the lift coefficient and the pitch moment coefficient are

$$C_L = \frac{2W}{\rho V^2 S} = 0.0865,$$

$$C_M = 0.$$

These expressions allow determination of an aircraft angle of attack of 0.55 degrees and an elevator deflection of −0.5 degrees.

In summary, the maximum airspeed for steady descending flight with a flight path angle of −2 degrees at an altitude of 10,000 ft is achieved using the following pilot inputs: elevator deflection of −0.5 degrees, full throttle, and zero bank angle. This maximum airspeed is approximately 1,005.86 ft/s.

Minimum Airspeed

The minimum airspeed of the executive jet aircraft in a steady descent must satisfy both the stall constraint and the maximum thrust constraint of the jet engine. The stall speed of the executive jet aircraft in steady descending flight with a flight path angle of −2 degrees occurs when the lift coefficient has its maximum value $C_{L_{max}} = 2.8$. The stall speed is

$$V_{stall} = \sqrt{\frac{2W}{\rho S C_{L_{max}}}} = 176.83 \text{ ft/s}.$$

The minimum airspeed due to the jet engine thrust constraint is 158.92 ft/s. In this case, the minimum airspeed due to the thrust constraint is less than the stall speed, so the stall constraint is active and the engine constraints are inactive. Consequently, the minimum airspeed of the aircraft is equal to the stall speed, namely 176.63 ft/s. At the minimum airspeed in steady descending flight, the lift coefficient and the pitch moment coefficient are

$$C_L = \frac{2W}{\rho V^2 S} = 2.8,$$

$$C_M = 0.$$

These expressions determine an aircraft angle of attack of 23.17 degrees and an elevator deflection of 14.04 degrees.

In summary, the minimum airspeed for steady descending flight with a flight path angle of −2 degrees at an altitude of 10,000 ft is achieved using the following pilot

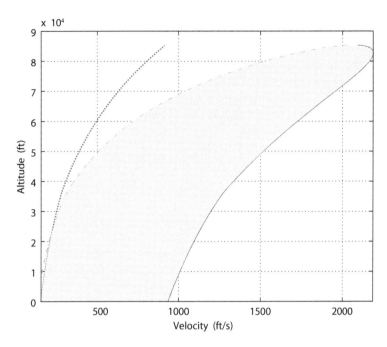

Figure 8.7. Steady longitudinal flight envelope: executive jet aircraft ($\gamma = -2$ degrees).

inputs: elevator deflection of 14.04 degrees, 81% throttle, and zero bank angle. The minimum airspeed in this steady descent is 176.63 ft/s.

Steady Descending Flight Envelope

The steady descending flight envelope of the executive jet aircraft is developed, assuming that the aircraft is in steady descending flight with a flight path angle of -2 degrees. This steady descending flight envelope consists of the set of all pairs of altitude and airspeed, satisfying the jet engine thrust constraints and the stall constraint, for which this steady descending flight condition can be maintained.

A graphical representation of the steady descending flight envelope, assuming a steady descent with a flight path angle of -2 degrees, is developed using the previous analysis. By following the same procedure described above, the maximum airspeed, the minimum airspeed due to the thrust constraint, and the stall speed can be computed at various altitudes. The previously defined Matlab function SCFJ is used to compute the maximum airspeed, the minimum airspeed due to the maximum thrust constraint, and the stall speed for a specified altitude. Then, the steady descending flight envelope is obtained, as shown in Figure 8.7. The inaccuracy of the drag assumptions in the transonic and supersonic range invalidates the high-speed part of this flight envelope; the physical significance of the high altitude part of this flight envelope is questionable.

8.7 Steady Longitudinal Flight Envelopes: Executive Jet Aircraft

The steady longitudinal flight envelope for the executive jet aircraft is defined in terms of the flight altitude, the airspeed of the aircraft, and the flight path angle of the aircraft. This steady longitudinal flight envelope is defined in terms of three flight variables and hence must be viewed as a set in three dimensions. This notion generalizes the steady level flight envelope and the steady climbing or descending flight envelope for a fixed flight path angle. In particular, the steady level flight envelope can be viewed as a two-dimensional cross section of the steady longitudinal flight envelope corresponding to a flight path angle of zero.

A steady flight surface can be computed that defines the boundary of the steady longitudinal flight envelope by evaluating the Matlab SCFJ function for various altitudes and flight path angles. This approach allows development of a graphical representation of the steady longitudinal flight envelope using the Matlab graphics features. The boundary surface of the steady longitudinal flight envelope is shown in Figure 8.8. It is obtained from the following Matlab m-file.

Matlab function 8.8 FigSCDFJ.m
```
gamma=linspace(-2*pi/180,5*pi/180,40);
for i=1:size(gamma,2)
    hmax(i)=fsolve(@(h) eqnFCCJ(h,gamma(i)),10000);
    h(i,:)=linspace(0,floor(hmax(i)),40);
    for j=1:size(h,2)
        [Vmax(i,j)VminTC(i,j)Vstall(i,j)]=SCFJ(h(i,j),gamma(i));
    end
end
Vmin=max(VminTC,Vstall);
surf(Vmax,gamma*180/pi,h,'LineStyle','none');
hold on;
surf(Vmin,gamma*180/pi,h,'LineStyle','none');
xlabel('Velocity (ft/s)');
ylabel('Flight path angle (deg)');
zlabel('Altitude (ft)');
```

Any triple of values for the flight altitude, aircraft airspeed, and aircraft flight path angle that lies within the steady longitudinal flight envelope illustrated in Figure 8.8 corresponds to a feasible steady flight condition for which the jet engine maximum thrust constraint and the stall constraint are satisfied. For each such feasible triple of values, methodology has been provided to compute the pilot inputs, namely the elevator deflection, the throttle setting, and zero bank angle, that give this steady flight condition. In addition, the methodology allows computation of the associated angle of attack, values of the lift and drag coefficients, and values of the lift and drag.

The steady level flight envelope in Figure 8.4 is a two-dimensional cross section of the three-dimensional flight envelope illustrated in Figure 8.8.

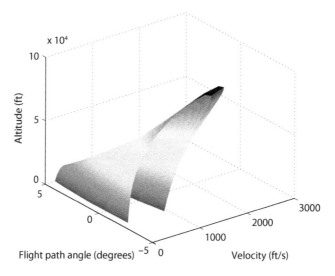

Figure 8.8. *Steady longitudinal flight envelope: executive jet aircraft.*

8.8 Steady Climbing Longitudinal Flight: General Aviation Aircraft

In this section, steady climbing longitudinal flight of the general aviation aircraft is analyzed. The weight of the aircraft is 2,900 lbs. Suppose that the aircraft has a rate of climb of 16 ft/s at an altitude of 10,000 ft; our analysis holds over a time period for which the change in air density is insignificant.

Power Required

Suppose that the airspeed in this steady climbing flight condition is 200 ft/s. The power required for this steady climbing flight condition is

$$P = W V_{climb} + \frac{1}{2}\rho V^3 S C_{D_0} + \frac{2K W^2}{\rho V S} = 93,131 \, \text{ft lbs/s} = 169.33 \, \text{hp}.$$

This computation verifies the obvious fact that the power required for a steady climb with a rate of climb of 16 ft/s exceeds the power required for steady level flight at the same altitude.

More generally, a graphical plot of the power required curve for this steady climbing flight condition, as it depends on the aircraft airspeed, is given in Figure 8.9. This curve is generated by the following Matlab m-file:

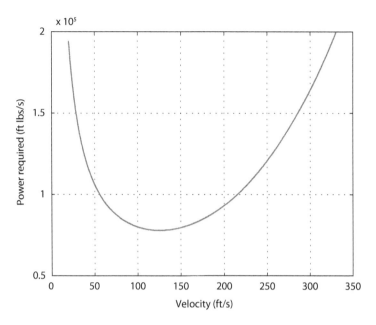

Figure 8.9. Power required for steady longitudinal flight: general aviation aircraft ($V_{climb} = 16$ ft/s).

Matlab function 8.9 PRSCFP.m

```
V=linspace(20,350,500);
W=2900; S=175; CD0=0.026; K=0.054; rho=1.7553e-3; Vclimb=16;
P=W*Vclimb+1/2*rho*V.^3*S*CD0+2*K*W^2/rho./V/S
plot(V,P);
xlabel('Velocity (ft/s)');
ylabel('Power required (ft lbs/s)');
grid on;
```

Thus Figure 8.9 shows the required power to achieve a steady climb with climb rate of 16 ft/s as a function of the airspeed. It has the typical form; high power is required at both low and high airspeeds.

Minimum Required Power

A steady climb with a rate of climb of 16 ft/s can be achieved by the general aviation aircraft for a range of different airspeeds. As seen in Figure 8.9, there is an airspeed for which the required power for a steady climb with a climb rate of 16 ft/s is a minimum. This flight condition is now analyzed.

The airspeed to achieve this minimum required power for this steady climb can be obtained graphically from Figure 8.9. This is also the airspeed at which the power required to balance the drag is a minimum, namely

$$V = \sqrt{\frac{2W}{\rho S} \sqrt{\frac{K}{3C_{D_0}}}} = 125.34 \text{ ft/s}.$$

At this airspeed, the lift coefficient and the pitch moment coefficient are given by

$$C_L = \frac{2W}{\rho V^2 S} = 1.21,$$
$$C_M = 0.$$

The aircraft angle of attack and the elevator deflection are determined according to

$$0.02 + 0.12\alpha = 1.21,$$
$$0.12 - 0.08\alpha + 0.075\delta_e = 0.$$

Thus, the angle of attack is 9.849 degrees and the elevator deflection is 8.91 degrees. The minimum power required for this steady climb is

$$P_{\min} = W V_{climb} + \frac{4}{3}\sqrt{\frac{2W^3}{\rho S} \sqrt{3K^3 C_{D_0}}} = 77{,}854 \text{ ft lbs/s} = 141.55 \text{ hp}.$$

The corresponding throttle setting for this steady climb is

$$\delta_t = \frac{P_{\min}}{\eta P_{\max}^s} \left(\frac{\rho^s}{\rho}\right)^m = 0.732.$$

In summary, the propeller aircraft can achieve a minimum power steady climb with a climb rate of 16 ft/s at an altitude of 10,000 ft using the following pilot inputs: elevator deflection of 8.91 degrees, throttle setting of 0.732, and zero bank angle. This steady climb condition occurs at an airspeed of 125.34 ft/s.

Maximum Airspeed

The maximum airspeed of the general aviation aircraft in a steady climb with a climb rate of 16 ft/s at an altitude of 10,000 ft is limited by the engine power. The maximum airspeed can be obtained graphically using Figure 8.10; the maximum airspeed occurs at the intersection of the power required curve and the curve defined by the maximum power that the engine and propeller can provide at this altitude.

The maximum airspeed can also be obtained by solving the equation

$$W V_{climb} + \frac{1}{2}\rho V^3 S C_{D_0} + \frac{2KW^2}{\rho V S} = \eta P_{\max}^s \left(\frac{\rho}{\rho^s}\right)^m.$$

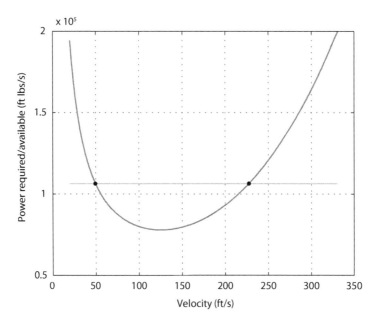

Figure 8.10. Maximum airspeed for steady longitudinal flight: general aviation aircraft (V_{climb} = 16 ft/s).

Multiplying the above equation by V, the following fourth degree polynomial equation is obtained:

$$\frac{1}{2}\rho S C_{D_0} V^4 + \left(W V_{climb} - \eta P_{max}^s \left(\frac{\rho}{\rho^s} \right)^m \right) V + \frac{2 K W^2}{\rho S} = 0.$$

This equation can be solved numerically using the Matlab function MMSSCFP as follows:

Matlab function 8.10 MMSSCFP.m
```
function [Vmin,Vmax]=MMSSCFP
%Output: Minimum air speed and maximum air speed in a climb
W=2900; S=175; CD0=0.026; K=0.054; Psmax=290*550; m=0.6; eta=0.8;
[Ts ps rhos]=StdAtpUS(0);
[Th ph rhoh]=StdAtpUS(h);
Vclimb=16;
[Vmin,Vmax]=sort(roots([1/2*rho*S*CD0 0 0 W*Vclimb-eta*Psmax*
(rho/rhos)^m 2*K*W^2/rho/S]));
```

Matlab computes two positive solution values of 227.43 ft/s and 49.70 ft/s. The high-speed solution is the maximum airspeed of the aircraft. At the given altitude, the speed of sound is 1,077.39 ft/s. Consequently, the corresponding Mach number of the aircraft in this steady climbing flight condition is 0.21. The maximum airspeed is much less than the speed of sound.

At this maximum airspeed, the lift coefficient and the pitch moment coefficient are

$$C_L = \frac{2W}{\rho V^2 S} = 0.365,$$

$$C_M = 0.$$

Thus, the aircraft angle of attack is 2.88 degrees and the elevator deflection is 1.47 degrees.

In summary, the maximum airspeed for the general aviation aircraft in a steady climb with a climb rate of 16 ft/s at an altitude of 10,000 ft is achieved with the following pilot inputs: elevator deflection of 1.47 degrees, full throttle, and zero bank angle. The maximum airspeed in this steady climbing flight condition is 227.43 ft/s.

Minimum Airspeed

The minimum airspeed for the general aviation aircraft in a steady climb with a climb rate of 16 ft/s is limited by the engine power constraint and by the stall constraint.

Since the condition for stall occurs at the maximum value of the lift coefficient $C_{L_{max}} = 2.4$, the airspeed at stall is

$$V_{stall} = \sqrt{\frac{2W}{\rho S C_{L_{max}}}} = 88.7 \, \text{ft/s}.$$

As seen in the prior section, the minimum airspeed due to the engine power constraint is 49.70 ft/s. In this case, the stall speed is larger than the minimum airspeed due to the maximum power constraint, so the stall constraint is active and the engine power constraint is not active. Consequently, the minimum airspeed in this steady climbing flight condition is 88.7 ft/s. At this minimum airspeed, the lift coefficient and the pitch moment coefficient are

$$C_L = 2.4,$$

$$C_M = 0.$$

This allows determination of an aircraft angle of attack of 19.83 degrees and an elevator deflection of 19.55 degrees.

In summary, the minimum airspeed for the general aviation aircraft in a steady climb with a climb rate of 16 ft/s at an altitude of 10,000 ft is achieved with the following pilot inputs: elevator deflection of 19.6 degrees, 34.5% throttle, and zero bank angle. The minimum airspeed in this steady climbing flight condition is 88.7 ft/s.

Maximum Climb Rate

The general aviation aircraft can maintain steady longitudinal flight at many different climb rates. Due to the flight constraints that must be satisfied, especially the limit on the maximum power that the engine and propeller can provide, there is a maximum climb rate. Flight conditions for a maximum climb rate in longitudinal flight are analyzed in this section.

The rate of climb is the ratio of the excess power and the weight, where the excess power is the difference between the power produced by the engine and propeller and the power required to overcome the drag and to climb. To maximize the climb rate, the power provided by the engine and propeller should be a maximum; that is, full throttle should be used. In addition, the airspeed should be selected to minimize the power to overcome the drag. The maximum climb rate in steady longitudinal flight at an altitude of 10,000 ft is

$$(V_{climb})_{max} = \frac{\eta P_{max}^s}{W} \left(\frac{\rho}{\rho^s}\right)^m - \frac{4}{3} \sqrt{\frac{2W}{\rho S}} \sqrt{3K^3 C_{D_0}} = 25.84 \, \text{ft/s}.$$

The airspeed for which the climb rate is maximum is

$$V = \sqrt{\frac{2W}{\rho S} \sqrt{\frac{K}{3C_{D_0}}}} = 125.34 \, \text{ft/s}.$$

At this flight condition for the maximum climb rate, the lift coefficient and the pitch moment coefficient are

$$C_L = \frac{2W}{\rho V^2 S} = 1.2,$$

$$C_M = 0.$$

The aircraft angle of attack is 9.85 degrees and the elevator deflection is 8.9 degrees.

In summary, the maximum climb rate of the general aviation aircraft at 10,000 ft is 25.84 ft/s corresponding to the following pilot inputs: elevator deflection of 8.9 degrees, full throttle, and zero bank angle.

Steady Climbing Flight Ceiling

The flight ceiling for the general aviation aircraft in a steady climbing flight condition with a 16 ft/s climb rate occurs when the power provided by the engine and propeller is maximum and equal to the power required for steady climbing flight. This flight ceiling is determined by solving the equation

$$W V_{climb} + \frac{4}{3} \sqrt{\frac{2W^3}{\rho S}} \sqrt{3K^3 C_{D_0}} - \eta P_{max}^s \left(\frac{\rho}{\rho^s}\right)^m = 0,$$

where ρ is a function of altitude according to the standard atmospheric model. The function on the left-hand side of this equation is defined by the Matlab function eqnFCCP.

Matlab function 8.11 eqnFCCP.m
```
function error=eqnFCCP(h,Vclimb)
%Input: altitude h (ft), climb rate Vclimb (ft/s)
%Output: error
W=2900; S=175; CD0=0.026; K=0.054; Psmax=290*550; m=0.6; eta=0.8;
[Ts ps rhos]=StdAtpUS(0);
[Th ph rhoh]=StdAtpUS(h);
error=W*Vclimb+4/3*sqrt(2*W^3/rhoh/S*sqrt(3*K^3*CD0))
-eta*Psmax*(rhoh/rhos)^m;
```

The following Matlab command computes the flight ceiling for which the value of the function defined in eqnFCCP is zero.

```
hmax=fsolve(@(h) eqnFCCP(h,16),10000)
```

where 10,000 is an initial guess of the flight ceiling. After a few iterations, the value of the flight ceiling of 21,675.4 ft is obtained.

In summary, the flight ceiling for the general aviation aircraft in a steady climb with a climb rate of 16ft/s is 21,675.4 ft.

Steady Climbing Flight Envelope

The steady climbing longitudinal flight envelope of the general aviation aircraft is developed, assuming a rate of climb of 16 ft/s. This steady climbing flight envelope consists of the set of all pairs of altitude and airspeed values, satisfying the power constraints and the stall constraint.

A graphical representation of the steady climbing flight envelope using the previous analysis is now developed. The maximum airspeed, the minimum airspeed due to the maximum power constraint, and the stall speed are computed for various altitudes. Define the Matlab function SCFP that computes the maximum airspeed, the minimum airspeed due to the maximum power constraint, and the stall speed for a specified altitude and climb rate.

Matlab function 8.12 SCFP.m
```
function [Vmax VminPC Vstall]=SCFP(h,Vclimb);
%Input: altitude h (ft)
%Output: max air speed Vmax (ft/s), min air speed due to thrust
constraint VminPC (ft/s),
%stall speed Vstall (ft/s)
W=2900; S=175; CD0=0.026; K=0.054; Psmax=290*550; m=0.6;
CLmax=2.4; eta=0.8;
[Ts ps rhos]=StdAtpUS(0);
[T p rho]=StdAtpUS(h);
tmp=sort(roots([1/2*rho*S*CD0 0 0 W*Vclimb-eta*Psmax*
(rho/rhos)^m
2*K*W^2/rho/S]));
```

(continued)

(continued)

```
Vmax=tmp(2);
VminPC=tmp(1);
Vstall=sqrt(2*W/rho/S/CLmax);
```

The steady climbing flight envelope is determined by evaluating the `SCFP` function at various altitudes for the fixed value of the climb rate. A graphical representation of this steady climbing flight envelope is obtained using the following Matlab m-file.

Matlab function 8.13 FigSCFP.m
```
h=linspace(0,21675,500);
for k=1:size(h,2)
    [Vmax(k) VminTC(k) Vstall(k)]=SCFP(h(k),16);
end
Vmin=max(VminTC,Vstall);
area([Vmin Vmax(end:-1:1)],[h h(end:-1:1)],...
    'FaceColor',[0.8 1 1],'LineStyle','none');
hold on;
plot(Vmax,h,VminTC,h,Vstall,h);
grid on;
xlim([0 400]);
xlabel('Velocity (ft/s)');
ylabel('Altitude (ft)');
```

The steady climbing flight envelope is shown in Figure 8.11, assuming that the aircraft is in steady climb with a climb rate of 16 ft/s.

8.9 Steady Descending Longitudinal Flight: General Aviation Aircraft

In this section, an analysis is made of the general aviation aircraft, with a weight of 2,900 lbs, in a steady descending flight condition with a climb rate of -10 ft/s at an altitude of 10,000 ft.

Power Required

Suppose that the airspeed in this steady descending flight condition is $V = 200$ ft/s. The power required for this steady descending flight condition is

$$P = W V_{climb} + \frac{1}{2}\rho V^3 S C_{D_0} + \frac{2KW^2}{\rho V S} = 17,731 \text{ lb ft/s} = 32.24 \text{ hp}.$$

This confirms the obvious fact that the power required for steady descending flight is smaller than the power required for steady level flight.

Next, the power required for this steady descending flight condition is computed for varying airspeeds. The power required curve for this steady descending flight, shown in Figure 8.12, is developed by the following Matlab m-file.

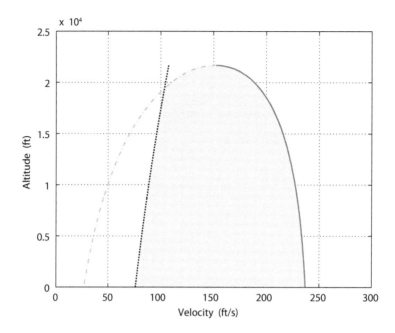

Figure 8.11. *Steady longitudinal flight envelope: general aviation aircraft* ($V_{climb} = 16$ ft/s).

Matlab function 8.14 PRSDFP.m
```
V=linspace(20,350,500);
W=2900; S=175; CD0=0.026; K=0.054; rho=1.7553e-3;
Vclimb=-10;
P=W*Vclimb+1/2*rho*V.^3*S*CD0+2*K*W^2/rho./V/S plot(V,P);
xlabel('Velocity (ft/s)');
ylabel('Power required (ft lbs/s)');
grid on;
```

The power required curve for steady descending flight shows the typical features: the required power for steady descending flight is high at both low and high airspeeds.

Minimum Required Power

The general aviation aircraft can achieve steady descent with a climb rate of -10 ft/s at a range of airspeeds. As seen in Figure 8.12, there is an airspeed at which the power required for steady descending flight at the specified negative flight path angle is a minimum. This minimum could be determined graphically from Figure 8.12.

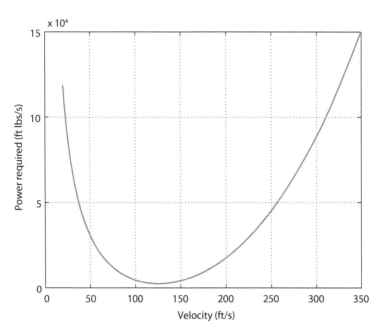

Figure 8.12. Power required for steady longitudinal flight: general aviation aircraft (V_{climb} = − 10 ft/s).

The minimum required power for steady descending flight occurs at the airspeed for which the power required to overcome the drag is a minimum. The value of the airspeed for which this power to overcome the drag is a minimum is

$$V = \sqrt{\frac{2W}{\rho S} \sqrt{\frac{K}{3 C_{D_0}}}} = 125.34 \, \text{ft/s}.$$

At this airspeed, the lift coefficient and the pitch moment coefficient are

$$C_L = \frac{2W}{\rho V^2 S} = 1.2,$$

$$C_M = 0.$$

These expressions allow determination of the aircraft angle of attack and the elevator deflection according to

$$0.02 + 0.12\alpha = 1.2,$$

$$0.12 - 0.08\alpha + 0.075\delta_e = 0.$$

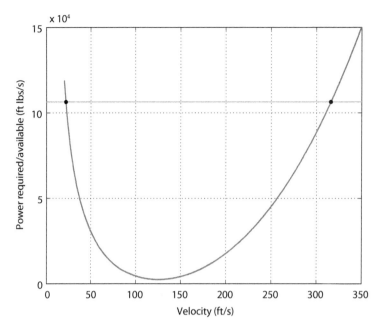

Figure 8.13. Maximum airspeed for steady longitudinal flight: general aviation aircraft (V_{climb} = − 10 ft/s).

Consequently, the angle of attack is 9.85 degrees and the elevator deflection is 8.91 degrees. The minimum power for this steady descending flight condition is

$$P_{min} = W V_{climb} + \frac{4}{3} \sqrt{\frac{2W^3}{\rho S}} \sqrt{3K^3 C_{D_0}} = 2{,}453.9 \text{ ft lbs/s} = 4.46 \text{ hp}.$$

The corresponding throttle setting is

$$\delta_t = \frac{P_{min}}{\eta P^s_{max}} \left(\frac{\rho^s}{\rho} \right)^m = 0.023.$$

In summary, the minimum required power for steady descending flight with a climb rate of −10 ft/s at an altitude of 10,000 ft is achieved using the following pilot inputs: elevator deflection of 8.91 degrees, throttle setting of 0.023, and zero bank angle. The value of the minimum required power is 4.46 hp.

Maximum Airspeed

As shown in Figure 8.13, the maximum airspeed of the general aviation aircraft for steady descending flight with a climb rate of −10 ft/s at an altitude of 10,000 ft occurs when the curve for the maximum power produced by the engine and propeller intersects the required power curve for steady descending flight.

The airspeed at which this steady descending flight condition holds can also be found by solving

$$W V_{climb} + \frac{1}{2}\rho V^3 S C_{D_0} + \frac{2KW^2}{\rho V S} = \eta P_{max}^s \left(\frac{\rho}{\rho^s}\right)^m.$$

Multiplying the above equation by V, the following fourth degree polynomial equation is obtained:

$$\frac{1}{2}\rho S C_{D_0} V^4 + \left(W V_{climb} - \eta P_{max}^s \left(\frac{\rho}{\rho^s}\right)^m\right) V + \frac{2KW^2}{\rho S} = 0.$$

This equation can be solved numerically using the Matlab function MMSSDFP as follows:

Matlab function 8.15 MMSSDFP.m
```
function [Vmin,Vmax]=MMSSDFP
%Output: Minimum air speed and maximum air speed in descent
W=2900; S=175; CD0=0.026; K=0.054; Psmax=290*550; m=0.6;
[Ts ps rhos]=StdAtpUS(0);
[Th ph rhoh]=StdAtpUS(h);
Vclimb=-10;
[Vmin,Vmax]=sort(roots([1/2*rho*S*CD0 0 0 W*Vclimb-eta*Psmax*
(rho/rhos)^m 2*K*W^2/rho/S]));
```

Matlab provides two positive values that solve this polynomial equation: 316.02 ft/s and 21.85 ft/s. The high-speed solution is the maximum airspeed of the aircraft in this descending flight condition. At the given altitude, the speed of sound is 1,077.39 ft/s. The corresponding Mach number of the general aviation aircraft in this steady descending flight condition is 0.293.

At this maximum airspeed in this steady descending flight conditon, the lift coefficient and the pitch moment coefficient are

$$C_L = \frac{2W}{\rho V^2 S} = 0.189,$$

$$C_M = 0.$$

These expressions allow determination of an aircraft angle of attack of 1.41 degrees and an elevator deflection of -0.09 degrees.

In summary, the maximum airspeed of the general aviation aircraft for steady descending flight with a climb rate of -10 ft/s at an altitude of 10,000 ft is achieved using the following pilot inputs: elevator deflection of -0.09 degrees, full throttle, and zero bank angle. This maximum airspeed is 316.02 ft/s.

Minimum Airspeed

The minimum airspeed of the general aviation aircraft in a steady descent must satisfy both the stall constraint and the maximum power constraint of the engine and propeller. The stall speed of the general aviation aircraft is

$$V_{\text{stall}} = \sqrt{\frac{2W}{\rho S C_{L_{\max}}}} = 88.70\,\text{ft/s}.$$

As seen in the previous section, the minimum airspeed due to the engine and propeller power constraints is 21.85 ft/s. In this case, the stall speed is larger than the minimum airspeed due to the power constraint, so the stall constraint is active and the engine power constraints are not active. Consequently, the minimum airspeed of the aircraft is equal to the stall speed, 88.7 ft/s. At the minimum airspeed in steady descending flight, the lift coefficient and the pitch moment coefficient are

$$C_L = \frac{2W}{\rho V^2 S} = 2.4,$$

$$C_M = 0.$$

These expressions determine an aircraft angle of attack of 19.83 degrees and an elevator deflection of 19.55 degrees.

In summary, the minimum airspeed for steady descending flight with a climb rate of -10 ft/s at an altitude of 10,000 ft is achieved using the following pilot inputs: elevator deflection of 19.6 degrees, 33.6% throttle, and zero bank angle. The minimum airspeed in this steady descent is 21.85 ft/s.

Steady Descending Longitudinal Flight Envelope

The steady descending longitudinal flight envelope of the general aviation aircraft is developed, assuming that the aircraft is in steady flight with a climb rate of -10 ft/s. This steady descending flight envelope consists of the set of all pairs of altitude and airspeed, satisfying the engine and propeller constraints and the stall constraint.

A graphical representation of the steady descending flight envelope, assuming a steady descent with a climb rate of -10 ft/s, is developed. The maximum airspeed, the minimum airspeed due to the power constraint, and the stall speed can be computed at various altitudes. The previously defined Matlab function SCFP is used to compute the maximum airspeed, the minimum airspeed due to the power constraints and the stall speed for a specified altitude. The steady descending flight envelope is shown in Figure 8.14.

8.10 Steady Longitudinal Flight Envelopes: General Aviation Aircraft

A steady longitudinal flight envelope of a general aviation aircraft is defined in terms of the flight altitude, the airspeed of the aircraft, and the climb rate of the aircraft. This steady

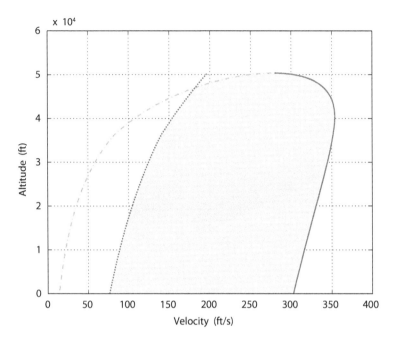

Figure 8.14. Steady Longitudinal flight envelope: general aviation aircraft ($V_{climb} = -10$ ft/s).

longitudinal flight envelope is defined in terms of three flight variables and hence must be viewed as a set in three dimensions. This notion generalizes the steady level flight envelope and the steady climbing or descending flight envelope for a fixed rate of climb. In particular, the steady level flight envelope can be viewed as a two-dimensional cross section of the steady longitudinal flight envelope corresponding to a climb rate of zero.

A steady flight surface is computed that defines the boundary of the steady longitudinal flight envelope by evaluating the Matlab SCFP function for various altitudes and climb rates. This approach allows development of a graphical representation of the steady longitudinal flight envelope using the Matlab graphics features. The boundary surface of the steady longitudinal flight envelope is shown in Figure 8.15. It is obtained from the following Matlab m-file.

Matlab function 8.16 FigSCDFP.m
```
Vclimb=linspace(-20,20,40);
for i=1:size(Vclimb,2)
    hmax(i)=fsolve(@(h) eqnFCCP(h,Vclimb(i)),10000);
    h(i,:)=linspace(0,floor(hmax(i)),40);
    for j=1:size(h,2)
        [Vmax(i,j)VminTC(i,j)Vstall(i,j)]=SCFP(h(i,j),Vclimb(i));
    end
end
```

(continued)

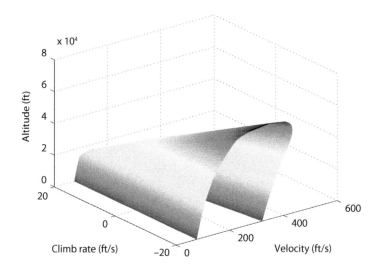

Figure 8.15. Steady longitudinal flight envelope: general aviation aircraft.

(continued)

```
Vmin=max(VminTC,Vstall);
surf(Vmax,Vclimb,h,'LineStyle','none');
hold on;
surf(Vmin,Vclimb,h,'LineStyle','none');
xlabel('Velocity (ft/s)');
ylabel('Climb rate (ft/s)');
zlabel('Altitude (ft)');
```

Any triple of values for the flight altitude, aircraft airspeed, and aircraft climb rate that lies within the steady longitudinal flight envelope illustrated in Figure 8.15 corresponds to a feasible steady flight condition for which the engine power constraints and the stall constraint are satisfied. For each such feasible triple of values, methodology has been provided to compute the pilot inputs, namely the elevator deflection, the throttle setting, and zero bank angle, that give this steady flight condition. In addition, the methodology also allows computation of the associated angle of attack, values of the lift and drag coefficients, and values of the lift and drag.

The steady longitudinal flight envelopes in Figures 8.11 and 8.14 are two-dimensional cross sections of the three-dimensional steady longitudinal flight envelope illustrated in Figure 8.15.

8.11 Conclusions

Conditions for steady climbing or descending longitudinal flight have been studied in detail. This analysis has led to the development of several important performance measures. These results have been illustrated in detail based on the two aircraft case studies.

The results in this chapter can be viewed as a generalization of the prior results for steady gliding flight in chapter 6; those prior steady gliding flight results are obtained as a special case by assuming zero thrust. The results in this chapter can also be viewed as a generalization of the prior results for steady level flight in chapter 7; in particular, those prior steady level flight results are obtained as a special case by assuming the flight path angle (or the climb rate) is zero. Subsequent chapters treat the case of steady turning flight.

8.12 Problems

8.1. Consider the executive jet aircraft in steady climbing longitudinal flight. Assume the aircraft weight is a constant 73,000 lbs.

   (a) What is the thrust required to maintain a steady flight path angle of 2 degrees and an airspeed of 330 ft/s at an altitude of 21,000 ft? What are the corresponding throttle setting and elevator deflection that maintain this steady climbing flight condition?

   (b) Assume the aircraft is in steady flight with a flight path angle of 2 degrees using 100 percent throttle at an altitude of 21,000 ft. Determine all possible airspeeds that maintain this steady climbing flight condition.

   (c) What is the minimum thrust that maintains a steady flight path angle of 2 degrees at an altitude of 21,000 ft? What are the corresponding airspeed, throttle setting, and elevator deflection that maintain this steady climbing flight condition?

   (d) What is the flight ceiling for steady climbing flight of the executive jet aircraft with a flight path angle of 2 degrees?

   (e) What is the flight envelope for steady climbing flight of the executive jet aircraft with a flight path angle of 2 degrees?

8.2. Consider the propeller-driven general aviation aircraft in steady climbing longitudinal flight. Assume the aircraft weight is a constant 2,900 lbs.

   (a) What is the power required to maintain a steady climb rate of 3.3 ft/s and an airspeed of 230 ft/s at an altitude of 12,000 ft? What are the corresponding throttle setting and elevator deflection that maintain this steady climbing flight condition?

   (b) Assume the aircraft is in steady flight with a climb rate of 3.3 ft/s using 100 percent throttle at an altitude of 12,000 ft. Determine all possible airspeeds that maintain this steady climbing flight condition.

   (c) What is the minimum power that maintains a steady climb rate of 3.3 ft/s at an altitude of 12,000 ft? What are the corresponding airspeed, throttle setting, and elevator deflection that maintain this steady climbing flight condition?

   (d) What is the flight ceiling for steady climbing flight of the general aviation aircraft with a climb rate of 3.3 ft/s?

   (e) What is the flight envelope for steady climbing flight of the general aviation aircraft with a climb rate of 3.3 ft/s?

8.3. Consider the UAV in steady climbing longitudinal flight. Assume the aircraft weight is a constant 45 lbs.

(a) What is the thrust required to maintain a steady flight path angle of 2.5 degrees and an airspeed of 88 ft/s at an altitude of 2,000 ft? What are the corresponding throttle setting and elevator deflection that maintain this steady climbing flight condition?

(b) Assume the aircraft is in steady flight with a flight path angle of 2.5 degrees using 100 percent throttle at an altitude of 2,000 ft. Determine all possible airspeeds that maintain this steady climbing flight condition.

(c) What is the minimum thrust that maintains a steady flight path angle of 2.5 degrees at an altitude of 2,000 ft? What are the corresponding airspeed, throttle setting, and elevator deflection that maintain this steady climbing flight condition?

(d) What is the flight ceiling for steady climbing flight of the UAV with a flight path angle of 2.5 degrees?

(e) What is the flight envelope for steady climbing flight of the UAV with a flight path angle of 2.5 degrees?

8.4. Consider the executive jet aircraft in steady descending longitudinal flight. Assume the aircraft weight is a constant 73,000 lbs.

(a) What is the thrust required to maintain a steady flight path angle of −2 degrees and an airspeed of 330 ft/s at an altitude of 21,000 ft? What are the corresponding throttle setting and elevator deflection that maintain this steady climbing flight condition?

(b) Assume the aircraft is in steady flight with a flight path angle of −2 degrees using 100 percent throttle at an altitude of 21,000 ft. Determine all possible airspeeds that maintain this steady climbing flight condition.

(c) What is the minimum thrust that maintains a steady flight path angle of −2 degrees at an altitude of 21, 000 ft? What are the corresponding airspeed, throttle setting, and elevator deflection that maintain this steady climbing flight condition?

(d) What is the flight ceiling for steady climbing flight of the executive jet aircraft with a flight path angle of −2 degrees?

(e) What is the flight envelope for steady climbing flight of the executive jet aircraft with a flight path angle of −2 degrees?

8.5. Consider the propeller-driven general aviation aircraft in steady descending longitudinal flight. Assume the aircraft weight is a constant 2,900 lbs.

(a) What is the power required to maintain a steady climb rate of −3.3 ft/s and an airspeed of 230 ft/s at an altitude of 12,000 ft? What are the corresponding throttle setting and elevator deflection that maintain this steady climbing flight condition?

(b) Assume the aircraft is in steady flight with a climb rate of -3.3 ft/s using 100 percent throttle at an altitude of 12,000 ft. Determine all possible airspeeds that maintain this steady climbing flight condition.

(c) What is the minimum power that maintains a steady climb rate of -3.3 ft/s at an altitude of 12,000 ft? What are the corresponding airspeed, throttle setting, and elevator deflection that maintain this steady climbing flight condition?

(d) What is the flight ceiling for steady climbing flight of the general aviation aircraft with a climb rate of -3.3 ft/s?

(e) What is the flight envelope for steady climbing flight of the general aviation aircraft with a climb rate of -3.3 ft/s?

8.6. Consider the UAV in steady descending longitudinal flight. Assume the aircraft weight is a constant 45 lbs.

(a) What is the thrust required to maintain a steady flight path angle of -2.5 degrees and an airspeed of 88 ft/s at an altitude of 2,000 ft? What are the corresponding throttle setting and elevator deflection that maintain this steady climbing flight condition?

(b) Assume the aircraft is in steady flight with a flight path angle of -2.5 degrees using 100 percent throttle at an altitude of 2,000 ft. Determine all possible airspeeds that maintain this steady climbing flight condition.

(c) What is the minimum thrust that maintains a steady flight path angle of -2.5 degrees at an altitude of 2,000 ft? What are the corresponding airspeed, throttle setting, and elevator deflection that maintain this steady climbing flight condition?

(d) What is the flight ceiling for steady climbing flight of the UAV with a flight path angle of -2.5 degrees?

(e) What is the flight envelope for steady climbing flight of the UAV with a flight path angle of -2.5 degrees?

8.7. Consider the executive jet aircraft in steady climbing or descending longitudinal flight. Assume the aircraft weight is a constant 73,000 lbs.

(a) Suppose the aircraft is to maintain a steady climb with a constant flight path angle of 2 degrees and a constant airspeed of 330 ft/s as it climbs from 21,000 ft to 32,000 ft. What is the thrust required to maintain this steady climbing flight condition? What are the corresponding throttle setting and elevator deflection? Express these as functions of the air density. How long does it take the aircraft to climb from 21,000 ft to 32,000 ft?

(b) Suppose the aircraft is to maintain a steady descent with a constant flight path angle of -2 degrees and a constant airspeed of 330 ft/s as it descends from 21,000 ft to 5,000 ft. What is the thrust required to maintain this steady descending flight condition? What are the corresponding throttle setting and elevator deflection? Express these as functions of the air density. How long does it take the aircraft to descend from 21,000 ft to 5,000 ft?

8.8. Consider the propeller-driven general aviation aircraft in steady climbing or descending longitudinal flight. Assume the aircraft weight is a constant 2,900 lbs.

(a) Suppose the aircraft is to maintain a steady climb with a constant climb rate of 3.3 ft/s and a constant airspeed of 230 ft/s as it climbs from 12,000 ft to 23,000 ft. What is the power required to maintain this steady climbing flight condition? What are the corresponding throttle setting and elevator deflection? Express these as functions of the air density. How long does it take the aircraft to climb from 12,000 ft to 23,000 ft?

(b) Suppose the aircraft is to maintain a steady descent with a constant descent rate of 3.3 ft/s and a constant airspeed of 230 ft/s as it descends from 12,000 ft to 3,000 ft. What is the power required to maintain this steady descending flight condition? What are the corresponding throttle setting and elevator deflection? Express these as functions of the air density. How long does it take the aircraft to descend from 12,000 ft to 3,000 ft?

8.9. Consider the executive jet aircraft in steady descending longitudinal flight at an altitude of 10,000 ft with a flight path angle of −5 degrees. Assume the aircraft weight is a constant 73,000 lbs.

(a) What is the required thrust as a function of the airspeed that maintains this steady descending flight condition?

(b) Taking the thrust constraints into account, what are the airspeeds at which this steady descending flight is feasible? What is the throttle setting as a function of the airspeed that maintains this steady descending flight condition over the feasible range of airspeeds?

(c) Based on these results, describe an interesting geometric feature of the steady descending flight envelope.

8.10. Consider the propeller-driven general aviation aircraft in steady descending longitudinal flight at an altitude of 10,000 ft with a climb rate of −12 ft/s. Assume the aircraft weight is a constant 2,900 lbs.

(a) What is the required power as a function of the airspeed that maintains this steady descending flight condition?

(b) Taking the power constraints into account, what are the airspeeds at which this steady descending flight is feasible? What is the throttle setting as a function of the airspeed that maintains this steady descending flight condition over the feasible range of airspeeds?

(c) Based on these results, describe an interesting geometric feature of the steady descending flight envelope.

8.11. Consider the executive jet aircraft in steady longitudinal flight at an altitude of 21,000 ft. Assume the aircraft weight is a constant 73,000 lbs.

    (a) What is the maximum steady flight path angle of the aircraft? What are the corresponding airspeed, throttle setting, and elevator deflection that maintain this steady climbing flight condition?

    (b) What is the minimum steady flight path angle of the aircraft? What are the corresponding airspeed, throttle setting, and elevator deflection that maintain this steady descending flight condition?

8.12. Consider the propeller-driven general aviation aircraft in steady longitudinal flight at an altitude of 12,000 ft. Assume the aircraft weight is a constant 2,900 lbs.

    (a) What is the maximum steady climb rate of the aircraft? What are the corresponding airspeed, throttle setting, and elevator deflection that maintain this steady climbing flight condition?

    (b) What is the maximum descent rate of the aircraft? What are the corresponding airspeed, throttle setting, and elevator deflection that maintain this steady descending flight condition?

8.13. Consider the UAV in steady longitudinal flight at an altitude of 2,000 ft. Assume the aircraft weight is a constant 45 lbs.

    (a) What is the maximum steady flight path angle of the aircraft? What are the corresponding airspeed, throttle setting, and elevator deflection that maintain this steady climbing flight condition?

    (b) What is the minimum steady flight path angle of the aircraft? What are the corresponding airspeed, throttle setting, and elevator deflection that maintain this steady descending flight condition?

8.14. Consider the executive jet aircraft in steady longitudinal flight. Is it possible that the aircraft can remain in steady longitudinal flight with an airspeed of 300 ft/s and a flight path angle of 4 degrees at an altitude of 15,000 ft? In other words, is this flight condition in the steady longitudinal flight envelope?

    If this flight condition is in the steady longitudinal flight envelope, determine the throttle setting and the elevator deflection required to maintain this flight condition.

    If this flight condition is not in the steady longitudinal flight envelope, which constraints are violated?

8.15. Consider the executive jet aircraft in steady longitudinal flight. Is it possible that the aircraft can remain in steady longitudinal flight with an airspeed of 500 ft/s and a flight path angle of 2 degrees at an altitude of 20,000 ft? In other words, is this flight condition in the steady longitudinal flight envelope?

    If this flight condition is in the steady longitudinal flight envelope, determine the throttle setting and the elevator deflection required to maintain this flight condition.

If this flight condition is not in the steady longitudinal flight envelope, which constraints are violated?

8.16. Consider the propeller-driven general aviation aircraft in steady longitudinal flight. Is it possible that the aircraft can remain in steady longitudinal flight with an airspeed of 95 ft/s and a climb rate of 18 ft/s at an altitude of 5,000 ft? In other words, is this flight condition in the steady longitudinal flight envelope?

If this flight condition is in the steady longitudinal flight envelope, determine the throttle setting and the elevator deflection required to maintain this flight condition.

If this flight condition is not in the steady longitudinal flight envelope, which constraints are violated?

8.17. Consider the propeller-driven general aviation aircraft in steady longitudinal flight. Is it possible that the aircraft can remain in steady longitudinal flight with an airspeed of 250 ft/s and a climb rate of 15 ft/s at an altitude of 5,000 ft? In other words, is this flight condition in the steady longitudinal flight envelope?

If this flight condition is in the steady longitudinal flight envelope, determine the throttle setting and the elevator deflection required to maintain this flight condition.

If this flight condition is not in the steady longitudinal flight envelope, which constraints are violated?

Aircraft Steady Level Turning Flight

This chapter treats steady level turns of an aircraft in flight through a stationary atmosphere. A brief description is given of steady turns by side-slipping, and the disadvantages of this method of accomplishing a steady turn are indicated. Then, the more common and efficient method for turning an aircraft by banking the aircraft is described in the context of steady level turns. The various flight conditions are analyzed, and turning flight performance is assessed.

9.1 Turns by Side-Slipping

One method for achieving a steady turn for an aircraft is to maintain a constant side-slip angle. This means that the velocity vector of the aircraft does not lie in the plane of mass symmetry of the aircraft. The velocity component of the aircraft along its aircraft-fixed y_A-axis results in a steady turn. This method for achieving a steady turn relies on using the thrust to turn; since the thrust is typically much smaller that the lift, this approach to steady turning flight is relatively ineffective. Side-slipping turns are utilized primarily by conventional missiles. The more common method for flight vehicles with a lifting surface makes use of banking the vehicle to turn.

9.2 Steady Level Banked Turning Flight

A steady level banked turn is achieved by rotating the lift vector about the velocity vector so that a horizontal component of lift causes the aircraft to turn. If the lift vector is rotated then the aircraft is in a steady turn. For most conventional aircraft, the lift force is much larger than the thrust that can be produced by the propulsion system, so this results in an effective turn strategy. A related treatment of steady level banked turning flight is given in Vinh (1995).

Free-Body Diagram

The free-body diagram in Figure 9.1 shows several views of the forces that act on an aircraft performing a steady level turn by banking. The center of mass of the aircraft is assumed to move along a horizontal circular path with constant airspeed V and constant turn radius r.

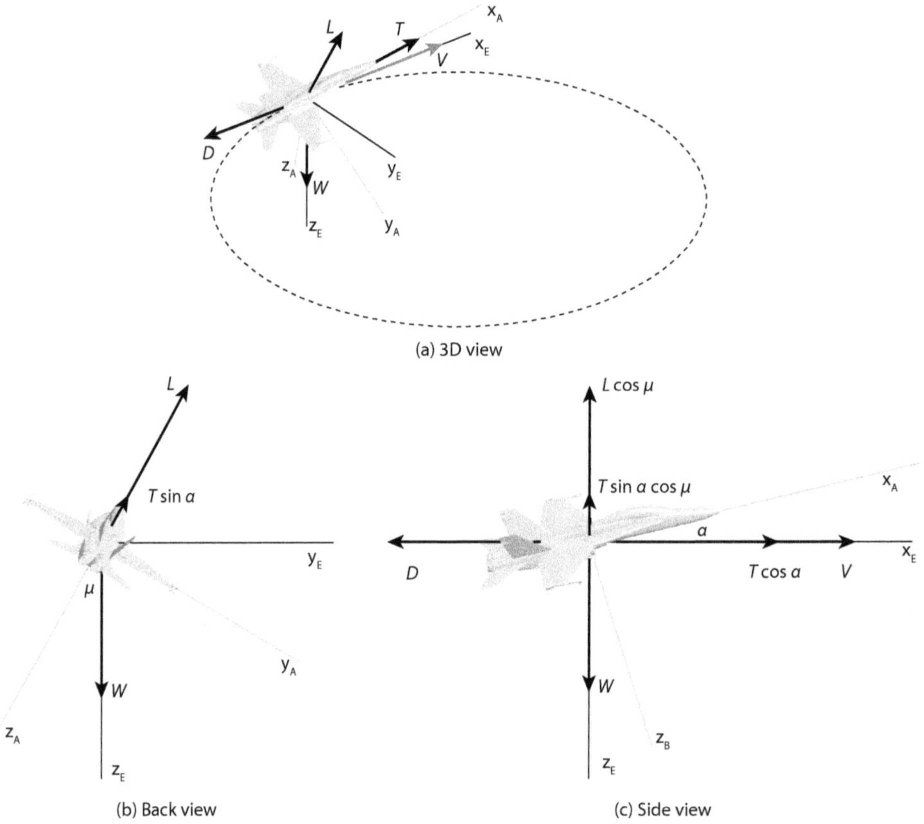

Figure 9.1. Free-body diagram of an aircraft in a steady level turn by banking.

Consequently, the aircraft has a constant centrifugal acceleration,

$$\frac{V^2}{r},$$

that is always directed in the horizontal plane toward the center of the circular path. Coordinated flight of the aircraft is also assumed; that is, the side-slip angle is zero.

The lift force on the aircraft acts normal to the velocity vector in the plane of aircraft symmetry; the lift force is rotated about the velocity vector from the vertical plane by the constant bank angle μ. The drag force on the aircraft acts opposite to the velocity vector. The thrust vector, with constant magnitude T, acts along the aircraft-fixed x_A-axis of the aircraft.

The free-body diagram in Figure 9.1 shows the forces acting on the aircraft in a steady level turn. This gives the three equations

$$-D + T\cos\alpha = 0, \tag{9.1}$$

$$\frac{W}{g}\frac{V^2}{r} = L \sin\mu + T \sin\alpha \sin\mu, \qquad (9.2)$$

$$W - L \cos\mu - T \sin\alpha \cos\mu = 0, \qquad (9.3)$$

and steady flight also requires that the pitch moment satisfy

$$M = 0. \qquad (9.4)$$

Equation (9.1) shows that the sum of the components of the forces along the velocity vector is zero since there is no aircraft acceleration in this direction; note that the projection of the thrust force in the direction of the velocity vector is given by $T \cos\alpha$. Equation (9.2) shows that the mass of the aircraft multiplied by the (horizontal) radial or centrifugal acceleration of the aircraft equals the sum of the (horizontal) radial components of the forces, namely the indicated components of the lift and the thrust. Equation (9.3) shows that the sum of the vertical components of the forces is zero since there is no aircraft acceleration in this direction; the vertical force components are the weight, the projection of the lift in this direction, and the projection of the thrust in this direction.

A positive bank angle corresponds to a steady right turn; a negative bank angle corresponds to a steady left turn; and zero bank angle corresponds to steady level longitudinal flight. The subsequent development holds for any bank angle, be it positive, negative, or zero.

As in chapter 7, a reasonable simplification is that the thrust magnitude is assumed to be much smaller than the lift, and the angle of attack is assumed to be a small angle, in radian measure, so that $\sin\alpha$ is approximated by α and $\cos\alpha$ is approximated by 1. The bank angle μ is *not* assumed to be a small angle. These assumptions lead to the following simplified equations for steady level turning flight:

$$-D + T = 0, \qquad (9.5)$$

$$\frac{W}{g}\frac{V^2}{r} = L \sin\mu, \qquad (9.6)$$

$$W - L \cos\mu = 0. \qquad (9.7)$$

These three equations form the basis for our subsequent analysis of steady level turning flight.

Aerodynamics

The equations that describe the lift force, the drag force, and the pitch moment from chapter 3 are

$$L = \frac{1}{2}\rho V^2 S C_L, \qquad (9.8)$$

$$D = \frac{1}{2}\rho V^2 S C_D, \qquad (9.9)$$

$$M = \frac{1}{2}\rho V^2 S c C_M. \tag{9.10}$$

The quadratic drag polar expression that relates the drag coefficient to the lift coefficient is given by

$$C_D = C_{D_0} + K C_L^2, \tag{9.11}$$

where $K = \frac{1}{\pi e AR}$. The lift coefficient is expressed as a linear function of the angle of attack as

$$C_L = C_{L_0} + C_{L_\alpha} \alpha. \tag{9.12}$$

The pitch moment coefficient is expressed as a linear function of the angle of attack and the elevator deflection as

$$C_M = C_{M_0} + C_{M_\alpha} \alpha + C_{M_{\delta_e}} \delta_e. \tag{9.13}$$

Flight Constraints

There are three types of important flight constraints that apply to steady level turning flight. These constraints play a central role in determining the performance properties of an aircraft in a steady level turn.

The first constraint arises from the aerodynamics assumption that the flight of the aircraft must avoid stalling. Based on the discussion in chapter 2, the stall constraint can be expressed as an inequality on the lift coefficient given by

$$C_L \leq C_{L_{\max}}. \tag{9.14}$$

The second set of constraints arises from limits on the operation of the propulsion system. For a jet aircraft, the thrust T required for a steady level turn must satisfy

$$0 \leq T \leq T_{\max}^s \left(\frac{\rho}{\rho^s}\right)^m. \tag{9.15}$$

The jet engine thrust constraints in expression (9.15) can also be given in terms of the throttle setting. The jet engine thrust constraints are equivalent to:

$$0 \leq \delta_t \leq 1, \tag{9.16}$$

where

$$\delta_t = \frac{T}{T_{\max}^s} \left(\frac{\rho^s}{\rho}\right)^m. \tag{9.17}$$

For a internal combustion engine and propeller-driven aircraft, the power P required for a steady level turn must satisfy

$$0 \leq P \leq \eta P_{max}^s \left(\frac{\rho}{\rho^s} \right)^m, \tag{9.18}$$

where η is the efficiency of the propeller. The engine power constraints in expression (9.18) can also be given in terms of the throttle setting. The engine power constraints are equivalent to the following:

$$0 \leq \delta_t \leq 1, \tag{9.19}$$

where

$$\delta_t = \frac{P}{\eta P_{max}^s} \left(\frac{\rho^s}{\rho} \right)^m. \tag{9.20}$$

The third constraint on a steady level turn is a wing loading constraint that is expressed as

$$L \leq n_{max} W, \tag{9.21}$$

where n_{max} is the maximum load factor. Define the load factor as the lift to weight ratio given by

$$n = \frac{L}{W}. \tag{9.22}$$

Thus the wing loading constraint can also be written as $n \leq n_{max}$. The wing loading constraint on the maximum lift arises from structural limits on the force that the wings can support. This structural limit is an important factor in determining the steady turning performance of an aircraft.

9.3 Steady Level Banked Turning Flight Analysis

Flight conditions for steady level turning flight are analyzed. Several turning flight relationships for the turn radius, the turn rate, and the required thrust and required power to maintain steady level turning flight are obtained.

Turn Radius and Turn Rate

Using equations (9.7) and (9.22), it is seen that the load factor for steady level turning flight is directly related to the bank angle as

$$n = \sec \mu. \tag{9.23}$$

Steady level flight, without turning, corresponds to zero bank angle or a load factor $n = 1$, while a steady level banked turn, either to the right or to the left, requires a load factor greater than 1. For a bank angle $\mu = 60$ degrees, the load factor is $n = 2$; such a turn is referred to as a *2 g* turn.

Divide equation (9.6) by equation (9.7) to obtain the kinematics relation

$$\tan \mu = \frac{V^2}{gr}. \tag{9.24}$$

Alternatively, the airspeed for steady level turning flight can be expressed in terms of the turn radius and the bank angle as

$$V = \sqrt{gr \tan \mu}, \tag{9.25}$$

and the turn radius for steady level turning flight can be expressed in terms of the airspeed and the bank angle as

$$r = \frac{V^2}{g \tan \mu}. \tag{9.26}$$

Another convenient expression for the turn radius can be developed. Using equation (9.7), the turn radius can be expressed in terms of the bank angle and the lift as

$$r = \frac{WV^2}{Lg \sin \mu}; \tag{9.27}$$

the turn radius can be expressed in terms of the bank angle and the lift coefficient as

$$r = \frac{2}{\rho g C_L} \frac{1}{\sin \mu} \frac{W}{S}. \tag{9.28}$$

By using the relation between the bank angle and the load factor, the turn radius can also be written as

$$r = \frac{2}{\rho g C_L} \frac{n}{\sqrt{n^2 - 1}} \frac{W}{S}. \tag{9.29}$$

This expression for the turn radius of an aircraft in a steady level turn shows that the turn radius depends on the lift coefficient, the bank angle or the load factor, the wing loading W/S, and the flight altitude.

The angular turn rate ω can be written as

$$\omega = \frac{V}{r} = \frac{g \tan \mu}{V}. \tag{9.30}$$

The turn period τ, the time required for the aircraft to make one revolution on the circular path, is given by

$$\tau = \frac{2\pi}{\omega} = \frac{2\pi V}{g \tan \mu}. \tag{9.31}$$

Thrust Required

Consider the conditions for a steady level turn supposing that V is the airspeed and μ is the bank angle. Expressing the lift in terms of the dynamic pressure and the lift coefficient and using equation (9.7), the required lift coefficient for steady level turning flight is

$$C_L = \frac{2W}{\rho V^2 S \cos \mu}. \tag{9.32}$$

Consequently, the required drag coefficient for steady level turning flight is

$$C_D = C_{D_0} + K \left(\frac{2W}{\rho V^2 S \cos \mu} \right)^2. \tag{9.33}$$

The drag on the aircraft in a steady level turn is

$$D = \frac{1}{2} \rho V^2 S C_{D_0} + \frac{2K W^2}{\rho V^2 S \cos^2 \mu}. \tag{9.34}$$

Hence the required thrust for steady level turning flight, expressed in terms of the airspeed and bank angle, is

$$T = \frac{1}{2} \rho V^2 S C_{D_0} + \frac{2K W^2}{\rho V^2 S \cos^2 \mu}. \tag{9.35}$$

The thrust required for a steady level turn necessarily exceeds the thrust required for steady level longitudinal flight given by equation (7.23). Equation (9.35) has the following interpretation. The first term on the right represents the thrust required to balance the lift independent part of the drag, and the second term represents the thrust required to balance the lift dependent part of the drag; this second term depends on the bank angle.

Equation (9.35) can be expressed in an alternative form. If the thrust T and the bank angle μ are assumed known, then equation (9.35) can be written as

$$\frac{1}{2} \rho V^4 S C_{D_0} - T V^2 + \frac{2K W^2}{\rho S \cos^2 \mu} = 0, \tag{9.36}$$

which can be viewed as an equation whose solution is the airspeed for steady level turning flight at the given values of thrust and bank angle. Solving this equation requires finding the roots of a fourth degree polynomial. Equation (9.36) has two real and positive solutions corresponding to a low-speed steady level turn and a high-speed steady level turn. The low-speed solution must be checked to guarantee that it satisfies the stall constraint, expressed as an inequality in the aircraft airspeed, given by

$$\sqrt{\frac{2W}{\rho S C_{L_{\max}} \cos \mu}} \leq V. \tag{9.37}$$

The angle of attack and the elevator deflection for an aircraft in a steady level turn can be determined based on the value of the lift coefficient computed from equation (9.32) and the fact that the pitch moment coefficient must be zero for steady level turning flight. The throttle setting is easily determined from the above analysis. Since the bank angle is to be maintained constant, there should be no deflection of the ailerons and no deflection of the rudder.

The thrust required for a steady level turn with a fixed bank angle has a minimum at some airspeed. This minimum value can be obtained using the methods of calculus: at the minimum the thrust required curve has zero slope; that is, the derivative of the thrust with respect to the airspeed is zero. In particular, the thrust is minimized when the turning drag is minimized. This occurs if the airspeed is

$$ V = \sqrt{\frac{2W}{\rho S \cos \mu} \sqrt{\frac{K}{C_{D_0}}}}. \tag{9.38} $$

The airspeed value at which the minimum thrust for steady level flight occurs depends on the air density and hence on the flight altitude according to the standard atmospheric model; this airspeed increases with an increase in the flight altitude.

The minimum value of the thrust required for steady level turning flight, expressed in terms of the bank angle, is obtained by substituting expression (9.38) into equation (9.35) to obtain

$$ T_{\min} = \frac{2W}{\cos \mu} \sqrt{K C_{D_0}}. \tag{9.39} $$

The minimum value of the thrust required for a steady level turn does not depend on the air density and hence does not depend on the flight altitude.

The above equations for steady level banked turning flight are expressed in terms of the aircraft thrust, the aircraft speed, and the bank angle. These can be rewritten in terms of the aircraft thrust, the aircraft speed, and the load factor, using equation (9.22). For a given thrust level, graphs showing how the load factor depends on the aircraft speed are commonly employed and are referred to as $V - n$ diagrams.

Power Required

The required power for steady level turning flight, expressed in terms of the airspeed and bank angle, is obtained from equation (9.35) as

$$ P = \frac{1}{2} \rho V^3 S C_{D_0} + \frac{2K W^2}{\rho V S \cos^2 \mu}. \tag{9.40} $$

Equation (9.40) has the following interpretation. The first term on the right represents the power required to balance the lift independent part of the drag and the second term on the right represents the power required to balance the lift dependent part of the drag; this second term depends on the bank angle.

Equation (9.40) can be expressed in an alternative form. If the power P and the bank angle μ are assumed known, then equation (9.40) can be written as

$$\frac{1}{2}\rho V^4 S C_{D_0} - PV + \frac{2KW^2}{\rho S \cos^2 \mu} = 0, \tag{9.41}$$

which can be viewed as an equation whose solution is the airspeed for steady level turning flight at the given value of power and bank angle. Equation (9.41) has two real and positive solutions corresponding to a low-speed steady level turn and a high-speed steady level turn. Solving this equation requires finding the roots of a fourth degree polynomial. The low-speed solution should be checked to see that the stall constraint is satisfied:

$$\sqrt{\frac{2W}{\rho S C_{L_{\max}} \cos \mu}} \leq V. \tag{9.42}$$

The angle of attack and the elevator deflection for an aircraft in a steady level turn can be determined based on the value of the lift coefficient computed from equation (10.32) and the fact that the pitch moment coefficient must be zero for steady level flight. The throttle setting is easily determined from the above analysis. Since the bank angle is to be maintained constant, there should be no deflection of the ailerons and no deflection of the rudder.

For a fixed value of the bank angle, the power required for a steady level turn has a minimum value at some airspeed. This minimum value can be obtained using the methods of calculus: at the minimum the power required curve has zero slope; that is, the derivative of the power with respect to the airspeed is zero. This occurs if the airspeed is

$$V = \sqrt{\frac{2W}{\rho S \cos \mu}\sqrt{\frac{K}{3C_{D_0}}}}. \tag{9.43}$$

The minimum value of the power for a steady level turn is obtained by substituting expression (9.43) into (9.40) to obtain

$$P_{\min} = \frac{4}{3}\sqrt{\frac{2W^3}{\rho S \cos^3 \mu}\sqrt{3K^3 C_{D_0}}}. \tag{9.44}$$

The minimum power required for a steady level turn with a fixed bank angle depends on the air density and hence on the flight altitude; the minimum power increases as the flight altitude increases.

The above equations for steady level banked turning flight are expressed in terms of the aircraft power, the aircraft speed, and the bank angle. These can be rewritten in terms of the aircraft power, the aircraft speed, and the load factor, using equation (9.22). For a given power level, graphs showing how the load factor depends on the aircraft speed are referred to as $V - n$ diagrams.

9.4 Jet Aircraft Steady Level Turning Flight Performance

Several different flight conditions that characterize the optimal performance of jet aircraft in steady level turns are studied. The following flight performance measures are analyzed: maximum airspeed, minimum airspeed, minimum turn radius, and flight ceiling. The steady level turning flight envelope is also introduced. These several flight performance measures are related to the issue of flight maneuverability or flight agility of the aircraft.

Minimum and Maximum Airspeeds

Flight conditions for minimum and maximum airspeeds of a jet aircraft are first analyzed for a steady level turn with a fixed bank angle. The maximum thrust that the engine can provide limits the minimum and the maximum possible airspeeds for which the aircraft can maintain a steady level turn. The minimum and maximum airspeeds are characterized by the fact that the thrust required for a steady level turn is equal to the maximum thrust that the engine can provide at full throttle; this leads to the fourth degree polynomial equation

$$\frac{1}{2}\rho V^4 S C_{D_0} - T_{\max}^s \left(\frac{\rho}{\rho^s}\right)^m V^2 + \frac{2KW^2}{\rho S \cos^2 \mu} = 0. \tag{9.45}$$

This equation has two positive solutions. The high-speed solution is the maximum possible airspeed of the aircraft in a steady level turn at the fixed bank angle; the low-speed solution may be the minimum possible airspeed of the aircraft in a steady level turn at the fixed bank angle.

Two important qualifications must be made about the maximum airspeed and the minimum airspeed obtained from this analysis. The computed minimum airspeed is the minimum airspeed if it satisfies the stall constraint given by inequality (9.37); otherwise the stall speed is the minimum airspeed for this steady level turn at the given thrust and bank angle. The computed maximum airspeed may be near to or exceed the speed of sound; if this occurs, the computed maximum airspeed is inaccurate due to supersonic aerodynamics effects.

Minimum Radius Turns

Flight conditions for a minimum radius steady level turn of a jet aircraft are now considered. Determining the minimum turn radius for a jet aircraft in a steady level turn is somewhat complicated; one approach for determining the flight conditions for a minimum turn radius is suggested.

There are several possible formulations for the minimum radius steady level turn as an optimization problem. One formulation is as follows: determine the values of airspeed V and bank angle μ for which the steady level turn radius

$$r = \frac{V^2}{g \tan \mu} \tag{9.46}$$

is a minimum, subject to satisfaction of the three inequalities defined by the stall constraint, the thrust constraint, and the wing loading constraint:

$$\frac{2W}{\rho V^2 S \cos \mu} \leq C_{Lmax}, \tag{9.47}$$

$$\frac{1}{2}\rho V^2 S C_{D_0} + \frac{2K W^2}{\rho V^2 S \cos^2 \mu} \leq \left(\frac{\rho}{\rho^s}\right)^m T^s_{max}, \tag{9.48}$$

$$\sec \mu \leq n_{max}. \tag{9.49}$$

This optimization problem is complicated by the fact that the active constraints and the inactive constraints are not a priori known. Based on equation (9.29), any one of the following situations is possible:

- Exactly one of the constraints is active.
  - The jet engine thrust constraint is active, but the stall constraint and the wing loading constraint are inactive.

- Exactly two of the constraints are active.
  - The stall constraint and the jet engine thrust constraint are active, but the wing loading constraint is inactive.
  - The stall constraint and the wing loading constraint are active, but the jet engine thrust constraint is inactive.
  - The wing loading constraint and the jet engine thrust constraint are active, but the stall constraint is inactive.

- All three of the constraints are active.

Since any one of these cases can characterize the minimum radius turn for a jet aircraft in steady level flight, in principle all cases must be analyzed and the corresponding flight condition and the corresponding steady turn radius computed for each case. The minimum radius turn for the jet aircraft in steady level flight is then determined by a direct comparison of the computed turn radius values for which all the constraint combinations are satisfied. In some instances, a priori physical knowledge can be used to make an educated hypothesis about which are the active constraints and the inactive constraints for the minimum radius turn.

An alternative graphical approach is based on a cross section of the steady level turning flight envelope (introduced in section 9.6 of this chapter) for a fixed flight altitude of the jet aircraft. In particular a cross section of all feasible steady level turning flight conditions can be graphically represented in two dimensions using the airspeed and bank angle variables, assuming a fixed flight altitude; these flight conditions must satisfy the stall constraint, the thrust constraint, and the wing loading constraint.

A graphical representation of this feasible set of airspeeds and bank angles, for a fixed flight altitude, can be obtained by defining boundary curves that correspond to one of the three constraints being active. The boundary curves for the stall constraint active and for the wing loading constraint active are simply described. If the maximum thrust constraint is active, there is a low-speed solution and a high-speed solution for each bank angle value as seen in the previous sections; this defines the boundary curve for this maximum thrust constraint being active. The feasible set of airspeed and bank angle values is simultaneously bounded by all three of these boundary curves.

Contour curves that correspond to constant values of the steady level turn radius are defined by equation (9.46). The desired minimum radius turn flight condition is characterized by the geometric fact that it lies on the smallest turn radius contour curve that intersects the feasible set of airspeed and bank angle values. Graphical tools can be used to plot the boundary curves that define this feasible set and to plot constant contour turn radius curves; the minimum radius steady level turning flight condition can be determined by inspection. Once the active constraints are clear, the optimal values of the airspeed and the bank angle, and the other flight variables, can be determined using the methods developed in this chapter.

Equation (9.29) also makes clear that the smallest possible turn radius is achieved, all other factors considered constant, at the lowest possible altitude. Further, the minimum turn radius is proportional to the wing loading W/S. That is, jet aircraft with small wing loading can achieve tighter turns than can jet aircraft with larger wing loading. The wing loading, a fundamental aircraft design parameter, is seen to be important in determining the maneuverability of the aircraft.

Flight Ceiling

The flight ceiling of a jet aircraft in a steady level turn with a fixed bank angle is determined. There is a maximum altitude, referred to as the steady level turning flight ceiling, at which a steady level turn can be maintained. The flight condition at this flight ceiling is characterized by the fact that the minimum thrust required for steady level turning flight, given in equation (9.39), is exactly equal to the maximum thrust that the engine can produce at this altitude. This leads to the equation

$$\frac{2W}{\cos \mu} \sqrt{K C_{D_0}} = T^s_{\max} \left(\frac{\rho}{\rho^s} \right)^m.$$ (9.50)

Since the air density depends on the altitude according to the standard atmospheric model, equation (9.50) can be viewed as an implicit equation in the altitude; the flight ceiling solves this equation.

Flight Envelope

The flight envelope of a jet aircraft in a steady level turn is described. The steady level turning flight envelope can be represented as a set in three dimensions defined by the altitude, airspeed, and bank angle variables that satisfy the steady level turning flight equations and constraints. Specific values of altitude, airspeed, and bank angle lie in the steady level turning flight envelope if the stall constraint, the engine thrust constraints, and the wing loading constraint are satisfied.

Since the prior analysis in this chapter provides the basis for computation of the thrust required, the lift coefficient, and the load factor for steady level turning flight, these constraints can easily be checked. The boundary of the steady level turning flight envelope is defined by surfaces that are obtained by making the thrust constraint active, by making the stall constraint active, or by making the wing loading constraint active.

One simplified visualization approach is to represent a cross section of the steady level turning flight envelope for a fixed value of the bank angle. Each such cross section defines a set of feasible flight conditions in two dimensions of altitude and airspeed. In fact the steady level longitudinal flight envelope is a cross section of the steady level turning flight envelope for the special case of zero bank angle. Similarly, cross sections of the steady level turning flight envelope can be represented for any positive or negative value of the bank angle.

A different visualization approach is to represent a cross section of the steady level turning flight envelope for a fixed value of the flight altitude. Each such cross section is a set of feasible flight conditions in two dimensions defined by the airspeed and bank angle. This is the cross section mentioned previously that is useful in determining the minimum radius steady level turn for a jet aircraft using graphical methods.

9.5 General Aviation Aircraft Steady Level Turning Flight Performance

Several different flight conditions that characterize the optimal performance of propeller-driven general aviation aircraft in steady level turns are studied. The following flight performance measures are analyzed: maximum airspeed, minimum airspeed, minimum turn radius, and flight ceiling. The steady level turning flight envelope is also introduced. These several flight performance measures are related to the issue of flight maneuverability or flight agility.

Minimum and Maximum Airspeeds

Flight conditions for minimum and maximum airspeeds of a propeller-driven general aviation aircraft are analyzed for a steady level turn with a fixed bank angle. The maximum power that the engine can provide limits the minimum and maximum possible airspeeds for which the aircraft can maintain a steady level turn. The minimum airspeed and the maximum airspeed are characterized by the fact that the power required for a steady level turn is equal to the maximum power that the engine can provide at full throttle; this leads to the fourth degree polynomial equation

$$\frac{1}{2}\rho V^4 S C_{D_0} - \eta P^s_{\max}\left(\frac{\rho}{\rho^s}\right)^m V + \frac{2KW^2}{\rho S \cos^2 \mu} = 0. \tag{9.51}$$

This equation typically has two positive solutions. The high-speed solution is the maximum possible airspeed of the aircraft at the fixed bank angle; the low-speed solution is the minimum possible airspeed of the aircraft at the fixed bank angle.

Two important qualifications must be made about the maximum airspeed and the minimum airspeed obtained from this analysis. This computed minimum airspeed is the minimum airspeed if it satisfies the stall constraint given by inequality (9.42); otherwise the stall speed is the minimum airspeed for the steady level turn at the given power and bank angle. The computed maximum airspeed may be near to or exceed the speed of sound; if this case occurs, the computed maximum airspeed is inaccurate.

Minimum Radius Turns

Flight conditions for a minimum radius steady level turn of a propeller-driven general aviation aircraft are now considered. The problem of determining the minimum turn radius in a steady level turn is somewhat complicated. There are several possible formulations for the minimum steady level radius turn as an optimization problem.

One formulation is as follows: determine the values of airspeed V and bank angle μ for which the steady level turn radius

$$r = \frac{V^2}{g \tan \mu} \tag{9.52}$$

is a minimum, subject to satisfaction of the three inequalities defined by the stall constraint, the power constraint, and the wing loading constraint:

$$\frac{2W}{\rho V^2 S \cos \mu} \leq C_{Lmax}, \tag{9.53}$$

$$\frac{1}{2}\rho V^3 S C_{D_0} + \frac{2K W^2}{\rho V S \cos^2 \mu} \leq \eta \left(\frac{\rho}{\rho^s}\right)^m P_{max}^s, \tag{9.54}$$

$$\sec \mu \leq n_{max}. \tag{9.55}$$

This optimization problem is complicated by the fact that the active constraints and the inactive constraints are not a priori known. Any one of the following situations is possible:

- One constraint is active.
    - The engine power constraint is active, but the stall constraint and the wing loading constraint are inactive.

- Exactly two of the constraints are active.
    - The stall constraint and the engine power constraint are active, but the wing loading constraint is inactive.
    - The stall constraint and the wing loading constraint are active, but the engine power constraint is inactive.
    - The wing loading constraint and the engine power constraint are active, but the stall constraint is inactive.

- All three of the constraints are active.

Since any one of these cases can characterize the minimum radius turn for a propeller-driven general aviation aircraft, in principle all cases can be analyzed and the corresponding flight condition and the corresponding steady turn radius computed for each case. The minimum radius turn for the propeller-driven general aviation aircraft in steady level flight is then determined by a direct comparison of the computed turn radius values for which all the constraint combinations are satisfied. In some instances, a priori physical knowledge can be used to make an educated hypothesis about which are the active constraints and the inactive constraints for the minimum radius turn.

A graphical representation of this feasible set of airspeeds and bank angles, for a fixed flight altitude, can be obtained by defining boundary curves that corresponding to a single one of the three constraint being active. The boundary curves for the stall constraint active and for the wing loading constraint active are easily described. If the maximum power constraint is active, there is a low-speed solution and a high-speed solution for each bank angle value as seen in the previous sections; this defines the boundary curve for this maximum power constraint being active. The feasible set of airspeed and bank angle values is simultaneously bounded by all three of these boundary curves.

Contour curves that correspond to constant values of the steady level turn radius are defined by equation (9.52). The desired minimum radius turn flight condition is characterized by the geometric fact that it lies on the smallest turn radius contour curve that intersects the feasible set of airspeed and bank angle values. Graphical tools can be used to plot the boundary curves that define this feasible set and to plot constant contour turn radius curves; the minimum radius steady level turning flight condition can be determined by inspection. Once the active constraints are identified, the optimal values of the airspeed and the bank angle, and the other flight variables, can be determined using the methods developed in this chapter.

Equation (9.29) also makes clear that the smallest possible turn radius is achieved, all other factors considered constant, at the lowest possible altitude. Further, the turn radius is proportional to the wing loading W/S. That is, propeller-driven general aviation aircraft with small wing loading can achieve tighter turns than can a propeller-driven general aviation aircraft with larger wing loading. The wing loading is a fundamental aircraft design parameter that is important in determining the maneuverability of the aircraft.

Flight Ceiling

The flight ceiling of a propeller-driven general aviation aircraft in a steady level turn with a fixed bank angle is analyzed. There is a maximum altitude, referred to as the steady level turning flight ceiling, at which steady level turning flight can be maintained. The flight condition at this flight ceiling is characterized by the fact that the minimum power required for a steady level turn, given in equation (9.44), is exactly equal to the maximum power that the engine can produce. This leads to the equation

$$\frac{4}{3}\sqrt{\frac{2W^3}{\rho S \cos^3 \mu}}\sqrt{3K^3 C_{D_0}} = \eta P_{\max}^s \left(\frac{\rho}{\rho^s}\right)^m. \tag{9.56}$$

Since the air density depends on the altitude according to the standard atmospheric model, equation (9.56) can be viewed as an implicit equation in the altitude; the flight ceiling solves this equation.

Flight Envelope

The flight envelope of a propeller-driven general aviation aircraft in a steady level turn is described. The steady level turning flight envelope is a set in three dimensions defined by the altitude, airspeed, and bank angle values that satisfy the steady level turning flight equations and constraints. Specific values of altitude, airspeed, and bank angle lie in the steady level turning flight envelope if the stall constraint, the engine power constraints, and the wing loading constraint are satisfied. Since the prior analysis in this chapter provides the basis for computation of the power required, the lift coefficient, and the load factor for steady level turning flight, these constraints can easily be checked. The boundary of the steady level turning flight envelope is defined by surfaces that are obtained by making the power constraint active, by making the stall constraint active, or by making the wing loading constraint active.

One simplified visualization approach is to represent a cross section of the steady level turning flight envelope for a fixed value of the bank angle. Each such cross section is a set of feasible flight conditions in two dimensions defined by the altitude and airspeed. In fact the steady level longitudinal flight envelope is a cross section of the steady level turning flight envelope for the special case of zero bank angle. Similarly, cross sections of the steady level turning flight envelope can be represented for any positive or negative value of the bank angle.

A different visualization approach is to represent a cross section of the steady level turning flight envelope for a fixed value of the flight altitude. Each such cross section is a set of feasible flight conditions in two dimensions defined by the airspeed and bank angle. This is the cross section mentioned previously that is useful in determining the minimum radius steady level turn for a propeller-driven general aviation aircraft using graphical methods.

9.6 Steady Level Turning Flight: Executive Jet Aircraft

Assume that the executive jet aircraft, with a weight of 73,000 lbs, is in a steady level turn at an altitude of 10,000 ft. The analysis in this section is based on the assumption that the bank angle is 30 degrees. This bank angle corresponds to a load factor of 1.15, which is significantly below the maximum load factor of 2.

Thrust Required

Suppose that the airspeed for a steady level turn with bank angle of 30 degrees is 500 ft/s. The thrust required for this steady level turn is

$$T = \frac{1}{2}\rho V^2 S C_{D_0} + \frac{2KW^2}{\rho V^2 S \cos^2 \mu} = 4,831 \text{ lbs}.$$

The thrust required curve, shown in Figure 9.2, is generated using the following Matlab m-file.

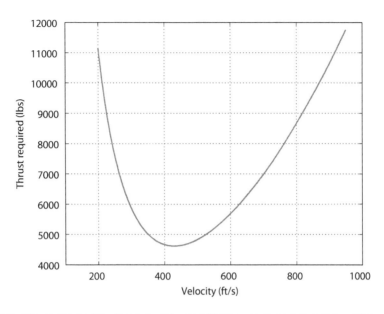

Figure 9.2. Thrust required for steady level turning flight: executive jet aircraft ($\mu = 30$ degrees).

Matlab function 9.1 TRSLTFJ.m
```
V=linspace(150,1000,500);
W=73000; S=950; CD0=0.015; K=0.05;
[T p rho]=StdAtpUS(10000);
mu=30* pi/180;
T=1/2*rho*V.^2*S*CD0+2*K*W^2/rho./V.^2/S/cos(mu)^2;
plot(V,T);
xlabel('Velocity (ft/s)');
ylabel('Thrust required (lbs)');
grid on;
```

As seen in Figure 9.2, the thrust required for a steady level turn with a bank angle of 30 degrees has the usual characteristics: the thrust is large for low and high airspeeds; the thrust has a minimum value at some intermediate airspeed.

Minimum Required Thrust

The minimum required thrust to achieve a steady level turn with a bank angle of 30 degrees can be obtained graphically from Figure 9.2.

The airspeed that achieves minimum required thrust for this steady level turn occurs when the drag is a minimum; this airspeed is given by

$$V = \sqrt{\frac{2W}{\rho S \cos \mu} \sqrt{\frac{K}{C_{D_0}}}} = 429.63 \text{ ft/s.}$$

At this steady turning flight condition, the lift coefficient and the pitch moment coefficient are

$$C_L = \frac{2W}{\rho V^2 S \cos \mu} = 0.548,$$

$$C_M = 0.$$

These equations can be expressed in terms of the aircraft angle of attack and the elevator deflection as

$$0.02 + 0.12\alpha = 0.548,$$

$$0.24 - 0.18\alpha + 0.28\delta_e = 0.$$

Thus, the angle of attack is 4.40 degrees and the elevator deflection is 1.97 degrees. The value of the minimum thrust for this steady level turn is given by

$$T_{min} = \frac{2W}{\cos \mu} \sqrt{KC_{D_0}} = 4,617 \, \text{lbs}.$$

The corresponding throttle setting is

$$\delta_t = \frac{T_{min}}{T_{max}^s} \left(\frac{\rho^s}{\rho}\right)^m = 0.443.$$

The turn radius corresponding to this minimum thrust steady level turning flight condition is

$$r = \frac{V^2}{g \tan \mu} = 9,928.7 \, \text{ft}.$$

In summary, the minimum thrust required for the executive jet aircraft to achieve a steady level turn with a bank angle of 30 degrees is 4,617 lbs and occurs with the following pilot inputs: elevator deflection of 1.97 degrees and a throttle setting of 0.443. The corresponding turn radius is 9,928.7 ft and the airspeed is 429.63 ft/s.

Maximum Airspeed

The maximum airspeed of the executive jet aircraft in a steady level turn with a bank angle of 30 degrees at an altitude of 10,000 ft is limited by the engine thrust. This maximum airspeed characterizes the intersection of the thrust required curve and the curve defined by the maximum thrust available from the jet engine at this altitude; this intersection is shown in Figure 9.3.

This maximum airspeed can also be found by solving the equation

$$\frac{1}{2}\rho V^2 S C_{D_0} + \frac{2KW^2}{\rho V^2 S \cos^2 \mu} = T_{max}^s \left(\frac{\rho}{\rho^s}\right)^m.$$

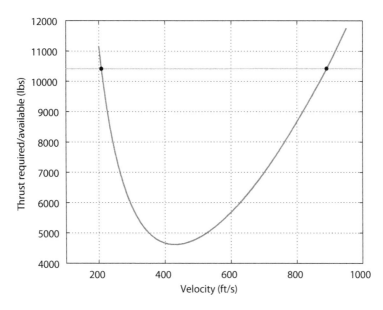

Figure 9.3. Maximum airspeed for steady level turning flight: executive jet aircraft ($\mu = 30$ degrees).

Multiplying the above equation by V^2, the following fourth degree polynomial equation is obtained:

$$\frac{1}{2}\rho S C_{D_0} V^4 - T_{\max}^s \left(\frac{\rho}{\rho^s}\right)^m V^2 + \frac{2K W^2}{\rho S \cos^2 \mu} = 0.$$

This equation can be solved numerically using the Matlab function MMSSLTFJ:

Matlab function 9.2 MMSSLTFJ.m
```
function [Vmin,Vmax]=MMSSLTFJ
%Output: Minimum air speed and maximum air speed in a
level turn
W=73000; S=950; CD0=0.015; K=0.05; Tsmax=12500; m=0.6;
[Ts ps rhos]=StdAtpUS(0);
[Th ph rhoh]=StdAtpUS(h);
gamma=3*pi/180;
[Vmin,Vmax]=sort(roots([1/2*rho*S*CD0 0 -
Tsmax*(rho/rhos)^m 0 2*K*W^2/rho/S/cos(mu)^2]));
```

Matlab provides two positive solutions, 888.9 ft/s and 207.65 ft/s. The high-speed solution is the maximum airspeed of the aircraft in a steady level turn with a 30 degree bank angle. At the given altitude, the speed of sound is 1,077.39 ft/s. Consequently, the Mach number for this maximum airspeed in steady turning flight is 0.82. At the maximum airspeed, the lift coefficient and the moment coefficient are

$$C_L = \frac{2W}{\rho V^2 S \cos \mu} = 0.128,$$

$$C_M = 0.$$

These expressions allow determination of the aircraft angle of attack as 0.9 degrees and the elevator deflection as −0.28 degrees. The turn radius corresponding to this maximum airspeed steady turn is

$$r = \frac{V^2}{g \tan \mu} = 42{,}502.1 \text{ ft.}$$

In summary, the maximum airspeed of the executive jet aircraft in a steady level turn at an altitude of 10,000 ft with a bank angle of 30 degrees is 888.9 ft/s and occurs with the following pilot inputs: elevator deflection of −0.28 degrees and with full throttle. The corresponding turn radius is 42,502.1 ft.

Minimum Airspeed

The minimum airspeed of the executive jet aircraft in a steady level turn at an altitude of 10,000 ft altitude with a bank angle of 30 degrees must satisfy both the stall constraint and the jet engine thrust constraint. The stall speed for steady level turning flight with a bank angle of 30 degrees is

$$V_{\text{stall}} = \sqrt{\frac{2W}{\rho S C_{L_{\text{max}}} \cos \mu}} = 190.02 \text{ ft/s.}$$

As seen previously, the minimum airspeed of the executive jet aircraft in a steady level turn due to the thrust constraint is 207.65 ft/s. In this case, the minimum airspeed due to the thrust constraint is larger than the stall speed, so the jet engine thrust constraint is active and the stall constraint is not active. Consequently, the minimum airspeed is 207.65 ft/s.

At the minimum airspeed in this steady level turn, the lift coefficient and pitch moment coefficient are

$$C_L = \frac{2W}{\rho V^2 S \cos \mu} = 2.345,$$

$$C_M = 0.$$

These equations determine the aircraft angle of attack as 19.37 degrees and the elevator deflection as 11.59 degrees. The turn radius corresponding to this minimum speed steady level turn is

$$r = \frac{V^2}{g \tan \mu} = 2{,}319.4 \text{ ft.}$$

In summary, the minimum airspeed of the executive jet aircraft in a steady level turn at an altitude of 10,000 ft with a bank angle of 30 degrees is 207.65 ft/s and occurs with the following pilot inputs: elevator deflection of 11.59 degrees and with full throttle. The corresponding turn radius is 2,319.4 ft.

Minimum Radius Steady Level Turns

The executive jet aircraft is to execute a steady level turn at an altitude of 10,000 ft that has the minimum possible turn radius. The analysis is based on a graphical two-dimensional representation of the feasible set for the airspeed and bank angle values that satisfy all of the constraints.

The stall constraint can be written as

$$V \geq \sqrt{\frac{2W}{\rho S C_{L_{\max}} \cos \mu}}.$$

The required thrust is less than or equal to the maximum thrust available from the jet engine:

$$\frac{1}{2}\rho V^2 S C_{D_0} + \frac{2K W^2}{\rho V^2 S \cos^2 \mu} \leq T_{\max}^s \left(\frac{\rho}{\rho^s}\right)^m.$$

The maximum load factor of 2 corresponds to a bank angle of 60 degrees. Thus,

$$\mu \leq 60 \text{ degrees}.$$

Matlab m-files are now presented that allow determination of the minimum turn radius. The maximum airspeed due to the jet engine thrust constraint, the minimum airspeed due to the jet engine thrust constraint, and the stall speed can be computed for various altitudes and bank angles. Define the Matlab function SLTFJ that computes the maximum airspeed, the minimum airspeed due to the thrust constraint, and the stall speed for a specified altitude and bank angle. This Matlab function can be used to confirm the numerical results in the previous section, and it is also useful for subsequent minimum radius turning flight and flight envelope computations for the executive jet aircraft.

Matlab function 9.3 SLTFJ.m
```
function [Vmax VminTC Vstall]=SLTFJ(h,mu);
%Input: altitude h (ft), bank angle mu (rad)
%Output: max air speed Vmax (ft/s), min air speed VminTC (ft/s),
stall speed Vstall (ft/s)
W=73000; S=950; CD0=0.015; K=0.05; Tsmax=12500; m=0.6; CLmax=2.8;
[Ts ps rhos]=StdAtpUS(0);
[T p rho]=StdAtpUS(h);
tmp=sort(roots([1/2*rho*S*CD0 0 -
Tsmax*(rho/rhos)^m 0 2*K*W^2/rho/S/cos(mu)^2]));
Vmax=tmp(4);
VminTC=tmp(3);
Vstall=sqrt(2*W/rho/S/cos(mu)/CLmax);
```

This Matlab m-file can be used to compute and visualize the boundary curves that define the feasible set for steady level turning flight at this altitude; by superimposing the curves of constant turn radius it is possible to identify the point in the feasible set that defines the minimum radius turn. The following Matlab m-file accomplishes this.

Matlab function 9.4 MinRLJ.m
```
g=32.2; h=10000;
nmax=2; n=100;
mus=linspace(0*pi/180,67.435*pi/180,n);
Vs=zeros(n,n);
for k=1:n
    [Vmax0 VminTC0 Vstall0]=SLTFJ(h,mus(k));
    Vmaxs(k)=Vmax0;
    VminTCs(k)=VminTC0;
    Vstalls(k)=Vstall0;
end
figure(1);
plot(VminTCs,mus*180/pi,'r',...
    Vmaxs,mus*180/pi,'r',...
    Vstalls,mus*180/pi,'b',...
    [100 1000],acos(1/nmax)*180/pi*ones(2,1),'g');
ylim([0,70]);
xlim([100,1000]);
xlabel('Velocity (ft/sec)');
ylabel('Bank angle (deg)');
figure(2);
plot(VminTCs,mus*180/pi,'r',...
    Vmaxs,mus*180/pi,'r',...
    Vstalls,mus*180/pi,'b',...
    [100 1000],acos(1/nmax)*180/pi*ones(2,1),'g');hold on;
muopt=7.4562e-01;
Vopt=2.4765e+02;
Ropt=2.0625e+03;
Vss=linspace(100,1000,n);
Rss=[0.2 0.5 1 2 3 4]*Ropt;
for j=1:size(Rss,2)
    for i=1:n
        muss(j,i)=atan(Vss(i)^2/g/Rss(j));
    end
    plot(Vss,muss(j,:)*180/pi,'k:');
end
plot(Vopt,muopt*180/pi,'ro');
ylim([0,70]);
xlim([100,1000]);
xlabel('Velocity (ft/sec)');
ylabel('Bank angle (deg)');
```

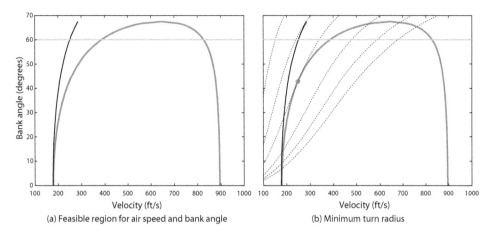

(a) Feasible region for air speed and bank angle (b) Minimum turn radius

Figure 9.4. Minimum radius steady level turn: executive jet aircraft.

The feasible set of airspeed and bank angle values that satisfy the above three constraints is shown in Figure 9.4(a), where the low-speed boundary curve indicates the stall limit, the inverted U curve indicates the boundary of the thrust constraint, and the horizontal boundary curve indicates the maximum load factor.

The minimum turn radius is obtained by following a graphical approach. In Figure 9.4(b), several contour lines for fixed values of the turn radius are given by equation (9.27). The minimum turn radius flight condition is denoted by the small circle in Figure 9.4(b); at this optimal turning flight condition only the thrust constraint is active; the stall constraint and the wing loading constraint are inactive.

The subsequent analysis of the minimum radius steady level turn for the jet aircraft is based on the fact that only the thrust constraint is active. This means that the minimum radius climbing turn corresponds to the values of airspeed V and bank angle μ that minimizes the turn radius

$$r = \frac{V^2}{g \tan \mu},$$

while satisfying the airspeed constraint

$$\frac{1}{2}\rho V^2 S C_{D_0} + \frac{2KW^2}{\rho S V^2 \cos^2 \mu} = \left(\frac{\rho}{\rho^s}\right)^m T_{max}^s.$$

This somewhat simpler optimization problem can be solved using the methods of calculus to obtain a solution given by the airspeed 247.65 ft/s and a bank angle of 42.72 degrees. These values can also be seen from Figure 9.4. The minimum value of the turn radius is computed as 2,062.6 ft. The corresponding lift coefficient $C_L = 1.943$; thus the angle of attack is 16.03 degrees and the elevator deflection is 9.45 degrees.

These results are summarized as follows. The minimum radius steady level turn of the executive jet aircraft at an altitude of 10,000 ft occurs with the following pilot inputs: elevator deflection of 9.45 degrees, full throttle, and with a bank angle of 42.72 degrees. The minimum value of the turn radius is 2,062.6 ft and the airspeed is 247.65 ft/s.

Steady Level Turning Flight Ceiling

The steady level flight ceiling of the executive jet aircraft is computed, assuming the aircraft is in a steady banked turn with a bank angle of 30 degrees.

The flight ceiling for this steady level turn with a 30 degree bank angle is determined by solving the following equation:

$$\frac{2W}{\cos \mu} \sqrt{K C_{D_0}} - T^s_{\max} \left(\frac{\rho}{\rho^s} \right)^m = 0,$$

where ρ is a function of altitude according to the standard atmospheric model. Define the Matlab function eqnFCLTJ whose value is zero at the steady level turning flight ceiling.

Matlab function 9.5 eqnFCLTJ.m

```
function error=eqnFCLTJ(h,mu)
%Input: altitude h (ft), bank angle mu (rad)
%Output: error
W=73000; CD0=0.015; K=0.05; Tsmax=12500; m=0.6;
[Ts ps rhos]=StdAtpUS(0);
[Th ph rhoh]=StdAtpUS(h);
error=2*W/cos(mu)*sqrt(K*CD0)-Tsmax*(rhoh/rhos)^m;
```

The following Matlab command computes the steady level turning flight ceiling for which the value of the function defined by eqnFCLTJ is zero.

```
hmax=fsolve(@(h) eqnFCLTJ(h,30*pi/180),10000)
```

where 10,000 is an initial guess of the flight ceiling. After a few iterations, the flight ceiling for a steady level turn with a 30 degree bank angle is obtained; the computed flight ceiling for this steady level turn is 45,373.5 ft.

In summary, the steady level flight ceiling of the executive jet aircraft is 45,373.5 ft, assuming the aircraft is in a steady banked turn with a bank angle of 30 degrees.

Steady Level Turning Flight Envelope

A steady level turning flight envelope of the executive jet aircraft is computed, assuming the aircraft is in a steady banked turn with a bank angle of 30 degrees; this steady level turning flight envelope is described in terms of the flight altitude and the airspeed.

A two-dimensional graphical representation of the steady level turning flight envelope of the executive jet aircraft at this fixed bank angle is developed using the previously defined Matlab function SLTFJ that computes the maximum and minimum airspeeds due to the thrust constraint and the stall speed. This steady level turning flight envelope, shown in Figure 9.5, is generated by the following Matlab m-file.

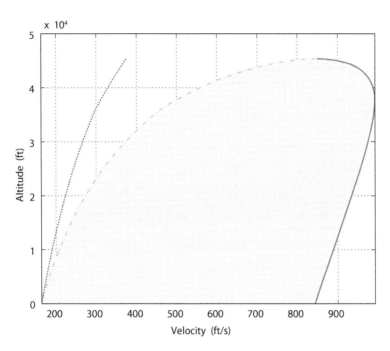

Figure 9.5. Steady level turning flight envelope: executive jet aircraft (μ = 30 degrees).

Matlab function 9.6 FigSLTFJ.m
```
mu=30*pi/180;
h=linspace(0,24400,500);
for k=1:size(h,2)
    [Vmax(k) VminTC(k) Vstall(k)]=SLTFJ(h(k),mu);
end
Vmin=max(VminTC,Vstall);
area([Vmin Vmax(end:-1:1)],[h h(end:-1:1)],...
    'FaceColor',[0.8 1 1],'LineStyle','none');
hold on;
plot(Vmax,h,VminTC,h,Vstall,h);
grid on;
xlim([0 1200]);
xlabel('Velocity (ft/s)');
ylabel('Altitude (ft)');
```

9.7 Steady Level Turning Flight Envelopes: Executive Jet Aircraft

A steady level turning flight envelope can be defined in terms of the flight altitude, the airspeed of the aircraft, and the bank angle of the aircraft. This steady level turning flight envelope is defined in terms of three flight variables and hence must be viewed as a feasible set in three dimensions. This generalizes the steady level turning flight envelope for a fixed bank angle. In particular, the steady level longitudinal flight envelope can be viewed as a two-dimensional cross section of the steady level turning flight envelope corresponding to zero bank angle.

A surface that defines the boundary of the steady level turning flight envelope can be computed and visualized using the Matlab graphics features. A boundary surface of the steady level turning flight envelope is shown in Figure 9.6. This boundary surface is obtained from the following Matlab m-file.

Matlab function 9.7 FigSLTFEJ.m
```
mu=linspace(0*pi/180,70*pi/180,40);
for i=1:size(mu,2)
    hmax(i)=fsolve(@(h) eqnFCLTJ(h,mu(i)),10000);
    h(i,:)=linspace(0,floor(hmax(i)),40);
    for j=1:size(h,2)
        [Vmax(i,j) VminTC(i,j) Vstall(i,j)]=SLTFJ(h(i,j),mu(i));
    end
end
Vmin=max(VminTC,Vstall);
surf(Vmax,mu*180/pi,h,'LineStyle','none');
hold on;
surf(Vmin,mu*180/pi,h,'LineStyle','none');
xlabel('Velocity (ft/s)');
ylabel('Bank angle (deg)');
zlabel('Altitude (ft)');
```

Any triple of values for the flight altitude, aircraft airspeed, and aircraft bank path angle that lies within the steady level turning flight envelope illustrated in Figure 9.6 corresponds to a feasible steady flight condition for which the thrust constraints, the stall constraint, and the wing loading constraint are satisfied. For each such feasible triple of values, methodology has been provided to compute the pilot inputs, namely the elevator deflection and the throttle setting that correspond to this steady flight condition. In addition, the methodology also allows computation of the associated angle of attack, values of the lift and drag coefficients, and values of the lift and drag.

The steady level turning flight envelope shown in Figure 9.5 is a two-dimensional cross section of the three-dimensional steady level turning flight envelope shown in Figure 9.6, where the cross section corresponds to the fixed bank angle of 30 degrees.

9.8 Steady Level Turning Flight: General Aviation Aircraft

Assume that the propeller-driven general aviation aircraft, with a weight of 2,900 lbs, is in a steady level turn at an altitude of 10,000 ft. The analysis in this section is based on the assumption that the bank angle is 25 degrees. This bank angle corresponds to a load factor of 1.1, which is significantly below the maximum load factor of 2; consequently the wing loading constraint is not active.

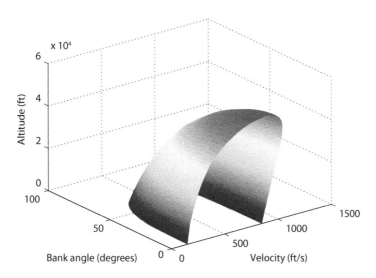

Figure 9.6. Steady level turning flight envelope: executive jet aircraft.

Power Required

Suppose that the airspeed for a steady level turn with a bank angle of 25 degrees is 200 ft/s. The power required for this steady level turn is

$$P = \frac{1}{2}\rho V^3 S C_{D_0} + \frac{2KW^2}{\rho V S \cos^2 \mu} = 49{,}945 \text{ ft lbs/s} = 90.81 \text{ hp}.$$

The pilot inputs for this steady turning flight condition can be computed following the methodology indicated previously. In particular, the required elevator deflection and the required throttle setting can be computed corresponding to the specified bank angle of 25 degrees.

The required power can be computed for various airspeeds. The power required curve, shown in Figure 9.7, is generated using the following Matlab m-file.

Matlab function 9.8 PRSLTFP.m
```
V=linspace(20,350,500);
W=2900; S=175; CD0=0.026; K=0.054; rho=1.7553e-3;
mu=25*pi/180;
P=1/2*rho*V.^3*S*CD0+2*K*W^2/rho./V/S/cos(mu)^2
plot(V,P);
xlabel('Velocity (ft/s)');
ylabel('Power required (ft lbs/s)');
grid on;
```

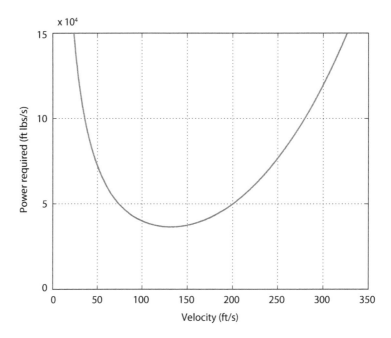

Figure 9.7. Power required for steady level turning flight: general aviation aircraft ($\mu = 25$ degrees).

As seen in Figure 9.7, the power required for a steady level turn with a bank angle of 25 degrees has the usual characteristics: it is large at low airspeeds, it is large at high airspeeds, and it has a minimum value at an intermediate airspeed.

Minimum Required Power

The minimum required power can be determined graphically using Figure 9.7.

The airspeed that achieves the minimum required power for this steady level turn occurs when the airspeed is given by

$$V = \sqrt{\frac{2W}{\rho S \cos \mu}} \sqrt{\frac{K}{3C_{D_0}}} = 131.66 \, \text{ft/s}.$$

At this steady turning flight condition, the lift coefficient and the pitch moment coefficient are

$$C_L = \frac{2W}{\rho V^2 S \cos \mu} = 1.2,$$

$$C_M = 0.$$

These equations can be expressed in terms of the aircraft angle of attack and the elevator deflection as

$$0.02 + 0.12\alpha = 1.2,$$
$$0.12 - 0.08\alpha + 0.075\delta_e = 0.$$

The results are an angle of attack of 9.85 degrees and an elevator deflection of 8.91 degrees. The value of the minimum required power for this steady level turn is given by

$$P_{\min} = \frac{4}{3}\sqrt{\frac{2W^3}{\rho S \cos^3 \mu}}\sqrt{3K^3 C_{D_0}} = 36,455 \text{ lb ft/s} = 66.28 \text{ hp}.$$

The corresponding throttle setting is

$$\delta_t = \frac{P_{\min}}{\eta P_{\max}^s}\left(\frac{\rho^s}{\rho}\right)^m = 0.343.$$

The turn radius corresponding to this minimum power steady level turn is

$$r = \frac{V^2}{g \tan \mu} = 1{,}154.5 \text{ ft}.$$

In summary, the minimum power required for the general aviation aircraft to achieve a steady level turn with a bank angle of 25 degrees is 66.28 hp and occurs with the following pilot inputs: elevator deflection of 8.91 degrees and a throttle setting of 0.343. The corresponding turn radius is 1,154.5 ft and the airspeed is 131.66 ft/s.

Maximum Airspeed

The maximum airspeed of the general aviation aircraft in a steady level turn with a bank angle of 25 degrees at an altitude 10,000 ft is limited by the engine and propeller constraint. This maximum airspeed characterizes the intersection of the power required curve and the curve defined by the maximum power available from the propeller and engine at this altitude; this intersection is shown in Figure 9.8. This maximum airspeed can also be determined by solving the equation

$$\frac{1}{2}\rho V^3 S C_{D_0} + \frac{2KW^2}{\rho V S \cos^2 \mu} = \eta P_{\max}^s \left(\frac{\rho}{\rho^s}\right)^m.$$

Multiplying the above equation by V and rearranging, the following fourth degree polynomial equation is obtained:

$$\frac{1}{2}\rho S C_{D_0} V^4 - \eta P_{\max}^s \left(\frac{\rho}{\rho^s}\right)^m V + \frac{2KW^2}{\rho S \cos^2 \mu} = 0.$$

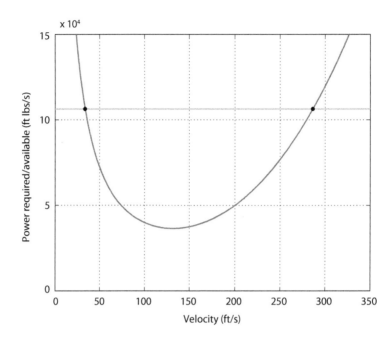

Figure 9.8. Maximum airspeed for steady level turning flight: general aviation aircraft ($\mu = 25$ degrees).

This equation can be solved numerically using the Matlab function MMSSLTP as follows:

Matlab function 9.9 MMSSLTP.m
```
function [Vmin,Vmax]=MMSSLFP
%Output: Minimum and maximum air speeds in a level turn
W=2900; S=175; CD0=0.026; K=0.054; Psmax=290*550; m=0.6; mu=25;
[Ts ps rhos]=StdAtpUS(0);
[Th ph rhoh]=StdAtpUS(h);
[Vmin,Vmax]=sort(roots([1/2*rho*S*CD0 0 0 -eta*Psmax*
(rho/rhos)^m 2*K*W^2/rho/S/cos(mu)^2]));
```

Matlab provides two positive solutions: 286.4 ft/s and 33.9 ft/s. The high-speed solution is the maximum airspeed of the aircraft in this steady level turn. At the given altitude, the speed of sound is 1,077.39 ft/s. Consequently, the Mach number for this maximum airspeed in a steady level turn is 0.266. At the maximum airspeed, the lift coefficient and the moment coefficient are

$$C_L = \frac{2W}{\rho V^2 S \cos \mu} = 0.254,$$
$$C_M = 0.$$

These expressions allow determination of the aircraft angle of attack of 1.95 degrees and the elevator deflection of 0.48 degrees. The turn radius corresponding to this maximum airspeed steady level turn is

$$r = \frac{V^2}{g \tan \mu} = 5{,}462.8 \text{ ft.}$$

In summary, the maximum airspeed of the propeller-driven general aviation aircraft in a steady level turn at 10,000 ft with a bank angle of 25 degrees is 286.4 ft/s and occurs with the following pilot inputs: elevator deflection of 0.48 degrees and with full throttle. The corresponding turn radius is 5,462.8 ft.

Minimum Airspeed

The minimum airspeed of the general aviation aircraft in a steady level turn at an altitude of 10,000 ft with a bank angle of 25 degrees must satisfy both the stall constraint and the engine and propeller constraints. The stall speed for steady level turning flight with a bank angle of 25 degrees is determined by the lift coefficient at stall

$$V_{\text{stall}} = \sqrt{\frac{2W}{\rho S C_{L_{\max}} \cos \mu}} = 93.17 \text{ ft/s.}$$

As seen previously, the minimum airspeed of the aircraft in a steady level turn due to the engine and propeller constraints is 33.9 ft/s. In this case, the stall speed is larger than the minimum airspeed due to the power constraint, so the stall constraint is active and the engine power constraint is not active. Consequently, the minimum airspeed in this steady level turn is 93.17 ft/s. At the minimum airspeed, the lift and the pitch moment coefficients are

$$C_L = \frac{2W}{\rho V^2 S \cos \mu} = 2.4,$$
$$C_M = 0.$$

These equations determine an aircraft angle of attack of 19.83 degrees and an elevator deflection of 19.55 degrees. The power required for this steady level turn is

$$P = \frac{1}{2}\rho V^3 S C_{D_0} + \frac{2KW^2}{\rho V S \cos^2 \mu} = 41{,}867 \text{ lb ft/s} = 76.12 \text{ hp.}$$

The corresponding throttle setting is

$$\delta_t = \frac{P}{\eta P_{\max}^s} \left(\frac{\rho^s}{\rho}\right)^m = 0.394.$$

The turn radius corresponding to this steady level turn is

$$r = \frac{V^2}{g \tan \mu} = 578.1 \text{ ft}.$$

In summary, the minimum airspeed of the general aviation aircraft in a steady level turn at an altitude of 10,000 ft with a bank angle of 25 degrees is 93.17 ft/s and occurs with the following pilot inputs: elevator deflection of 19.55 degrees and throttle setting of 0.394. The corresponding turn radius is 578.1 ft.

Minimum Radius Steady Level Turn

The general aviation aircraft is to execute a steady level turn with a bank angle of 25 degrees at an altitude of 10,000 ft that has the minimum possible turn radius. The analysis is based on a graphical two-dimensional representation of the feasible set for the airspeed and bank angle values that satisfy all of the constraints.

The stall constraint is given by

$$V \geq \sqrt{\frac{2W}{\rho S C_{L_{\max}} \cos \mu}}.$$

The required power is less than the maximum power available from the propeller and engine:

$$\frac{1}{2} \rho V^3 S C_{D_0} + \frac{2 K W^2}{\rho V S \cos^2 \mu} \leq \eta P^s_{\max} \left(\frac{\rho}{\rho^s} \right)^m.$$

The maximum load factor is 2, which corresponds to a bank angle of 60 degrees. Thus the wing loading constraint is satisfied.

Matlab m-files are presented that allow determination of the minimum turn radius. The maximum airspeed due to the engine and propeller constraints, the minimum airspeed due to the engine and propeller constraints, and the stall speed are computed for various altitudes and bank angles. Define the Matlab function SLTFP that computes the maximum airspeed, the minimum airspeed due to the power constraint, and the stall speed for a specified altitude and bank angle. This Matlab function can be used to confirm the numerical results in the previous section, and it is also useful for subsequent minimum radius turning flight and flight envelope computations for the general aviation aircraft.

Matlab function 9.10 SLTFP.m
```
function [Vmax VminPC Vstall]=SLTFP(h,mu);
%Input: altitude h (ft), bank angle mu (rad)
%Output: max air speed Vmax (ft/s), min air speed due to the
power constraint VminPC (ft/s),
%stall speed Vstall (ft/s)
```

(continued)

(continued)

```
W=2900; S=175; CD0=0.026; K=0.054; Psmax=290*550; m=0.6;
CLmax=2.4; eta=0.8;
[Ts ps rhos]=StdAtpUS(0);
[T p rho]=StdAtpUS(h);
tmp=sort(roots([1/2*rho*S*CD0 0 0 -eta*Psmax*(rho/rhos)^m
2*K*W^2/rho/S/cos(mu)^2]));
Vmax=tmp(2);
VminPC=tmp(1);
Vstall=sqrt(2*W/rho/S/cos(mu)/CLmax);
```

This Matlab m-file can be used to compute and visualize the boundary curves that define the feasible set for steady level turning flight at this altitude; by superimposing the curves of constant turn radius it is possible to identify the point in the feasible set that defines the minimum radius turn. The following Matlab m-file accomplishes this.

Matlab function 9.11 MinRLP.m
```
g=32.2; h=10000;
nmax=2; n=100;
mus=linspace(0*pi/180,64*pi/180,n);
Vs=zeros(n,n);
for k=1:n
    [Vmax0 VminTC0 Vstall0]=SLTFP(h,mus(k));
    Vmaxs(k)=Vmax0;
    VminTCs(k)=VminTC0;
    Vstalls(k)=Vstall0;
end
figure(1);
plot(VminTCs,mus*180/pi,'r',...
    Vmaxs,mus*180/pi,'r',...
    Vstalls,mus*180/pi,'b',...
    [0 300],acos(1/nmax)*180/pi*ones(2,1),'g');
ylim([0,70]);
xlim([0,300]);
xlabel('Velocity (ft/sec)');
ylabel('Bank angle (deg)');
figure(2);
plot(VminTCs,mus*180/pi,'r',...
    Vmaxs,mus*180/pi,'r',...
    Vstalls,mus*180/pi,'b',...
    [0 300],acos(1/nmax)*180/pi*ones(2,1),'g');hold on;
mu=60*pi/180;
[Vmax VminTC Vstall]=SLTFP(h,mu);
R=Vstall^2/g/tan(mu);
muopt=mu;
```

(continued)

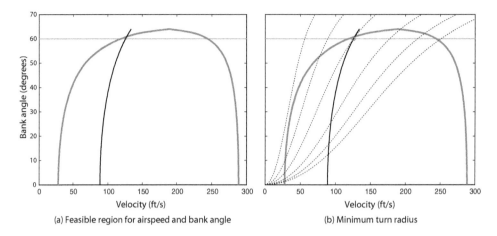

(a) Feasible region for airspeed and bank angle   (b) Minimum turn radius

Figure 9.9. *Minimum radius steady level turn: general aviation aircraft.*

(continued)

```
Vopt=Vstall;
Ropt=R;
Vss=linspace(0,400,n);
Rss=[0.2 0.5 1 2 3 4]*Ropt;
for j=1:size(Rss,2)
    for i=1:n
        muss(j,i)=atan(Vss(i)^2/g/Rss(j));
    end
    plot(Vss,muss(j,:)*180/pi,'k:');
end
plot(Vopt,muopt*180/pi,'ro');
ylim([0,70]);
xlim([0,300]);
xlabel('Velocity (ft/sec)');
ylabel('Bank angle (deg)');
```

The feasible set of the airspeed and bank angle values that satisfy the three constraints are shown in Figure 9.9(a), where the boundary curve indicating the stall limit is shown, the inverted U boundary curve indicates the power limit, and the horizontal boundary curve indicates the wing loading limit.

The minimum turn radius is obtained by following a graphical approach. In Figure 9.9(b), several contour lines for fixed values of the turn radius are shown. The minimum turn radius is denoted by the small circle in Figure 9.9(b); at this optimal turning flight condition the wing loading constraint and the stall constraint are active; the power constraint is not active although the margin is small.

The subsequent analysis of the minimum radius steady level turn for the general aviation aircraft is based on the fact that the wing loading constraint and the stall constraint are active; the engine power constraint is then evaluated and shown to be satisfied.

Since the wing loading constraint is active, the load factor is the maximum load factor of 2, which corresponds to a bank angle of 60 degrees. Since the stall constraint is active, this implies that the lift coefficient is a maximum. This allows determination of the airspeed

$$V = \sqrt{\frac{2W}{\rho S C_L \cos \mu}} = 125.43 \text{ ft/s}.$$

The turn radius corresponding to the bank angle of 60 degrees at the airspeed of 125.43 ft/s is

$$r = \frac{V^2}{g \tan \mu} = 282.12 \text{ ft}.$$

Now the engine and propeller power constraints are checked to make sure that they are satisfied. The required power is

$$P = W V_{climb} + \frac{1}{2} \rho V^3 S C_{D_0} + \frac{2 K W^2}{\rho V S \cos^2 \mu} = 1.0217 \times 10^5 \text{ ft lbs/s} = 185.76 \text{ hp}.$$

The corresponding throttle setting is

$$\delta_t = \frac{P}{\eta P_{max}^s} \left(\frac{\rho^s}{\rho}\right)^m = 0.96,$$

demonstrating that the required power does not exceed the maximum power that the propeller and engine can provide. The fact that the power constraints are satisfied with only a small margin is consistent with the graphical illustration in Figure 9.9(b). The expression for the lift coefficient at stall and the fact that the pitch moment coefficient is zero allows determination of the aircraft angle of attack of 19.8 degrees and the elevator deflection of 19.6 degrees.

These results are summarized as follows. The minimum radius steady level turn of the general aviation aircraft at an altitude of $10,000$ ft occurs with the following pilot inputs: elevator deflection of 19.6 degrees, throttle setting of 0.96, and with a bank angle of 60 degrees. The minimum value of the turn radius is 282.12 ft and the airspeed is 125.43 ft/s.

Steady Level Turning Flight Ceiling

The steady level flight ceiling of the general aviation aircraft is computed, assuming the aircraft is in a steady level banked turn with a bank angle of 25 degrees. The flight ceiling for this steady level turn is determined by solving the following equation:

$$\frac{4}{3} \sqrt{\frac{2W^3}{\rho S \cos^3 \mu}} \sqrt{3 K^3 C_{D_0}} - \eta P_{max}^s \left(\frac{\rho}{\rho^s}\right)^m = 0,$$

where ρ is a function of altitude according to the standard atmospheric model. Define the Matlab function eqnFCLTP whose value is zero at the steady level turning flight ceiling.

Matlab function 9.12 eqnFCLTP.m
```
function error=eqnFCLTP(h,mu)
W=2900; S=175; CD0=0.026; K=0.054; Psmax=290*550; m=0.6; eta=0.8;
[Ts ps rhos]=StdAtpUS(0);
[Th ph rhoh]=StdAtpUS(h);
error=4/3*sqrt(2*W^3/rhoh/S/cos(mu)^3*sqrt(3*K^3*CD0))
-eta*Psmax*(rhoh/rhos)^m;
```

The following Matlab command computes the steady level turning flight ceiling for which the value of eqnFCLTP is zero.

```
hmax=fsolve(@(h) eqnFCLTP(h,25*pi/180),10000)
```

where 10,000 is an initial guess of the flight ceiling. After a few iterations, the flight ceiling for a steady level turn with a bank angle of 25 degrees is obtained; the computed flight ceiling for this steady level turn is 37,398.96 ft.

In summary, the steady level flight ceiling of the general aviation aircraft is 37,398.96 ft, assuming the aircraft is in a steady level banked turn with a bank angle of 25 degrees.

Steady Level Turning Flight Envelope

A steady level flight envelope of the propeller-driven general aviation aircraft is computed, assuming the aircraft is in a steady level banked turn with a bank angle of 25 degrees; this steady level turning flight envelope is described in terms of the flight altitude and the airspeed.

A two-dimensional graphical representation of the steady level turning flight envelope at this fixed bank angle is developed using the previously defined Matlab function SLTFP that computes the maximum airspeed, the minimum airspeed due to the power constraint, and the stall speed for a specified altitude.

This steady level turning flight envelope, for a fixed bank angle of 25 degrees, is shown in Figure 9.10; it is generated by the following Matlab m-file.

Matlab function 9.13 FigSLTFP.m
```
mu=25*pi/180;
h=linspace(0,37398,500);
for k=1:size(h,2)
    [Vmax(k) VminTC(k) Vstall(k)]=SLTFP(h(k),mu);
end
Vmin=max(VminTC,Vstall);
area([Vmin Vmax(end:-1:1)],[h h(end:-1:1)],
    'FaceColor',[0.8 1 1],'LineStyle','none');
```

(continued)

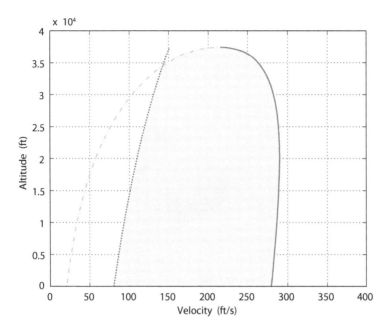

Figure 9.10. Steady level turning flight envelope: general aviation aircraft (μ = 25 degrees).

(continued)

```
hold on;
plot(Vmax,h,VminTC,h,Vstall,h);
grid on;
xlim([0 400]);
xlabel('Velocity (ft/s)');
ylabel('Altitude (ft)');
```

9.9 Steady Level Turning Flight Envelopes: General Aviation Aircraft

A steady level turning flight envelope for the general aviation aircraft can be defined in terms of the flight altitude, the airspeed of the aircraft, and the bank angle of the general aviation aircraft. This steady level turning flight envelope is defined in terms of three flight variables and hence must be viewed as a feasible set in three dimensions. This generalizes the steady level turning flight envelope for a fixed bank angle. In particular, the steady level longitudinal flight envelope can be viewed as a two-dimensional cross section of the steady level turning flight envelope corresponding to zero bank angle.

A surface that defines the boundary of the steady level turning flight envelope can be computed and visualized using the Matlab graphics features. A boundary surface of the steady level turning flight envelope is shown in Figure 9.11. This boundary surface is obtained from the following Matlab m-file.

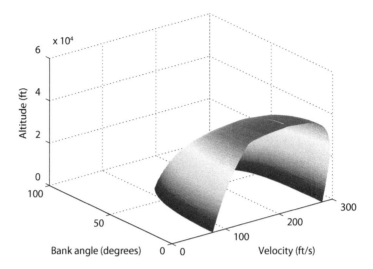

Figure 9.11. *Steady level turning flight envelope: general aviation aircraft.*

Matlab function 9.14 FigSLTFEP.m
```
mu=linspace(0,65*pi/180,40);
for i=1:size(mu,2)
    hmax(i)=fsolve(@(h) eqnFCLTP(h,mu(i)),10000);
    h(i,:)=linspace(0,floor(hmax(i)),40);
    for j=1:size(h,2)
        [Vmax(i,j) VminTC(i,j) Vstall(i,j)]=SLTFP(h(i,j),mu(i));
    end
end
Vmin=max(VminTC,Vstall);
surf(Vmax,mu*180/pi,h,'LineStyle','none');
hold on;
surf(Vmin,mu*180/pi,h,'LineStyle','none');
xlabel('Velocity (ft/s)');
ylabel('Bank angle (deg)');
zlabel('Altitude (ft)');
```

Any triple of values for the flight altitude, aircraft airspeed, and aircraft bank angle that lies within the steady level turning flight envelope illustrated in Figure 9.11 corresponds to a feasible steady flight condition for which the engine and propeller constraints, the stall constraint, and the wing loading constraint are satisfied. For each such feasible triple of values, methodology has been provided to compute the pilot inputs, namely the elevator deflection and the throttle setting corresponding to this steady flight condition. In addition, the methodology also allows computation of the associated angle of attack, values of the lift and drag coefficients, and values of the lift and drag.

The steady level turning flight envelope in Figure 9.10 is a two-dimensional cross section of the three-dimensional steady level turning flight envelope illustrated in Figure 9.11, where the cross section corresponds to the bank angle of 25 degrees.

9.10 Conclusions

This chapter has developed results for steady level turning flight and related performance measures. The results in this chapter can be viewed as a generalization of the prior results for steady level longitudinal flight in chapter 7; in particular, the steady level longitudinal flight results are obtained as a special case for zero bank angle. The next chapter generalizes the results of the current chapter to include the case of steady climbing or descending and turning flight.

9.11 Problems

9.1. Consider the executive jet aircraft in a steady level banked turn at an altitude of 33,000 ft. Assume the aircraft weight is a constant 73,000 lbs.

  (a) Assume the aircraft performs a horizontal banked turn at an airspeed of 660 ft/s with a bank angle of 15 degrees. What is the radius of the turn? What is the turn rate? What is the load factor? What are the required throttle and elevator deflection to maintain this steady level turn?

  (b) Assume the aircraft performs a horizontal banked turn at an airspeed of 660 ft/s with a bank angle of 45 degrees. What is the radius of the turn? What is the turn rate? What is the load factor? What are the required throttle and elevator deflection to maintain this steady level turn?

  (c) Assume the aircraft performs a horizontal banked turn with 100 percent throttle and a bank angle of 45 degrees. Determine all possible flight conditions for this coordinated turn. For each flight condition, determine the airspeed, the radius of the turn, the turn rate, and the load factor. What is the required elevator deflection to maintain this steady level turn?

9.2. Consider the executive jet aircraft in a steady level banked turn at an altitude of 3,000 ft. Assume the aircraft weight is a constant 73,000 lbs.

  (a) Assume the aircraft performs a horizontal banked turn at an airspeed of 660 ft/s with a bank angle of 15 degrees. What is the radius of the turn? What is the turn rate? What is the load factor? What are the required throttle and elevator deflection to maintain this steady level turn?

  (b) Assume the aircraft performs a horizontal banked turn at an airspeed of 660 ft/s with a bank angle of 45 degrees. What is the radius of the turn? What is the turn rate? What is the load factor? What are the required throttle and elevator deflection to maintain this steady level turn?

  (c) Assume the aircraft performs a horizontal banked turn with 100 percent throttle and a bank angle of 45 degrees. Determine all possible flight conditions for this coordinated turn. For each flight condition, determine the airspeed, the radius of the turn, the turn rate, and the load factor. What is the required elevator deflection to maintain this steady level turn?

9.3. Consider the propeller-driven general aviation aircraft in a steady level banked turn at an altitude of 18,000 ft. Assume the aircraft weight is a constant 2,900 lbs.

(a) Assume the aircraft performs a horizontal banked turn at an airspeed of 260 ft/s with a bank angle of 22 degrees. What is the radius of the turn? What is the turn rate? What is the load factor? What are the required throttle and elevator deflection to maintain this steady level turn?

(b) Assume the aircraft performs a horizontal banked turn at an airspeed of 260 ft/s with a bank angle of 40 degrees. What is the radius of the turn? What is the turn rate? What is the load factor? What are the required throttle and elevator deflection to maintain this steady level turn?

(c) Assume the aircraft performs a horizontal banked turn with 100 percent throttle and a bank angle of 22 degrees. Determine all possible flight conditions for this coordinated turn. For each flight condition, determine the airspeed, the radius of the turn, the turn rate, and the load factor. What is the required elevator deflection to maintain this steady level turn?

9.4. Consider the propeller-driven general aviation aircraft in a steady level banked turn at an altitude of 2,000 ft. Assume the aircraft weight is a constant 2,900 lbs.

(a) Assume the aircraft performs a horizontal banked turn at an airspeed of 260 ft/s with a bank angle of 22 degrees. What is the radius of the turn? What is the turn rate? What is the load factor? What are the required throttle and elevator deflection to maintain this steady level turn?

(b) Assume the aircraft performs a horizontal banked turn at an airspeed of 260 ft/s with a bank angle of 40 degrees. What is the radius of the turn? What is the turn rate? What is the load factor? What are the required throttle and elevator deflection to maintain this steady level turn?

(c) Assume the aircraft performs a horizontal banked turn with 100 percent throttle and a bank angle of 22 degrees. Determine all possible flight conditions for this coordinated turn. For each flight condition, determine the airspeed, the radius of the turn, the turn rate, and the load factor. What is the required elevator deflection to maintain this steady level turn?

9.5. Consider the UAV in a steady level banked turn at an altitude of 8,000 ft. Assume the aircraft weight is a constant 45 lbs.

(a) Assume the aircraft performs a horizontal banked turn at an airspeed of 92 ft/s with a bank angle of 28 degrees. What is the radius of the turn? What is the turn rate? What is the load factor? What are the required throttle and elevator deflection to maintain this steady level turn?

(b) Assume the aircraft performs a horizontal banked turn at an airspeed of 92 ft/s with a bank angle of 50 degrees. What is the radius of the turn? What is the turn rate? What is the load factor? What are the required throttle and elevator deflection to maintain this steady level turn?

(c) Assume the aircraft performs a horizontal banked turn with 100 percent throttle and a bank angle of 28 degrees. Determine all possible flight conditions for this coordinated turn. For each flight condition, determine the airspeed, the radius of the turn, the turn rate, and the load factor. What is the required elevator deflection to maintain this steady level turn?

9.6. Consider the UAV in a steady level banked turn at an altitude of 2,000 ft. Assume the aircraft weight is a constant 45 lbs.

(a) Assume the aircraft performs a horizontal banked turn at an airspeed of 92 ft/s with a bank angle of 28 degrees. What is the radius of the turn? What is the turn rate? What is the load factor? What are the required throttle and elevator deflection to maintain this steady level turn?

(b) Assume the aircraft performs a horizontal banked turn at an airspeed of 92 ft/s with a bank angle of 50 degrees. What is the radius of the turn? What is the turn rate? What is the load factor? What are the required throttle and elevator deflection to maintain this steady level turn?

(c) Assume the aircraft performs a horizontal banked turn with 100 percent throttle and a bank angle of 28 degrees. Determine all possible flight conditions for this coordinated turn. For each flight condition, determine the airspeed, the radius of the turn, the turn rate, and the load factor. What is the required elevator deflection to maintain this steady level turn?

9.7. Consider the executive jet aircraft in a steady level banked turn at an altitude of 33,000 ft. Assume the aircraft weight is a constant 73,000 lbs. It is desired to carry out a minimum radius steady level turn. For this minimum radius steady level turn:

(a) What is the minimum value of the steady level turn radius?
(b) What is the turn rate?
(c) What is the required airspeed?
(d) What are the required bank angle and load factor?
(e) What are the required throttle setting and elevator deflection?

9.8. Consider the executive jet aircraft in a steady level banked turn at an altitude of 3,000 ft. Assume the aircraft weight is a constant 73,000 lbs. It is desired to carry out a minimum radius steady level turn. For this minimum radius steady level turn:

(a) What is the minimum value of the steady level turn radius?
(b) What is the turn rate?
(c) What is the required airspeed?
(d) What are the required bank angle and load factor?
(e) What are the required throttle setting and elevator deflection?

9.9. Consider the propeller-driven general aviation aircraft in a steady level banked turn at an altitude of 18,000 ft. Assume the aircraft weight is a constant 2,900 lbs. It is desired to carry out a minimum radius steady level turn. For this minimum radius steady level turn:

(a) What is the minimum value of the steady level turn radius?
(b) What is the turn rate?
(c) What is the required airspeed?
(d) What are the required bank angle and load factor?
(e) What are the required throttle setting and elevator deflection?

9.10. Consider the propeller-driven general aviation aircraft in a steady level banked turn at an altitude of 2,000 ft. Assume the aircraft weight is a constant 2,900 lbs. It is desired to carry out a minimum radius steady level turn. For this minimum radius steady level turn:

(a) What is the minimum value of the steady level turn radius?
(b) What is the turn rate?
(c) What is the required airspeed?
(d) What are the required bank angle and load factor?
(e) What are the required throttle setting and elevator deflection?

9.11. Consider the executive jet aircraft in steady level turning flight. Is it possible that the aircraft can remain in steady level turning flight with an airspeed of 260 ft/s and a turn radius of 2,800 ft at an altitude of 12,000 ft? In other words, is this flight condition in the steady level turning flight envelope?

If this flight condition is in the steady level turning flight envelope, determine the throttle setting, the elevator deflection, and the bank angle required to maintain this flight condition.

If this flight condition is not in the steady level turning flight envelope, which constraints are violated?

9.12. Consider the executive jet aircraft in steady level turning flight. Is it possible that the aircraft can remain in steady level turning flight with an airspeed 500 ft/s and a turn radius of 4,500 ft at an altitude of 9,000 ft? In other words, is this flight condition in the steady level turning flight envelope?

If this flight condition is in the steady level turning flight envelope, determine the throttle setting, the elevator deflection, and the bank angle required to maintain this flight condition.

If this flight condition is not in the steady level turning flight envelope, which constraints are violated?

9.13. Consider the propeller-driven general aviation aircraft in steady level turning flight. Is it possible that the aircraft can remain in steady level turning flight with an airspeed of 95 ft/s and a turn radius of 1,200 ft at an altitude of 5,000 ft? In other words, is this flight condition in the steady level turning flight envelope?

If this flight condition is in the steady level turning flight envelope, determine the throttle setting, the elevator deflection, and the bank angle required to maintain this flight condition.

If this flight condition is not in the steady level turning flight envelope, which constraints are violated?

9.14. Consider the propeller-driven general aviation aircraft in steady level turning flight. Is it possible that the aircraft can remain in steady level turning flight with an airspeed of 250 ft/s and a turn radius of 2,000 ft at an altitude of 7,000 ft? In other words, is this flight condition in the steady level turning flight envelope?

If this flight condition is in the steady level turning flight envelope, determine the throttle setting, the elevator deflection, and the bank angle required to maintain this flight condition.

If this flight condition is not in the steady level turning flight envelope, which constraints are violated?

10

Aircraft Steady Turning Flight

This chapter treats aircraft flight, through a stationary atmosphere, in a steady climbing or descending turn. Banked turns during a steady climb or a steady descent are studied. Various flight conditions for steady turning flight are analyzed, and turning flight performance measures are introduced and assessed.

Steady turning and climbing or descending flight is the most general type of steady flight. It corresponds to motion of the center of mass of the aircraft along a helical path that has a vertical axis. One special case, when the turn radius is infinite, is steady longitudinal flight for which the center of mass of the aircraft moves along a straight path as in chapter 8. Another special case occurs when the flight path angle is zero so that the center of mass of the aircraft moves along a horizontal circular path as in chapter 9.

10.1 Steady Banked Turns

A banked climbing or descending turn is achieved by rotating the lift vector about the velocity vector so that it has a horizontal component that gives rise to centrifugal acceleration, thereby causing the aircraft to turn. If the lift vector is rotated then the aircraft is in a climbing or descending turn. For most conventional aircraft, the lift force is much larger than the thrust that can be produced by the propulsion system, so this results in an effective turn strategy. Banked steady turning flight is also studied in Vinh (1995).

Free-Body Diagram

The free-body diagram in Figure 10.1 shows several views of the forces that act on an aircraft performing a steady turn by banking while the aircraft is climbing. The center of mass of the aircraft moves along a helical path, with constant airspeed V, constant flight path angle γ or climb rate V_{climb}, and constant turn radius r. Consequently, the aircraft has a constant centrifugal acceleration of

$$\frac{V^2 \cos^2 \gamma}{r}$$

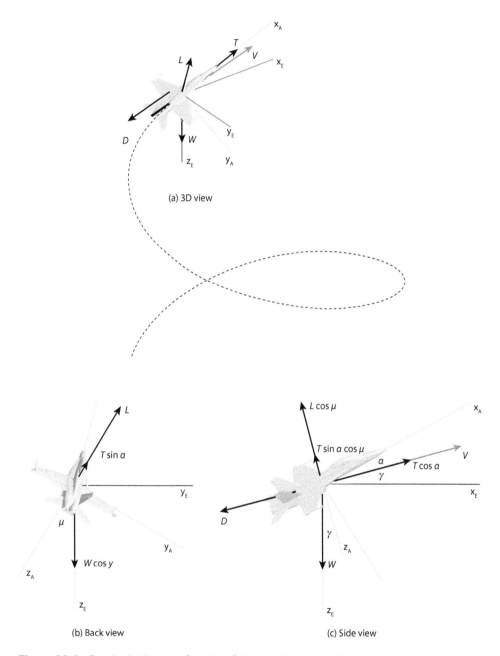

Figure 10.1. Free body diagram of an aircraft in a steady banked climbing turn.

that is always directed in the horizontal plane toward the vertical axis of the helical path. Coordinated flight of the aircraft is assumed; that is, the side-slip angle is zero.

The lift force on the aircraft acts normal to the velocity vector in the plane of mass symmetry of the aircraft; the lift force is rotated around the velocity vector from the vertical plane by the bank angle μ. The drag force on the aircraft acts opposite to the velocity vector. The thrust vector, with magnitude T, acts along the aircraft-fixed x_A-axis.

The free-body diagram, as shown in Figure 10.1, is somewhat complicated since the aircraft is in a steady turn and climb in three dimensions. Figure 10.1 should be carefully examined to see that the several perspectives of the free-body diagrams are consistent and they lead to the following algebraic equations that arise from Newton's law:

$$-W \sin \gamma - D + T \cos \alpha = 0, \tag{10.1}$$

$$\frac{W}{g} \frac{V^2 \cos^2 \gamma}{r} = L \sin \mu + T \sin \alpha \sin \mu, \tag{10.2}$$

$$W \cos \gamma - L \cos \mu - T \sin \alpha \cos \mu = 0. \tag{10.3}$$

Equation (10.1) states that the sum of the components of the forces in the direction of the velocity vector is zero since there is no aircraft acceleration along the velocity vector. Equation (10.2) states that the mass of the aircraft multiplied by the centrifugal acceleration is the sum of the force components in the (horizontal) radial direction. Equation (10.3) states that the sum of the forces, projected onto the direction normal to the velocity vector in the vertical plane, is zero since there is no aircraft acceleration in this direction.

In addition, the pitch moment is zero in steady flight:

$$M = 0. \tag{10.4}$$

A positive flight path angle corresponds to a steady turn while climbing; a negative flight path angle corresponds to a steady turn while descending. The case of steady level flight, for which the flight path angle is zero, reduces to the case of steady level turning flight studied in chapter 9. A positive bank angle corresponds to a steady right turn while climbing or descending; a negative bank angle corresponds to a steady left turn while climbing or descending. The subsequent development holds for any flight path angle, be it positive, negative, or zero, and any bank angle, be it positive, negative, or zero.

As in chapter 7, the thrust magnitude is assumed to be much smaller than the lift, and the flight path angle and the angle of attack are assumed to be small angles in radian measure so that $\sin \gamma$ is approximated by γ, $\sin \alpha$ is approximated by α, $\cos \gamma$ is approximated by 1, and $\cos \alpha$ is approximated by 1 in the equations (10.1)–(10.3). Since the bank angle can be relatively large for many turning maneuvers, a similar approximation is not made for the terms involving the bank angle. The above equations are approximated as

$$T = W \gamma + D, \tag{10.5}$$

$$\frac{W}{g} \frac{V^2}{r} = L \sin \mu, \tag{10.6}$$

$$L \cos \mu = W. \tag{10.7}$$

These equations form the basis for our subsequent analysis of steady turning and climbing or descending flight.

Aerodynamics

Recall the equations that describe the lift force, the drag force, and the pitch moment from chapter 3 are

$$L = \frac{1}{2}\rho V^2 S C_L, \tag{10.8}$$

$$D = \frac{1}{2}\rho V^2 S C_D, \tag{10.9}$$

$$M = \frac{1}{2}\rho V^2 S c C_M. \tag{10.10}$$

The quadratic drag polar expression that relates the drag coefficient to the lift coefficient is given by

$$C_D = C_{D_0} + K C_L^2, \tag{10.11}$$

where $K = \frac{1}{\pi e AR}$. The lift coefficient is expressed as a linear function of the angle of attack as

$$C_L = C_{L_0} + C_{L_\alpha} \alpha. \tag{10.12}$$

The pitch moment coefficient is expressed as a linear function of the angle of attack and the elevator deflection as

$$C_M = C_{M_0} + C_{M_\alpha} \alpha + C_{M_{\delta_e}} \delta_e. \tag{10.13}$$

Flight Constraints

As in chapter 9 there are three types of important flight constraints that apply to steady climbing or descending turning flight. These constraints are important in determining the performance properties of an aircraft in a steady climbing or descending turn.

The first constraint arises from the aerodynamics assumption that the flight of the aircraft must avoid stalling. Based on the discussion in chapter 2, the stall constraint is expressed by the inequality

$$C_L \le C_{L_{\max}}. \tag{10.14}$$

The second set of constraints arises from limits on the operation of the propulsion system. For a jet aircraft, the thrust T required for a steady climbing or descending turn must satisfy

$$0 \le T \le T_{\max}^s \left(\frac{\rho}{\rho^s}\right)^m. \tag{10.15}$$

The jet engine thrust constraints in expression (10.15) can also be expressed as constraints on the throttle setting

$$0 \leq \delta_t \leq 1, \tag{10.16}$$

where

$$\delta_t = \frac{T}{T_{\max}^s} \left(\frac{\rho^s}{\rho} \right)^m. \tag{10.17}$$

For an internal combustion engine and a propellor-driven aircraft, the power P required for a steady climbing or descending turn must satisfy

$$0 \leq P \leq \eta P_{\max}^s \left(\frac{\rho}{\rho^s} \right)^m, \tag{10.18}$$

where η is the efficiency of the propeller. The engine power constraints in expression (10.18) can also be expressed as constraints on the throttle setting

$$0 \leq \delta_t \leq 1, \tag{10.19}$$

where

$$\delta_t = \frac{P}{\eta P_{\max}^s} \left(\frac{\rho^s}{\rho} \right)^m. \tag{10.20}$$

The third constraint on a steady climbing or descending turn is the wing loading constraint that is expressed as

$$L \leq n_{max} W, \tag{10.21}$$

where n_{max} is the maximum load factor.

10.2 Steady Banked Turning Flight Analysis

Flight conditions for steady climbing or descending turning flight are analyzed. Several turning flight relationships are first obtained. Equations are obtained that describe the turn radius, the turn rate, the required thrust or power, the flight path angle or climb rate, the airspeed, and the bank angle for steady climbing or descending turns.

We first use equations (10.7) and (10.8) to obtain an expression for the airspeed of the aircraft:

$$V = \sqrt{\frac{2W}{\rho S C_L \cos \mu}}. \tag{10.22}$$

This expression for the airspeed holds for any steady climbing or descending flight condition.

Turn Radius and Turn Rate

Using equation (10.7), it is seen that the load factor $n = \frac{L}{W}$ for steady climbing or descending turning flight is directly related to the bank angle as

$$n = \sec \mu. \tag{10.23}$$

Divide equation (10.6) by equation (10.7) to obtain

$$\tan \mu = \frac{V^2}{gr}. \tag{10.24}$$

Alternatively, the airspeed for steady climbing or descending flight can be expressed in terms of the turn radius and the bank angle as

$$V = \sqrt{gr \tan \mu}. \tag{10.25}$$

Equation (10.24) can be rewritten so that the turn radius for steady climbing or descending flight is expressed in terms of the airspeed and bank angle:

$$r = \frac{V^2}{g \tan \mu}. \tag{10.26}$$

Another expression for the turn radius can be given in terms of the bank angle using equations (10.7), (10.8), and (10.26) to obtain

$$r = \frac{2}{\rho g C_L \sin \mu} \frac{W}{S}. \tag{10.27}$$

The turn radius can also be expressed in terms of the load factor as

$$r = \frac{2}{\rho g C_L} \frac{n}{\sqrt{n^2 - 1}} \frac{W}{S}. \tag{10.28}$$

These expressions for the turn radius of an aircraft in a steady climbing or descending turn show that the turn radius depends on the lift coefficient, the bank angle or the load factor, the wing loading W/S, and the flight altitude.

The angular turn rate can be written as

$$\omega = \frac{V}{r} = \frac{g \tan \mu}{V}. \tag{10.29}$$

The turn period, the time required for the aircraft to make one revolution on the helical path, is given by

$$\tau = \frac{2\pi}{\omega} = \frac{2\pi V}{g \tan \mu}. \tag{10.30}$$

The equations in this section for the load factor, the turn radius, the angular turn rate, and the turn period are exactly the same as those for a steady level turn. In other words, based on the assumption of a small flight path angle, these expressions are not influenced by the fact that the aircraft is climbing or descending.

Thrust Required

Using equations (10.5) and (10.7) and the relation $\frac{D}{L} = \frac{C_D}{C_L}$, the required thrust for a steady turn is

$$T = \left(\gamma + \frac{C_D}{C_L \cos \mu}\right) W. \tag{10.31}$$

This equation gives the thrust required for a steady climbing or descending turn as it depends on the flight path angle, the bank angle, and the aerodynamic parameters.

Other expressions for the thrust for a steady climbing or descending turn can be obtained. Expressing the lift in terms of the dynamic pressure and lift coefficient and using equation (10.7), the required lift coefficient for steady turning and climbing or descending flight is

$$C_L = \frac{2W}{\rho V^2 S \cos \mu}. \tag{10.32}$$

Consequently, the required drag coefficient for steady turning and climbing or descending flight is obtained from the quadratic drag polar expression as

$$C_D = C_{D_0} + K \left(\frac{2W}{\rho V^2 S \cos \mu}\right)^2. \tag{10.33}$$

The drag on the aircraft in steady turning and climbing or descending flight is

$$D = \frac{1}{2}\rho V^2 S C_{D_0} + \frac{2K W^2}{\rho V^2 S \cos^2 \mu}. \tag{10.34}$$

Hence the required thrust for an aircraft in steady turning flight in a climb or descent, expressed in terms of the airspeed, bank angle, and flight path angle, is obtained using equation (10.5):

$$T = W\gamma + \frac{1}{2}\rho V^2 S C_{D_0} + \frac{2K W^2}{\rho V^2 S \cos^2 \mu}. \tag{10.35}$$

The thrust required for a steady climbing or descending turn can also be expressed in terms of the airspeed, bank angle, and climb rate as

$$T = W \left(\frac{V_{climb}}{V} \right) + \frac{1}{2} \rho V^2 S C_{D_0} + \frac{2KW^2}{\rho V^2 S \cos^2 \mu}. \tag{10.36}$$

There are three terms in expressions (10.35) and (10.36) for the required thrust. The first term on the right of the equations describes the part of the thrust that is required for the aircraft to climb or descend; this thrust term is positive for climbing flight and negative for descending flight. The second term represents the drag component that is due to aerodynamic friction and is independent of the lift and the bank angle; this drag term increases as the airspeed increases. The third term is the part of the drag that is induced by the lift; this drag term increases as the airspeed decreases or the bank angle increases. The total thrust required to maintain a steady climbing or descending turn is the sum of all three terms.

Equation (10.35) can be rewritten to express the flight path angle in terms of the thrust, the airspeed, and the bank angle as

$$\gamma = \frac{T}{W} - \frac{1}{2} \frac{\rho V^2 S C_{D_0}}{W} - \frac{2KW}{\rho V^2 S \cos^2 \mu}. \tag{10.37}$$

Equation (10.36) can be rewritten to express the climb rate in terms of the thrust, the airspeed, and the bank angle as

$$V_{climb} = \frac{TV}{W} - \frac{1}{2} \frac{\rho V^3 S C_{D_0}}{W} - \frac{2KW}{\rho V S \cos^2 \mu}. \tag{10.38}$$

Equation (10.35) can be expressed in an alternative form. If the thrust, the flight path angle, and the bank angle are assumed known, then equation (10.35) can be written as

$$\frac{1}{2} \rho V^4 S C_{D_0} - (T - W\gamma) V^2 + \frac{2KW^2}{\rho S \cos^2 \mu} = 0, \tag{10.39}$$

which can be viewed as an equation whose solution is the airspeed for a steady climbing or descending turn at the given value of thrust, flight path angle, and bank angle. Equation (10.39) has two real and positive solutions corresponding to a low-speed, steady climbing or descending turn and a high-speed, steady climbing or descending turn. Solving this equation requires finding the roots of a fourth degree polynomial. The low-speed solution must be checked to see that the stall constraint is satisfied:

$$\sqrt{\frac{2W}{\rho S C_{L_{max}} \cos \mu}} \leq V. \tag{10.40}$$

The angle of attack and the elevator deflection for an aircraft in a steady climbing or descending turn can be determined based on the value of the lift coefficient computed from equation (10.32) and the fact that the pitch moment coefficient is zero for steady flight.

The throttle setting is easily determined from the above analysis. Since the bank angle is to be maintained constant throughout the maneuver, there should be no differential ailerons deflection and no rudder deflection.

The thrust required for a steady climbing or descending turn, with a fixed flight path angle and a fixed bank angle, has a minimum value. This minimum value can be obtained using the methods of calculus: at the minimum the thrust required curve has zero slope; that is, the derivative of the thrust with respect to the airspeed is zero. In particular, the thrust is minimized when the drag is minimized. This occurs if the airspeed is

$$V = \sqrt{\frac{2W}{\rho S \cos \mu} \sqrt{\frac{K}{C_{D_0}}}}. \tag{10.41}$$

The airspeed value at which the thrust required for a steady climbing or descending turn is a minimum depends on the air density and hence on the flight altitude; this airspeed increases with an increase in the flight altitude.

The minimum value of the thrust required for steady climbing or descending turning flight, expressed in terms of the flight path angle and the bank angle, is obtained by substituting expression (10.41) into equation (10.35) to obtain

$$T_{\min} = W\gamma + \frac{2W}{\cos \mu} \sqrt{K C_{D_0}}. \tag{10.42}$$

The minimum value of the thrust required for a steady climbing or descending turn, for fixed values of the flight path angle and bank angle, does not depend on the air density and hence does not depend on the flight altitude. The minimum value of the thrust required for steady climbing or descending turning flight, expressed in terms of the climb rate and the bank angle, is obtained by substituting expression (10.41) into equation (10.36) to obtain

$$T_{\min} = W V_{climb} \sqrt{\frac{\rho S \cos \mu}{2W} \sqrt{\frac{C_{D0}}{K}}} + \frac{2W}{\cos \mu} \sqrt{K C_{D_0}}. \tag{10.43}$$

The minimum value of the thrust required for a steady climbing or descending turn, for fixed values of the climb rate and bank angle, does depend on the air density and hence on the flight altitude.

For a jet aircraft, the thrust constraints given in (10.15) must be satisfied. In particular, for a positive flight path angle the most important thrust constraint is typically the upper thrust limit, while for a negative flight path angle the thrust constraint defined by the lower limit of zero thrust must also be considered.

The equations for a banked steady climbing or descending turn in this section are expressed in terms of the aircraft thrust, the aircraft speed, the flight path angle or the climb rate, and the bank angle. These can be rewritten in terms of the aircraft thrust, the aircraft speed, the flight path angle, and the load factor, using equation (10.23). For given values of thrust and flight path angle, a $V - n$ diagram shows how the load factor depends on the aircraft airspeed.

Power Required

The power required for steady climbing or descending turning flight, expressed in terms of the airspeed, flight path angle, and bank angle, is obtained by multiplying equation (10.35) by the airspeed to obtain

$$P = WV\gamma + \frac{1}{2}\rho V^3 SC_{D_0} + \frac{2KW^2}{\rho VS \cos^2 \mu}. \tag{10.44}$$

The power required for steady climbing or descending turning flight is also expressed in terms of the airspeed, the climb rate, and the bank angle as

$$P = WV_{climb} + \frac{1}{2}\rho V^3 SC_{D_0} + \frac{2KW^2}{\rho VS \cos^2 \mu}. \tag{10.45}$$

Equations (10.44) and (10.45) have the following interpretation. The first term on the right represents the power required to climb or descend, the second term on the right represents the power required to overcome the lift independent part of the drag, and the third term represents the power required to overcome the lift dependent part of the drag; this third term depends on the bank angle.

Equation (10.44) can be rewritten to express the flight path angle in terms of the power, the bank angle, and the airspeed as

$$\gamma = \frac{P}{WV} - \frac{1}{2}\frac{\rho V^2 SC_{D_0}}{W} - \frac{2KW}{\rho V^2 S \cos^2 \mu}. \tag{10.46}$$

Equation (10.45) can be rewritten to express the climb rate in terms of the power, the bank angle, and the airspeed as

$$V_{climb} = \frac{P}{W} - \frac{1}{2}\frac{\rho V^3 SC_{D_0}}{W} - \frac{2KW}{\rho VS \cos^2 \mu}. \tag{10.47}$$

Equation (10.45) can be expressed in an alternative form. If the power, the climb rate, and the bank angle are assumed known, then equation (10.45) can be written as

$$\frac{1}{2}\rho V^4 SC_{D_0} - (P - WV_{climb}) V + \frac{2KW^2}{\rho S \cos^2 \mu} = 0, \tag{10.48}$$

which can be viewed as an equation whose solution is the airspeed for steady climbing or descending turning flight at the given value of power, climb rate, and bank angle. Equation (10.48) has two real and positive solutions corresponding to a low-speed steady climbing or descending turn and a high-speed steady climbing or descending turn. Solving this equation requires finding the roots of a fourth degree polynomial. The low-speed solution should be checked to see that the stall constraint is satisfied:

$$\sqrt{\frac{2W}{\rho S C_{L_{\max}} \cos \mu}} \leq V. \tag{10.49}$$

For a fixed value of the flight path angle or climb rate and a fixed value of the bank angle, the power required for a steady climbing or descending turn has a minimum value at some airspeed. This minimum value can be obtained using the methods of calculus: at the minimum the power required curve has zero slope; that is, the derivative of the power with respect to the airspeed is zero. This occurs if the airspeed is

$$V = \sqrt{\frac{2W}{\rho S \cos \mu} \sqrt{\frac{K}{3 C_{D_0}}}}. \tag{10.50}$$

The airspeed at which the minimum power for a steady climbing or descending turn occurs depends on the air density and hence on the flight altitude; this airspeed increases with an increase in the flight altitude.

The minimum power for a steady climbing or descending turn, expressed in terms of the flight path angle and the bank angle, is obtained by substituting expression (10.50) into equation (10.44) to obtain

$$P_{\min} = W\gamma \sqrt{\frac{2W}{\rho S \cos \mu} \sqrt{\frac{K}{3 C_{D_0}}}} + \frac{4}{3} \sqrt{\frac{2W^3}{\rho S \cos^3 \mu}} \sqrt{3 K^3 C_{D_0}}. \tag{10.51}$$

The minimum value of the power required for a steady climbing or descending turn, for a fixed flight path angle and a fixed bank angle, depends on the air density and hence on the flight altitude.

The minimum value of the power for a steady climbing or descending turn, expressed in terms of the climb rate and the bank angle, is obtained by substituting expression (10.50) into equation (10.45) to obtain

$$P_{\min} = W V_{climb} + \frac{4}{3} \sqrt{\frac{2W^3}{\rho S \cos^3 \mu}} \sqrt{3 K^3 C_{D_0}}. \tag{10.52}$$

The minimum value of the power required for a steady climbing or descending turn, for a fixed flight path angle and a fixed bank angle, depends on the air density and hence on the flight altitude; the minimum power increases as the flight altitude increases.

For an aircraft powered by an internal combustion engine with a propeller, the power constraints given by the inequalities in (10.18) must be satisfied. In particular, for a positive climb rate the most important power constraint is typically the upper power limit, while for a negative climb rate the power constraint defined by the lower limit of zero power must also be considered.

The equations in this section for banked steady climbing or descending turning flight are expressed in terms of the aircraft power, the aircraft airspeed, the rate of climb, and the bank angle. These can be rewritten in terms of the aircraft power, the aircraft airspeed, the rate of climb, and the load factor, using equation (10.23). For given values of power and rate of climb, a $V - n$ diagram shows how the load factor depends on the aircraft airspeed.

10.3 Jet Aircraft Steady Turning Flight Performance

Flight conditions that characterize the optimal performance of jet aircraft in steady climbing or descending turns are studied. The following flight performance measures are analyzed: maximum airspeed for a given flight path angle and bank angle, minimum airspeed for a given flight path angle and bank angle, maximum flight path angle, minimum turn radius, and flight ceiling. The steady climbing or descending turning flight envelope is also introduced. These flight performance measures characterize the flight maneuverability or flight agility of the aircraft.

Minimum and Maximum Airspeeds

Flight conditions for minimum and maximum airspeeds of a jet aircraft are analyzed for a steady climbing or descending turn with a fixed bank angle and a fixed flight path angle. The maximum thrust that the jet engine can provide limits the minimum and maximum airspeeds for which the aircraft can maintain a steady climbing or descending turn. In particular, the minimum and maximum airspeeds are characterized by the fact that the thrust required is equal to the maximum thrust that the jet engine can provide at full throttle; this leads to the equation

$$W\gamma + \frac{1}{2}\rho V^2 S C_{D_0} + \frac{2KW^2}{\rho V^2 S \cos^2 \mu} = T^s_{\max}\left(\frac{\rho}{\rho^s}\right)^m. \qquad (10.53)$$

This equation has two positive solutions. The high-speed solution is the maximum possible airspeed of the aircraft in a steady climbing or descending turn at the fixed flight path angle and the fixed bank angle; the low-speed solution is the minimum possible airspeed of the aircraft in a steady climbing or descending turn at the fixed flight path angle and the fixed bank angle. Equation (10.53) can be expressed as a fourth degree polynomial equation.

Two important qualifications must be made about the maximum airspeed and the minimum airspeed obtained from this analysis. This computed minimum airspeed is the minimum airspeed if it satisfies the stall constraint given by inequality (10.40); otherwise the stall speed is the minimum airspeed for steady climbing or descending turn at the given thrust, flight path angle, and bank angle. If the computed maximum airspeed is near to or exceeds the speed of sound, the computed maximum airspeed is inaccurate.

Maximum Flight Path Angle

Flight conditions for a jet aircraft to have a maximum flight path angle, when the aircraft is in a steady turn with a fixed bank angle, are studied. Assume the bank angle is selected so that the wing loading constraint is satisfied. The flight path angle is the excess thrust, the thrust provided by the engine minus the drag, divided by the weight of the aircraft. The maximum flight path angle occurs if the thrust provided by the jet engine is maximum, that is, full throttle, and if the drag is a minimum. The drag depends on the bank angle and the drag is a minimum when the airspeed is given by

$$V = \sqrt{\frac{2W}{\rho S \cos \mu} \sqrt{\frac{K}{C_{D_0}}}}. \tag{10.54}$$

Consequently, the maximum flight path angle is given by

$$\gamma_{max} = \frac{T_{max}^s}{W} \left(\frac{\rho}{\rho^s} \right)^m - 2 \cos \mu \sqrt{K C_{D_0}}. \tag{10.55}$$

This flight condition for the maximum flight path angle of an aircraft turning by banking should be checked to make sure that the stall constraint given by inequality (10.40) is satisfied.

Although we do not develop the results in detail, there is also a minimum flight path angle for steady turning flight with a given bank angle. This occurs when the thrust provided by the engine is minimum (that is, the throttle setting is zero) and the drag is maximum; a local maximum of the drag occurs at the stall speed.

Minimum Radius Turns

Flight conditions for a minimum radius steady climbing or descending turn of a jet aircraft with a fixed flight path angle are now considered. The problem of determining the minimum turn radius for a jet aircraft in a steady climbing or descending turn is complicated; one approach for determining the flight conditions for a minimum turn radius is suggested.

There are several possible formulations for the steady climbing or descending minimum radius turn as an optimization problem. One formulation is as follows: determine the values of airspeed V and bank angle μ for which the steady climbing or descending turn radius

$$r = \frac{V^2}{g \tan \mu} \tag{10.56}$$

is a minimum, subject to satisfaction of the three inequalities defined by the stall constraint, the thrust constraints, and the wing loading constraint:

$$\frac{2W}{\rho V^2 S \cos \mu} \leq C_{Lmax}, \tag{10.57}$$

$$W\gamma + \frac{1}{2}\rho V^2 S C_{D_0} + \frac{2K W^2}{\rho V^2 S \cos^2 \mu} \leq \left(\frac{\rho}{\rho^s} \right)^m T_{max}^s, \tag{10.58}$$

$$\sec \mu \leq n_{max}. \tag{10.59}$$

This minimum radius turn optimization problem differs from the minimum radius turn optimization problem for steady level flight of a jet aircraft in chapter 9 only in the specification of the thrust constraint, which includes an additional thrust term required for the aircraft to climb or descend. This optimization problem is complicated by the fact that the active constraints and the inactive constraints are not a priori known. Any one of the following situations is possible:

- One constraint is active.
  - The jet engine thrust constraint is active, but the stall constraint and the wing loading constraint are inactive.

- Exactly two of the constraints are active.
  - The stall constraint and the jet engine thrust constraint are active, but the wing loading constraint is inactive.
  - The stall constraint and the wing loading constraint are active, but the jet engine thrust constraint is inactive.
  - The wing loading constraint and the jet engine thrust constraint are active, but the stall constraint is inactive.

- All three of the constraints are active.

Since any one of these cases can characterize the minimum radius turn for a jet aircraft in steady climbing or descending flight, in principle all cases can be analyzed and the corresponding flight condition and the corresponding steady turn radius computed for each case. The minimum radius turn for the jet aircraft in steady climbing or descending flight is then determined by a direct comparison of the computed turn radius values for which all the constraints are satisfied. In some instances, a priori physical knowledge can be used to make an educated hypothesis about which are the active constraints and the inactive constraints for the minimum radius turn.

A graphical representation of this feasible set of airspeeds and bank angles, for a fixed flight altitude and a fixed flight path angle, can be obtained by defining boundary curves that correspond to one of the three constraints being active. The boundary curves for the stall constraint active and for the wing loading constraint active are easily obtained. If the maximum thrust constraint is active, there is a low-speed solution and a high-speed solution for each bank angle value as seen in the previous sections; this defines the boundary curve for this maximum thrust constraint being active. The feasible set of airspeed and bank angle values is simultaneously bounded by all three of these boundary curves.

Contour curves that correspond to constant values of the steady level turn radius are defined by equation (10.56). The flight conditions for the desired minimum radius turn are characterized by the intersection of the smallest turn radius contour curve with the feasible set of airspeed and bank angle values. Graphical tools can be used to plot the boundary curves that define this feasible set and to plot constant contour turn radius curves; the minimum radius steady turning flight condition can be determined by inspection. Once the active constraints are clear, the optimal values of the airspeed and the bank angle, and the other flight variables, can be determined using the methods developed in this chapter.

Equation (10.27) makes clear that the smallest possible turn radius for a jet aircraft is achieved, all other factors considered constant, at the lowest possible altitude. Further, the minimum turn radius is proportional to the wing loading W/S. That is, jet aircraft with small wing loading can achieve tighter turns than can jet aircraft with larger wing loading. The wing loading is an important design parameter that determines the maneuverability of the aircraft.

Flight Ceiling

For fixed values of the flight path angle and bank angle, there is a maximum altitude, referred to as the steady climbing or descending turning flight ceiling, at which a steady climbing or descending turn can be maintained. The flight condition at this flight ceiling is characterized by the fact that the minimum thrust required for a steady climbing or descending turn, given by equation (10.42), equals the maximum thrust that the jet engine can produce at this altitude. This leads to the equation

$$W\gamma + \frac{2W}{\cos\mu}\sqrt{KC_{D_0}} = T^s_{\max}\left(\frac{\rho}{\rho^s}\right)^m.$$ (10.60)

Since the air density depends on the altitude according to the standard atmospheric model, equation (10.60) can be viewed as an implicit equation in the altitude; the flight ceiling solves this equation.

Flight Envelope

For a fixed value of the flight path angle, the flight envelope is the feasible set in three dimensions defined by the altitude, airspeed, and bank angle variables that satisfy the steady flight equations and constraints for the fixed value of the flight path angle. The steady level turning flight envelope is obtained if the flight path angle is zero.

Another possible three-dimensional visualization approach is to represent the flight envelope for a fixed value of the bank angle. For that fixed value of the bank angle, the flight envelope is the feasible set in three dimensions defined by the altitude, airspeed, and flight path angle variables that satisfy the steady flight equations and constraints for the fixed value of the bank angle. The steady longitudinal flight envelope is obtained if the bank angle is zero.

A simplified two-dimensional visualization approach is to represent the flight envelope for a fixed value of the flight path angle and a fixed value of the bank angle. Each such flight envelope is the feasible set in two dimensions, defined by the altitude and airspeed, that satisfy the steady flight equations and constraints. The steady level flight envelope is obtained for the special case of zero flight path angle and zero bank angle. Similarly, various flight envelopes can be represented for any positive or negative value of the flight path angle and for any positive or negative value of the bank angle.

A different two-dimensional visualization approach is to represent the flight envelope for a fixed value of the flight altitude and a fixed value of the flight path angle. Each such flight envelope is a feasible set in two dimensions, defined by the airspeed and bank angle, that satisfy the steady flight equations and constraints. Such flight envelopes are useful in determining the minimum radius steady turn for a jet aircraft.

10.4 General Aviation Aircraft Steady Turning Flight Performance

Flight conditions that characterize the optimal performance of propeller-driven general aviation aircraft in steady climbing or descending turns are studied. The following flight performance measures are analyzed: maximum airspeed, minimum airspeed, maximum climb rate, minimum turn radius, and flight ceiling. Steady climbing or descending turning flight envelopes are also introduced. These flight performance measures characterize the flight maneuverability or flight agility of the aircraft.

Minimum and Maximum Airspeeds

Flight conditions for minimum and maximum airspeeds of a propeller-driven general aviation aircraft are analyzed for a steady climbing or descending turn with a fixed bank angle and a fixed climb rate. The maximum power that the engine and propeller can provide limits the minimum and maximum airspeeds for which the aircraft can maintain a steady climbing or descending turn. The minimum and maximum airspeeds are characterized by the fact that the power required for a steady climbing or descending turn is equal to the maximum power that the propeller and engine can provide at full throttle; this leads to the equation

$$ W V_{climb} + \frac{1}{2}\rho V^3 S C_{D_0} + \frac{2KW^2}{\rho V S \cos^2 \mu} = \eta P_{max}^s \left(\frac{\rho}{\rho^s} \right)^m . \qquad (10.61) $$

This equation typically has two positive solutions. The high-speed solution is the maximum airspeed of the aircraft in a steady climbing or descending turn at the fixed climb rate and bank angle; the low-speed solution is the minimum airspeed of the aircraft in a steady climbing or descending turn at the fixed climb rate and bank angle. Equation (10.61) can be expressed in terms of a fourth degree polynomial.

This computed minimum airspeed is the minimum airspeed if it satisfies the stall constraint given by inequality (10.49); otherwise the stall speed is the minimum airspeed for steady climbing or descending turn at the given power, climb rate, and bank angle. If the computed maximum airspeed is near to or exceeds the speed of sound, the computed maximum airspeed is inaccurate.

Maximum Rate of Climb

Flight conditions for a propeller-driven general aviation aircraft to have a maximum rate of climb, when the aircraft is in a steady turn with a fixed bank angle, are studied. Assume the bank angle is selected so that the wing loading constraint is satisfied. Since the climb rate is the ratio of the excess power and the weight, maximization of the climb rate requires maximization of the excess power. This implies that the power provided by the propeller and engine is a maximum, that is, full throttle, and the power to overcome the drag is a minimum. The airspeed for which the climb rate is maximum is identical to the airspeed for which the power to overcome the drag is a minimum. According to equation (10.50), this airspeed is

$$V = \sqrt{\frac{2W}{\rho S \cos \mu} \sqrt{\frac{K}{3C_{D_0}}}}. \tag{10.62}$$

Using the results in inequality (10.18) for the maximum engine power and in equation (10.51) for the minimum value of the power to overcome the drag, the maximum climb rate is

$$(V_{climb})_{max} = \frac{\eta P^s_{max}}{W} \left(\frac{\rho}{\rho^s}\right)^m - \frac{4}{3} \sqrt{\frac{2W}{\rho S \cos^3 \mu} \sqrt{3K^3 C_{D_0}}}. \tag{10.63}$$

This expression shows that the maximum climb rate decreases as the flight altitude increases. This flight condition for a propeller-driven general aviation aircraft should be checked to make sure that the stall constraint is satisfied.

Although we do not develop the results, there is also a minimum climb rate for steady turning flight with a given bank angle. This occurs when the power provided by the engine is minimum (that is, the throttle setting is zero) and the power due to the drag is maximum; a local maximum of the power due to the drag occurs at the stall speed.

Minimum Radius Turns

Flight conditions for a minimum radius steady climbing or descending turn of a propeller-driven general aviation aircraft with a fixed climb rate are now considered. The problem of determining the minimum turn radius for a propeller-driven general aviation aircraft in a steady climbing or descending turn is complicated; one approach to determine the flight conditions for a minimum turn radius is suggested.

There are several possible formulations of the minimum radius steady climbing or descending turn as an optimization problem. One formulation is as follows: determine the values of airspeed and bank angle for which the steady climbing or descending turn radius

$$r = \frac{V^2}{g \tan \mu} \tag{10.64}$$

is a minimum, subject to satisfaction of the three inequalities defined by the stall constraint, the engine and propeller power constraints, and the wing loading constraint:

$$\frac{2W}{\rho V^2 S \cos \mu} \leq C_{Lmax}, \tag{10.65}$$

$$W V_{climb} + \frac{1}{2} \rho V^3 S C_{D_0} + \frac{2KW^2}{\rho V S \cos^2 \mu} \leq \eta \left(\frac{\rho}{\rho^s}\right)^m P^s_{max}, \tag{10.66}$$

$$\sec \mu \leq n_{max}. \tag{10.67}$$

This minimum radius turn problem differs from the minimum radius turn problem for steady level flight of a propeller-driven general aviation aircraft discussed in chapter 9 only in the specification of the power constraints, which includes an additional power term required for the aircraft to climb or descend.

This optimization problem is complicated by the fact that the active constraints and the inactive constraints are not a priori known; any one of the following situations is possible:

- One constraint is active.
  - The engine power constraint is active, but the stall constraint and the wing loading constraint are inactive.

- Exactly two of the constraints are active.
  - The stall constraint and the engine power constraint are active, but the wing loading constraint is inactive.
  - The stall constraint and the wing loading constraint are active, but the engine power constraint is inactive.
  - The wing loading constraint and the engine power constraint are active, but the stall constraint is inactive.

- All three of the constraints are active.

Since any one of these cases can characterize the minimum radius turn for a propeller-driven general aviation aircraft in steady climbing or descending flight, in principle all cases can be analyzed and the corresponding flight condition and the corresponding steady turn radius computed for each case. The minimum radius turn for the propeller-driven general aviation aircraft in steady climbing or descending flight is then determined by a direct comparison of the computed turn radius values for which all the constraints are satisfied. In some instances, a priori physical knowledge can be used to make an educated hypothesis about which are the active constraints and the inactive constraints for the minimum radius turn.

A graphical representation of this feasible set of airspeeds and bank angles, for a fixed flight altitude and a fixed climb rate, can be obtained by defining boundary curves that correspond to a single one of the three constraints being active. The boundary curves for the stall constraint active and for the wing loading constraint active are easily obtained. If the maximum power constraint is active, there is a low-speed solution and a high-speed solution for each bank angle value as seen in the previous sections; this defines the boundary curve for this maximum power constraint being active. The feasible set of airspeed and bank angle values is simultaneously bounded by all three of these boundary curves.

Contour curves that correspond to constant values of the steady level turn radius are defined by equation (10.64). The desired minimum radius turn flight condition is characterized by the geometric fact that it lies on the smallest turn radius contour curve that intersects the feasible set of airspeed and bank angle values. Graphical tools can be used to plot the boundary curves that define this feasible set and to plot constant contour turn radius curves; the minimum radius steady level turning flight condition can be determined by inspection. Once the active constraints are identified, the optimal values of the airspeed and the bank angle, and the other flight variables, can be determined using the methods developed in this chapter.

Equation (10.27) makes clear that the smallest possible turn radius for a propeller-driven general aviation aircraft is achieved, all other factors considered constant, at the lowest possible altitude. Further, the minimum turn radius is proportional to the wing loading W/S. That is, propeller-driven general aviation aircraft with small wing loading can achieve tighter turns than can such aircraft with larger wing loading. The wing loading, a fundamental aircraft design parameter, is important in determining the maneuverability of the aircraft.

Flight Ceiling

The flight ceiling of a propeller-driven general aviation aircraft in a steady climbing or descending turn with a fixed bank angle and a fixed climb rate is analyzed. There is a maximum altitude, referred to as the steady climbing or descending and turning flight ceiling, at which steady climbing or descending turning flight can be maintained. The flight condition at this flight ceiling is characterized by the fact that the minimum power required for the steady climbing or descending turn, given in equation (10.43), is exactly equal to the maximum power that the propeller and engine can produce. This leads to the equation

$$W V_{climb} + \frac{4}{3} \sqrt{\frac{2W^3}{\rho S \cos^3 \mu}} \sqrt{3K^3 C_{D_0}} = \eta P_{\max}^s \left(\frac{\rho}{\rho^s} \right)^m . \tag{10.68}$$

Since the air density depends on the altitude according to the standard atmospheric model, equation (10.68) can be viewed as an implicit equation in the altitude; the flight ceiling solves this equation.

Flight Envelope

For a fixed value of the climb rate, the flight envelope is the feasible set in three dimensions defined by the altitude, airspeed, and bank angle variables that satisfy the steady flight equations and constraints for the fixed value of the climb rate. The steady level turning flight envelope is obtained if the climb rate is zero.

Another possible visualization approach is to represent the flight envelope for a fixed value of the bank angle. For that fixed value of the bank angle, the flight envelope is the feasible set in three dimensions defined by the altitude, airspeed, and climb rate variables that satisfy the steady flight equations and constraints for the fixed value of the bank angle. The steady longitudinal flight envelope is obtained if the bank angle is zero.

A simplified two-dimensional visualization approach is to represent the flight envelope for a fixed value of the climb rate and a fixed value of the bank angle. Each such flight envelope is the feasible set in two dimensions defined by the altitude and airspeed that satisfy the steady flight equations and constraints. The steady level flight envelope is obtained for the special case of zero climb rate and zero bank angle. Similarly, various two-dimensional flight envelopes can be represented for any positive or negative value of the climb rate and for any positive or negative value of the bank angle.

A different two-dimensional visualization approach is to represent the flight envelope for a fixed value of the flight altitude and a fixed value of the climb rate. Each such flight envelope of this type is a feasible set in two dimensions, defined by the airspeed and bank angle, that satisfy the steady flight equations and constraints. Such flight envelopes are useful in determining the minimum radius steady turn for a propeller-driven aircraft.

10.5 Steady Climbing and Turning Flight: Executive Jet Aircraft

Assume that the executive jet aircraft, with a weight of 73,000 lbs, is in steady climbing and turning flight with a bank angle of 15 degrees and a flight path angle of 3 degrees at an altitude of 10,000 ft. This bank angle corresponds to a load factor of 1.15, which is significantly below the maximum load factor of 2; thus the wing loading constraint is not active.

Thrust Required

Suppose that the airspeed in this steady turning and climbing flight condition is $V = 500$ ft/s. The thrust required to maintain this steady turning and climbing flight condition is

$$T = W\gamma + \frac{1}{2}\rho V^2 SC_{D_0} + \frac{2KW^2}{\rho V^2 S \cos^2 \mu} = 8{,}319 \text{ lbs.}$$

The pilot inputs for this steady turning and climbing flight condition can be computed following the methodology indicated previously.

The required thrust can be computed for various airspeeds. The thrust required curve, shown in Figure 10.2, is developed by the following Matlab m-file.

Matlab function 10.1 TRSCTJ.m
```
V=linspace(150,1000,500);
W=73000; S=950; CD0=0.015; K=0.05;
[T p rho]=StdAtpUS(10000);
mu=15*pi/180; gamma=3*pi/180
T=W*gamma+1/2*rho*V.^2*S*CD0+2*K*W^2/rho./V.^2/S/cos(mu)^2;
plot(V,T);
xlabel('Velocity (ft/s)');
ylabel('Thrust required (lbs)');
grid on;
```

This thrust required curve for steady turning and climbing flight has the typical features: the thrust required for steady turning and climbing flight is large at both low and high airspeeds, and it has a minimum at an intermediate airspeed.

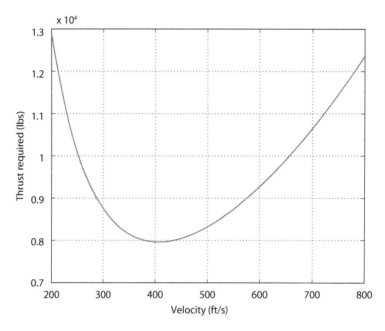

Figure 10.2. Thrust required for steady turning flight: executive jet aircraft ($\gamma = 3$ degrees, $\mu = 15$ degrees).

Minimum Required Thrust

As seen in Figure 10.2, there is a minimum required thrust to achieve a steady climbing turn with a bank angle of 15 degrees and a flight path angle of 3 degrees. This minimum required thrust can be determined by recognizing that the corresponding airspeed minimizes the aircraft drag. The airspeed that achieves the minimum required drag for this steady turning and climbing flight condition is

$$V = \sqrt{\frac{2W}{\rho S \cos \mu} \sqrt{\frac{K}{C_{D_0}}}} = 406.87 \text{ ft/s}.$$

At this steady climbing and turning flight condition, the lift coefficient and the pitch moment coefficient are

$$C_L = \frac{2W}{\rho V^2 S \cos \mu} = 0.548,$$

$$C_M = 0.$$

These expressions determine the aircraft angle of attack α and the elevator deflection δ_e which satisfy

$$0.02 + 0.12\alpha = 0.548,$$
$$0.24 - 0.18\alpha + 0.28\delta_e = 0.$$

Thus, the angle of attack is 4.40 degrees and the elevator deflection is 1.97 degrees. The value of the minimum thrust required for this steady turning and climbing flight condition is given by

$$T_{\min} = W\gamma + \frac{2W}{\cos \mu} \sqrt{K C_{D_0}} = 7,961.7 \, \text{lbs}.$$

The corresponding throttle setting is

$$\delta_t = \frac{T_{\min}}{T_{\max}^s} \left(\frac{\rho^s}{\rho} \right)^m = 0.764.$$

The turn radius corresponding to this minimum thrust steady turning and climbing flight condition is

$$r = \frac{V^2}{g \tan \mu} = 19,186 \, \text{ft}.$$

In summary, the minimum thrust required for the executive jet aircraft to achieve a steady turn with a bank angle of 15 degrees and with a flight path angle of 3 degrees is 7,961.7 lbs and occurs with the following pilot inputs: elevator deflection of 1.97 degrees and a throttle setting of 0.764. The corresponding turn radius is 19,186 ft.

Maximum Aircraft Airspeed

The maximum airspeed of the executive jet aircraft in a steady climbing turn with a bank angle of 15 degrees and with a flight path angle of 3 degrees occurs at full throttle; that is, the jet engine thrust is maximum. This maximum airspeed is found by solving the equation

$$W\gamma + \frac{1}{2}\rho V^2 S C_{D_0} + \frac{2K W^2}{\rho V^2 S \cos^2 \mu} = T_{\max}^s \left(\frac{\rho}{\rho^s} \right)^m.$$

Multiplying the above equation by V^2, the following fourth degree polynomial equation is obtained:

$$\frac{1}{2}\rho S C_{D_0} V^4 + \left[W\gamma - T_{\max}^s \left(\frac{\rho}{\rho^s} \right)^m \right] V^2 + \frac{2K W^2}{\rho S \cos^2 \mu} = 0.$$

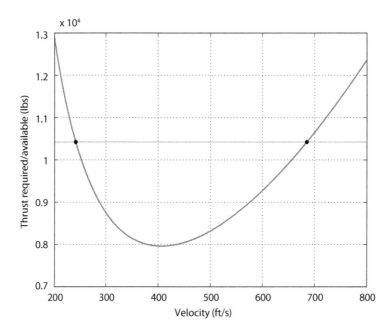

Figure 10.3. Maximum airspeed for steady turning flight: executive jet aircraft ($\gamma = 3$ degrees, $\mu = 15$ degrees).

This equation can be solved numerically using the Matlab function MMSSCTFJ:

Matlab function 10.2 MMSSCTFJ.m
```
function[Vmin,Vmax]=MMSSCTFJ
%Output: Minimum air speed and maximum air speed in a
climbing turn
W=73000; S=950; CD0=0.015; K=0.05; Tsmax=12500; m=0.6;
[Ts ps rhos]=StdAtpUS(0);
[Th ph rhoh]=StdAtpUS(h);
gamma=3*pi/180; mu=15*pi/180;
[Vmin,Vmax]=sort(roots([1/2*rho*S*CD0 0 W*gamma-Tsmax*
(rho/rhos)^m 0 2*K*W^2/rho/S/cos(mu)^2]));
```

Matlab provides two positive solutions of 685.03 ft/s and 241.58 ft/s. These two solutions can also be obtained from the intersections of the thrust required curve and the curve that defines the maximum thrust that the jet engines can provide as shown in Figure 10.3.

The high-speed solution is the maximum airspeed of the aircraft in a steady climbing turn with a bank angle of 15 degrees and with a flight path angle of 3 degrees. At the given altitude, the speed of sound is 1,077.39 ft/s. Consequently, the aircraft Mach number for this steady turning and climbing flight condition is 0.64.

At this maximum airspeed, the lift coefficient and the pitch moment coefficient are

$$C_L = \frac{2W}{\rho V^2 S \cos \mu} = 0.193,$$

$$C_M = 0.$$

These expressions allow determination of the aircraft angle of attack of 1.44 degrees and the elevator deflection of 0.07 degrees. The turn radius corresponding to this maximum airspeed steady turning and climbing flight condition is

$$r = \frac{V^2}{g \tan \mu} = 54{,}389 \text{ ft.}$$

In summary, the maximum airspeed of the executive jet aircraft in a steady turning and climbing flight condition at 10,000 ft with a bank angle of 15 degrees and a flight path angle of 3 degrees occurs with the following pilot inputs: elevator deflection of 0.07 degrees and with full throttle. The corresponding maximum airspeed is 685.03 ft/s and the turn radius is 54,389 ft.

Minimum Airspeed

The minimum airspeed of the executive jet aircraft in a steady climbing turn at an altitude of 10,000 ft with a bank angle of 15 degrees and a flight path angle of 3 degrees must satisfy both the stall constraint and the maximum thrust constraint of the jet engine. The stall speed for a steady turning and climbing flight condition with a bank angle of 15 degrees and a flight path angle of 3 degrees is

$$V_{\text{stall}} = \sqrt{\frac{2W}{\rho S C_{L_{\text{max}}} \cos \mu}} = 179.92 \text{ ft/s.}$$

As seen previously, the minimum airspeed of the executive jet aircraft in a steady climbing turn due to the thrust constraint is 241.58 ft/s. In this case, the minimum airspeed due to the thrust constraint is larger than the stall speed, so the maximum thrust constraint is active and the stall constraint is not active. Consequently, full throttle is required to achieve the minimum airspeed 241.58 ft/s. At the minimum airspeed in this steady turning and climbing flight condition, the lift coefficient and the pitch moment coefficient are

$$C_L = \frac{2W}{\rho V^2 S \cos \mu} = 1.553,$$

$$C_M = 0.$$

These equations determine the aircraft angle of attack as 12.77 degrees and the elevator deflection as 7.36 degrees. The turn radius corresponding to this minimum speed steady turning and climbing flight condition is

$$r = \frac{V^2}{g \tan \mu} = 6,764.2 \text{ ft.}$$

In summary, the minimum airspeed of the executive jet aircraft in a steady turning and climbing flight condition at an altitude of 10,000 ft with a bank angle of 15 degrees and a flight path angle of 3 degrees is 241.58 ft/s and occurs with the following pilot inputs: elevator deflection of 7.36 degrees and with full throttle. The corresponding turn radius is 6,764.2 ft.

Maximum Flight Path Angle in Turning Flight

The executive jet aircraft can maintain a steady turn with a bank angle of 15 degrees at many different flight path angles. Due to the flight constraints that must be satisfied, especially the limit on the maximum thrust that the jet engines can provide, there is a maximum flight path angle. Flight conditions for a maximum flight path angle in turning flight are analyzed in this section.

The flight path angle is the ratio of the excess thrust and the weight, where the excess thrust is the difference between the engine thrust and the drag. To maximize the flight path angle, it is clear that the thrust provided by the engines should be a maximum; that is, full throttle should be used. In addition, the airspeed should be selected to minimize the drag.

Thus the maximum flight path angle in steady turning flight at an altitude of 10,000 ft is

$$\gamma_{max} = \frac{T^s_{max}}{W} \left(\frac{\rho}{\rho^s}\right)^m - \frac{2}{\cos \mu} \sqrt{K C_{D_0}} = 0.086 \text{ rad} = 4.93 \text{ degrees.}$$

The airspeed for which the flight path angle is maximum is the airspeed that minimizes the drag in the steady turn, namely

$$V = \sqrt{\frac{2W}{\rho S \cos \mu} \sqrt{\frac{K}{C_{D_0}}}} = 416.81 \text{ ft/s.}$$

At this flight condition for maximum flight path angle, the lift coefficient and the pitch moment coefficient are

$$C_L = \frac{2W}{\rho V^2 S \cos \mu} = 0.529,$$

$$C_M = 0.$$

From these two values, it follows that the aircraft angle of attack is 4.24 degrees and the elevator deflection is 1.87 degrees. The turn radius corresponding to this steady climbing turn is

$$r = \frac{V^2}{g \tan \mu} = 19,180.9 \, \text{ft}.$$

In summary, the maximum flight path angle of the executive jet aircraft turning with a bank angle of 15 degrees at 10,000 ft is 4.93 degrees corresponding to the following pilot inputs: elevator deflection of 1.87 degrees with full throttle. The corresponding turn radius is 19,180.9 ft.

Minimum Radius Steady Climbing Turn

The executive jet aircraft is to execute a steady climbing turn at an altitude of 10,000 ft with a flight path angle of 3 degrees that has minimum turn radius. The analysis is based on a graphical two-dimensional representation of boundary curves that define the feasible set for the airspeed and bank angle values.

Matlab m-files are presented that determine the minimum turn radius. The maximum airspeed and the minimum airspeed due to the jet engine thrust constraint and the stall speed can be computed for various altitudes, flight path angles, and bank angles. Define the Matlab function SCTFJ that computes the maximum airspeed, the minimum airspeed due to the thrust constraint, and the stall speed for a specified altitude, flight path angle, and bank angle. This Matlab function can be used to confirm the numerical results in the previous section, and it is also useful for subsequent minimum radius turning flight and flight envelope computations for the executive jet aircraft.

Matlab function 10.3 SCTFJ.m
```
function [Vmax VminTC Vstall]=SCTFJ(h,gamma,mu);
%Input: altitude h (ft), flight path angle gamma (rad),
bank angle mu (rad)
%Output: max airspeed Vmax (ft/s), min airspeed VminTC (ft/s),
stall speed Vstall (ft/s)
W=73000; S=950; CD0=0.015; K=0.05; Tsmax=12500; m=0.6; CLmax=2.8;
[Ts ps rhos]=StdAtpUS(0);
[T p rho]=StdAtpUS(h);
tmp=sort(roots([1/2*rho*S*CD0 0 W*gamma-Tsmax*(rho/rhos)^m 0
2*K*W^2/rho/S/cos(mu)^2]));
Vmax=tmp(4);
VminTC=tmp(3);
Vstall=sqrt(2*W/rho/S/cos(mu)/CLmax);
```

This Matlab m-file can be used to compute and visualize the boundary curves that define the feasible set for the steady climbing turn at this altitude; by superimposing the curves of constant turn radius it is possible to identify the point in the feasible set that defines the minimum radius turn. The following Matlab m-file accomplishes this.

Matlab function 10.4 MinRCJ.m

```
g=32.2; h=10000; nmax=2; gamma=3*pi/180;
n=100;
mus=linspace(0*pi/180,52.9*pi/180,n);
Vs=zeros(n,n);
for k=1:n
    [Vmax0 VminTC0 Vstall0]=SCTFJ(h,gamma,mus(k));
    Vmaxs(k)=Vmax0;
    VminTCs(k)=VminTC0;
    Vstalls(k)=Vstall0;
end
figure(1);
plot(VminTCs,mus*180/pi,'r',...
    Vmaxs,mus*180/pi,'r',...
    Vstalls,mus*180/pi,'b',...
    [100 1000],acos(1/nmax)*180/pi*ones(2,1),'g');
ylim([0,70]);
xlim([100,800]);
xlabel('Velocity (ft/sec)');
ylabel('Bank angle (deg)');
figure(2);
plot(VminTCs,mus*180/pi,'r',...
    Vmaxs,mus*180/pi,'r',...
    Vstalls,mus*180/pi,'b',...
    [100 800],acos(1/nmax)*180/pi*ones(2,1),'g');hold on;
Vopt=3.1122e+02;
muopt=6.7200e-01;
Ropt=3.7811e+03;
Vss=linspace(100,1000,n);
Rss=[0.2 0.5 1 2 3 4]*Ropt;
for j=1:size(Rss,2)
    for i=1:n
        muss(j,i)=atan(Vss(i)^2/g/Rss(j));
    end
    plot(Vss,muss(j,:)*180/pi,'k:');
end
plot(Vopt,muopt*180/pi,'ro');
ylim([0,70]);
xlim([100,800]);
xlabel('Velocity (ft/sec)');
ylabel('Bank angle (deg)');
```

The feasible set consists of the airspeed and the bank angle values that satisfy the three constraints; this feasible set is shown in Figure 10.4(a), where the low-speed boundary curve indicates the stall limit, the inverted U boundary curve indicates the maximum thrust limit, and the horizontal boundary curve indicates the wing loading limit. Several contour

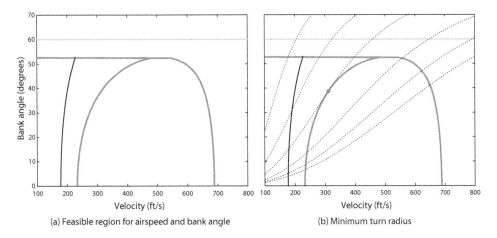

(a) Feasible region for airspeed and bank angle   (b) Minimum turn radius

Figure 10.4. Minimum radius steady turn: executive jet aircraft ($\gamma = 3$ degrees).

lines for fixed values of the turn radius are shown in Figure 10.4(b). The flight conditions for the minimum radius turn are characterized by the small circle in Figure 10.4(b). It is seen that, at the minimum radius turn, only the thrust constraint is active, while the stall constraint and the wing loading constraint are not active.

The subsequent analysis of the minimum radius steady climbing turn for the jet aircraft is based on the fact that only the thrust constraint is active. This means that the minimum radius climbing turn corresponds to the values of airspeed and bank angle that minimize the turn radius

$$r = \frac{V^2}{g \tan \mu},$$

while satisfying the maximum thrust constraint that is expressed in terms of the airspeed and bank angle:

$$W\gamma + \frac{1}{2}\rho V^2 SC_{D_0} + \frac{2K W^2}{\rho S V^2 \cos^2 \mu} = \left(\frac{\rho}{\rho^s}\right)^m T_{max}^s.$$

The thrust value in this expression is computed from the assumption that the maximum thrust constraint is active; that is, the jet engine operates at full throttle.

This simpler optimization problem can be solved using the methods of calculus to obtain a solution given by an airspeed of 311.22 ft/s and a bank angle of 38.5 degrees. The minimum value of the turn radius is computed as 3,781.2 ft. The corresponding lift coefficient and pitch moment are

$$C_L = 1.1551$$
$$C_M = 0,$$

so that the angle of attack is 9.46 degrees and the elevator deflection is 5.22 degrees.

These results are summarized as follows. The minimum radius steady climbing turn of the executive jet aircraft at an altitude of 10,000 ft with a flight path of angle of 3 degrees occurs with the following pilot inputs: elevator deflection of 5.22 degrees, full throttle, and with a bank angle of 38.5 degrees. The minimum value of the turn radius is 3,781.2 ft.

Steady Climbing and Turning Flight Ceiling

The steady climbing and turning flight ceiling of the executive jet aircraft is computed, assuming the aircraft is in a steady banked turn with a bank angle of 15 degrees and a flight path angle of 3 degrees.

The flight ceiling for this steady climbing turn with a bank angle of 15 degrees and a flight path angle of 3 degrees is determined by solving the following equation:

$$W\gamma + \frac{2W}{\cos\mu}\sqrt{KC_{D_0}} - T_{max}^s\left(\frac{\rho}{\rho^s}\right)^m = 0,$$

where ρ is a function of altitude according to the standard atmospheric model. Define the Matlab function eqnFCCTJ whose value is zero at the steady climbing turn flight ceiling.

Matlab function 10.5 eqnFCCTJ.m
```
function error=eqnFCCTJ(h,gamma,mu)
%Input: altitude h (ft), flight path angle gamma (rad),
bank angle mu (rad)
%Output: error
W=73000; CD0=0.015; K=0.05; Tsmax=12500; m=0.6;
[Ts ps rhos]=StdAtpUS(0);
[Th ph rhoh]=StdAtpUS(h);
error=W*gamma+2*W/cos(mu)*sqrt(K*CD0)-Tsmax*(rhoh/rhos)^m;
```

The following Matlab command computes the steady climbing turn flight ceiling for which the value of the function defined in eqnFCCTJ is zero.

```
hmax=fsolve(@(h) eqnFCCTJ(h,3*pi/180,15*pi/180),10000)
```

where 10,000 is an initial guess of the flight ceiling. After a few iterations, the flight ceiling for this steady turning and climbing flight condition is determined to be 23,551.3 ft.

In summary, the flight ceiling of the executive jet aircraft is 23,551.3 ft, assuming the aircraft is in a steady climbing turn with a bank angle of 10 degrees and a flight path angle of 3 degrees.

Steady Climbing and Turning Flight Envelope

A two-dimensional flight envelope of the executive jet aircraft is computed, assuming the aircraft is in a steady climbing turn with a bank angle of 15 degrees and a flight path angle

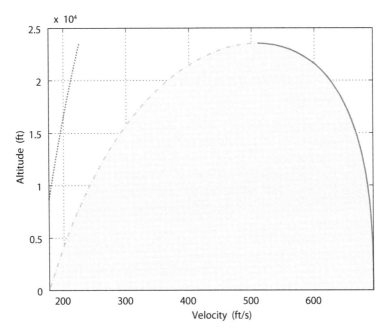

Figure 10.5. Steady turning flight envelope: executive jet aircraft (γ = 3 degrees, μ = 15 degrees).

of 3 degrees; this flight envelope is the set of feasible values of the flight altitude and the airspeed.

A two-dimensional graphical representation of this steady climbing and turning flight envelope for the executive jet aircraft at this fixed bank angle and fixed flight path angle is developed using the previously defined Matlab function SCTFJ. This particular flight envelope, as shown in Figure 10.5, is generated by the following Matlab m-file.

Matlab function 10.6 FigSCTFJ.m

```
gamma=3*pi/180; mu=15*pi/180;
h=linspace(0,23551,500);
for k=1:size(h,2)
    [Vmax(k) VminTC(k) Vstall(k)]=SCTFJ(h(k),gamma,mu);
end
Vmin=max(VminTC,Vstall);
area([Vmin Vmax(end:-1:1)],[h h(end:-1:1)],...
    'FaceColor',[0.8 1 1],'LineStyle','none');
hold on;
plot(Vmax,h,VminTC,h,Vstall,h);
grid on;
xlim([0 1200]);
xlabel('Velocity (ft/s)');
ylabel('Altitude (ft)');
```

In summary, all of the values of airspeed and altitude that lie in the flight envelope shown in Figure 10.5 imply that the executive jet aircraft can maintain a steady climbing turn with a bank angle of 15 degrees and a flight path angle of 3 degrees at that value of airspeed and altitude.

10.6 Steady Descending and Turning Flight: Executive Jet Aircraft

Steady descending turns for the executive jet aircraft are analyzed in this section. Assume that the executive jet aircraft, with a weight of 73,000 lbs, is in a steady descending and turning flight with a bank angle of 15 degrees and a flight path angle of -2 degrees at an altitude of 10,000 ft. This bank angle corresponds to a load factor of 1.15, which is significantly below the maximum load factor of 2; thus the wing loading constraint is not active.

Thrust Required

Suppose that the airspeed in this steady turning and descending flight condition is 500 ft/s. The thrust required to maintain this steady turning and descending flight condition is

$$T = W\gamma + \frac{1}{2}\rho V^2 S C_{D_0} + \frac{2KW^2}{\rho V^2 S \cos^2 \mu} = 1,948.5 \text{ lbs.}$$

The required thrust for this steady turning and descending flight condition can be computed for various airspeeds. The thrust required curve, shown in Figure 9.2, is developed from the following Matlab m-file.

Matlab function 10.7 TRSDTJ.m
```
V=linspace(150,1000,500);
W=73000; S=950; CD0=0.015; K=0.05;
[T p rho]=StdAtpUS(10000);
mu=15*pi/180; gamma=-2*pi/180
T=W*gamma+1/2*rho*V.^2*S*CD0+2*K*W^2/rho./V.^2/S/cos(mu)^2;
plot(V,T);
xlabel('Velocity (ft/s)');
ylabel('Thrust required (lbs)');
grid on;
```

This thrust required curve for steady turning and descending flight has the typical features: the thrust required is large at low airspeeds, the thrust required is large at high airspeeds, and the thrust required has a minimum value at an intermediate airspeed.

Minimum Required Thrust

The executive jet aircraft can maintain a steady descending turn with a flight path angle of -2 ft/s and a bank angle of 15 degrees at various airspeeds. As seen in Figure 10.6, there is an airspeed at which the thrust required for this steady descending turn is a minimum.

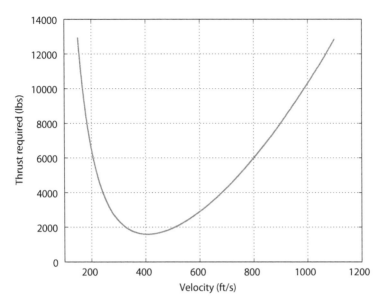

Figure 10.6. Thrust required for steady turning flight: executive jet aircraft ($\gamma = -2$ degrees, $\mu = 15$ degrees).

This minimum required thrust can be determined by recognizing that the corresponding airspeed minimizes the aircraft drag. The airspeed that achieves minimum required drag for this steady turning and descending flight condition is given by

$$V = \sqrt{\frac{2W}{\rho S \cos \mu} \sqrt{\frac{K}{C_{D_0}}}} = 406.87 \, \text{ft/s}.$$

The value of the minimum thrust required for this steady turning and descending flight condition is given by

$$T_{\min} = W\gamma + \frac{2W}{\cos \mu} \sqrt{K C_{D_0}} = 1{,}591.4 \, \text{lbs}.$$

The corresponding throttle setting is

$$\delta_t = \frac{T_{\min}}{T^s_{\max}} \left(\frac{\rho^s}{\rho} \right)^m = 0.153.$$

The turn radius corresponding to this minimum thrust steady turning and descending flight condition is computed from

$$r = \frac{V^2}{g \tan \mu} = 19{,}186.8 \, \text{ft}.$$

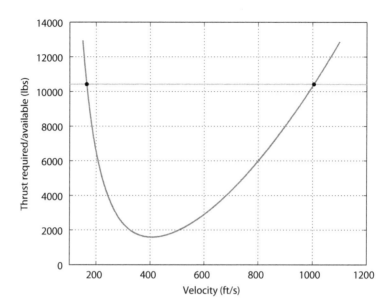

Figure 10.7. Maximum airspeed for steady turning flight: executive jet aircraft ($\gamma = -2$ degrees, $\mu = 15$ degrees).

In summary, the minimum thrust required for the executive jet aircraft to achieve a steady turn with a bank angle of 15 degrees, and a flight path angle of -2 degrees is 1,591.4 lbs and occurs with the following pilot inputs: elevator deflection of 1.97 degrees and throttle setting of 0.153. The corresponding turn radius is 19,186.8 ft.

Maximum Airspeed

The maximum airspeed of the executive jet aircraft in a steady turn with a flight path angle of -2 degrees and a bank angle of 15 degrees is characterized by the intersection of the thrust required curve and the curve defined by the maximum thrust that the jet engine can provide at this altitude. This intersection is shown in Figure 10.7.

This maximum airspeed can also be determined by solving the equation

$$W\gamma + \frac{1}{2}\rho V^2 S C_{D_0} + \frac{2K W^2}{\rho V^2 S \cos^2 \mu} = T^s_{\max} \left(\frac{\rho}{\rho^s} \right)^m.$$

Multiplying the above equation by V^2 and rearranging, the following fourth degree polynomial equation is obtained:

$$\frac{1}{2}\rho S C_{D_0} V^4 + \left(W\gamma - T^s_{\max} \left(\frac{\rho}{\rho^s} \right)^m \right) V^2 + \frac{2K W^2}{\rho S \cos^2 \mu} = 0.$$

This equation can be solved numerically using the Matlab function MMSSDTFJ:

Matlab function 10.8 MMSSDTFJ.m

```
function [Vmin,Vmax]=MMSSDTFJ
%Output: Minimum air speed and maximum air speed in a
climbing turn
W=73000; S=950; CD0=0.015; K=0.05; Tsmax=12500; m=0.6;
[Ts ps rhos]=StdAtpUS(0);
[Th ph rhoh]=StdAtpUS(h);
gamma=-2*pi/180; mu=15*pi/180;
[Vmin,Vmax]=sort(roots([1/2*rho*S*CD0 0 W*gamma-Tsmax*
(rho/rhos)^m 0 2*K*W^2/rho/S/cos(mu)^2]));
```

Matlab provides two positive solutions of 1,004.93 ft/s and 164.68 ft/s. These two solutions can also be obtained from the intersections of the thrust required curve and the curve that defines the maximum thrust that the jet engines can provide as shown in Figure 10.7.

The high-speed solution is the maximum airspeed of the aircraft in steady descending and turning flight with a bank angle of 15 degrees and with a flight path angle of −2 degrees. The corresponding Mach number of the executive jet aircraft in this steady turning and descending flight condition is 0.93. This maximum airspeed is close to the speed of sound so that the compressibility effects are important and the computed maximum airspeed may not be accurate.

At the maximum airspeed, the lift coefficient and the pitch moment coefficient are

$$C_L = \frac{2W}{\rho V^2 S \cos \mu} = 0.090,$$

$$C_M = 0.$$

These expressions allow determination of the aircraft angle of attack as 0.58 degrees and the elevator deflection as −0.48 degrees. The turn radius corresponding to this maximum speed steady turning and descending flight condition is

$$r = \frac{V^2}{g \tan \mu} = 117{,}047.8 \text{ ft.}$$

In summary, the maximum airspeed of the executive jet aircraft in a steady descending turn at an altitude of 10,000 ft with a bank angle of 15 degrees and a flight path angle of −2 degree is 1,004.93 ft/s and occurs with the following pilot inputs: elevator deflection of −0.48 degrees and with full throttle. The corresponding turn radius is 117,047.8 ft.

Minimum Airspeed

The minimum airspeed of the executive jet aircraft in a steady descending turn at an altitude of 10,000 ft with a bank angle of 15 degrees and a flight path angle of −2 degrees must satisfy both the stall constraint and the jet engine maximum thrust constraint. The stall speed for steady turning and climbing flight with a bank angle of 15 degrees and a flight path angle of -2 degrees is

$$V_{\text{stall}} = \sqrt{\frac{2W}{\rho S C_{L_{\max}} \cos \mu}} = 179.92 \text{ ft/s.}$$

As seen previously, the minimum airspeed of the executive jet aircraft in a steady descending turn due to the thrust constraint is 164.68 ft/s. In this case, the minimum airspeed due to the thrust constraint is less than the stall speed, so the stall constraint is active and the engine constraint is not active. Consequently, the minimum airspeed at this flight condition is equal to the stall speed, namely 179.92 ft/s. At the minimum airspeed in this steady turning and descending flight condition, the lift coefficient $C_L = 2.8$ and the pitch moment coefficient $C_M = 0$. This determines the aircraft angle of attack as 23.17 degrees and the elevator deflection as 14.04 degrees.

The value of the thrust required for this steady turning and descending flight condition is given by

$$T = W\gamma + \frac{1}{2}\rho V^2 S C_{D_0} + \frac{2KW^2}{\rho V^2 S \cos^2 \mu} = 8,437.21 \text{ lbs.}$$

The corresponding throttle setting is

$$\delta_t = \frac{T}{T_{\max}^s}\left(\frac{\rho^s}{\rho}\right)^m = 0.81.$$

The turn radius corresponding to this maximum speed steady turning and descending flight condition is

$$r = \frac{V^2}{g \tan \mu} = 3,751.9 \text{ ft.}$$

In summary, the minimum airspeed of the executive jet aircraft in a steady turning and descending flight condition at an altitude of 10,000 ft with a bank angle of 15 degrees and a flight path angle of -2 degrees is 179.92 ft/s and occurs with the following pilot inputs: elevator deflection of 14.04 degrees and throttle setting of 0.81. The corresponding turn radius is 3,751.9 ft.

Minimum Radius Steady Descending Turn

The executive jet aircraft is to execute a steady descending turn at an altitude of 10,000 ft with a flight path angle of -12 degrees that has minimum turn radius.

The Matlab m-file SCTFJ can be used to compute and visualize the boundary curves that define the feasible set for the steady climbing turn at this altitude; by superimposing the curves of constant turn radius it is possible to identify the point in the feasible set that defines the minimum radius turn. The following Matlab m-file accomplishes this.

Matlab function 10.9 MinRDJ.m

```
g=32.2; h=10000; nmax=2; gamma=-12*pi/180;
n=100;
mus=linspace(0*pi/180,81.45*pi/180,n);
Vs=zeros(n,n);
for k=1:n
    [Vmax0 VminTC0 Vstall0]=SCTFJ(h,gamma,mus(k));
    Vmaxs(k)=Vmax0;
    VminTCs(k)=VminTC0;
    Vstalls(k)=Vstall0;
end
figure(1);
plot(VminTCs,mus*180/pi,'r',...
    Vmaxs,mus*180/pi,'r',...
    Vstalls,mus*180/pi,'b',...
    [100 1500],acos(1/nmax)*180/pi*ones(2,1),'g');
ylim([0,90]);
xlim([100,1500]);
xlabel('Velocity (ft/sec)');
ylabel('Bank angle (deg)');
figure(2);
plot(VminTCs,mus*180/pi,'r',...
    Vmaxs,mus*180/pi,'r',...
    Vstalls,mus*180/pi,'b',...
    [100 1500],acos(1/nmax)*180/pi*ones(2,1),'g');hold on;
mu=60*pi/180;
[Vmax VminTC Vstall]=SCTFJ(10000,gamma,mu);
Ropt=Vstall^2/g/tan(mu);
Vopt=Vstall;
muopt=mu;
Vss=linspace(100,1400,n);
Rss=[0.2 0.5 1 2 3 4]*Ropt;
for j=1:size(Rss,2)
    for i=1:n
        muss(j,i)=atan(Vss(i)^2/g/Rss(j));
    end
    plot(Vss,muss(j,:)*180/pi,'k:');
end
plot(Vopt,muopt*180/pi,'ro');
ylim([40,80]);
xlim([100,400]);
xlabel('Velocity (ft/sec)');
ylabel('Bank angle (deg)');
```

The feasible set consists of the airspeed and bank angle values that satisfy the three constraints; the boundary of this feasible set is shown in Figure 10.8(a), where the

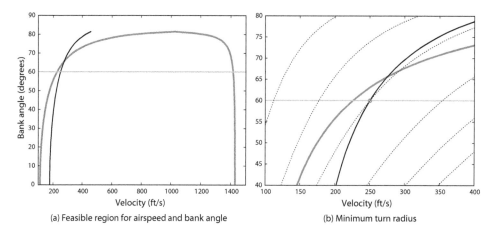

(a) Feasible region for airspeed and bank angle

(b) Minimum turn radius

Figure 10.8. Minimum radius steady turn: executive jet aircraft ($\gamma = -12$ degrees).

low-speed boundary curve indicates the stall limit, the inverted ∪ boundary curve characterizes the maximum thrust limit, and the horizontal boundary curve indicates the wing loading limit. In Figure 10.8(b), several contour lines for fixed values of the turn radius are shown. It is seen that, at the minimum turn radius, the wing loading constraint and the stall constraint are active; the jet engine thrust constraint is not active.

Since the wing loading constraint is active, the load factor is the maximum load factor of 2, which corresponds to a bank angle of 60 degrees. Since the stall constraint is active, the lift coefficient is a maximum. This allows determination of the airspeed according to

$$V = \sqrt{\frac{2W}{\rho S C_{L_{\max}} \cos \mu}} = 250.1 \text{ ft/s}.$$

The turn radius corresponding to full throttle, a bank angle of 60 degrees, and a flight path angle of -12 degrees, at an airspeed of 250.1 ft/s, is

$$r = \frac{V^2}{g \tan \mu} = 1,121.3 \text{ ft}.$$

The required thrust for this steady descending turn is

$$T = W\gamma + \frac{1}{2}\rho V^2 S C_{D_0} + \frac{2KW^2}{\rho V^2 S \cos^2 \mu} = 5,933.1 \text{ lbs}.$$

The corresponding throttle setting is

$$\delta_t = \frac{T}{T^s_{\max}} \left(\frac{\rho^s}{\rho}\right)^m = 0.569.$$

Hence the jet engine thrust constraints, the wing loading constraint, and the stall constraint are satisfied.

The value of the lift coefficient at stall and the fact that the pitch moment coefficient is zero allows determination of the aircraft angle of attack as 23.17 degrees and the elevator deflection as 14.04 degrees.

These results are summarized as follows. The minimum radius steady descending turn of the executive jet aircraft at an altitude of 10,000 ft with a flight path angle of −12 degrees occurs with the following pilot inputs: elevator deflection of 14.04 degrees, throttle setting of 0.569, and with a bank angle of 60 degrees. The minimum value of the turn radius is 1,121.3 ft.

Steady Descending and Turning Flight Ceiling

The steady descending and turning flight ceiling of the executive jet aircraft is computed, assuming the aircraft is in a steady banked turn with a bank angle of 15 degrees and a flight path angle of −2 degrees. The flight ceiling for this steady turning and descending flight condition is determined by solving the following equation:

$$W\gamma + \frac{2W}{\cos\mu} \sqrt{KC_{D_0}} - T^s_{max} \left(\frac{\rho}{\rho^s}\right)^m = 0,$$

where ρ is a function of altitude according to the standard atmospheric model. We use the Matlab function eqnFCCTJ, whose value is zero at the steady descending turning flight ceiling. The following Matlab command computes the steady descending turning flight ceiling such that the value of the function defined in eqnFCCTJ is zero.

```
hmax=fsolve(@(h) eqnFCCTJ(h,-2*pi/180,15*pi/180),10000)
```

where 10,000 is an initial guess of this flight ceiling. After a few iterations, the flight ceiling for this steady turning and descending flight condition is determined to be 82,022.98 ft.

In summary, the flight ceiling of the executive jet aircraft is 82,003 ft, assuming the aircraft is in a steady descending turn with a bank angle of 10 degrees and a flight path angle of −2 degrees. This performance result is of doubtful practical importance.

Steady Descending and Turning Flight Envelope

A two-dimensional flight envelope of the executive jet aircraft is computed, assuming the aircraft is in a steady descending turn with a bank angle of 15 degrees and a flight path angle of −2 degrees; this flight envelope is the set of feasible values of the flight altitude and the airspeed.

A graphical representation of this flight envelope of the executive jet aircraft at this fixed bank angle and fixed flight path angle is developed using the previous analysis. The Matlab function SCTFJ is used to compute the maximum and minimum airspeeds due to the engine thrust constraint and the stall speed. This particular flight envelope, as shown in Figure 10.9, is generated by the following Matlab m-file.

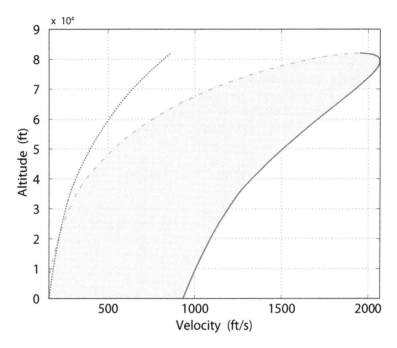

Figure 10.9. Steady turning flight envelope: executive jet aircraft ($\gamma = -2$ degrees, $\mu = 15$ degrees).

Matlab function 10.10 FigSDTFJ.m

```
gamma=-2*pi/180; mu=15*pi/180;
h=linspace(0,23551,500);
for k=1:size(h,2)
    [Vmax(k) VminTC(k) Vstall(k)]=SCTFJ(h(k),gamma,mu);
end
Vmin=max(VminTC,Vstall);
area([Vmin Vmax(end:-1:1)],[h h(end:-1:1)],...
    'FaceColor',[0.8 1 1],'LineStyle','none');
hold on;
plot(Vmax,h,VminTC,h,Vstall,h);
grid on;
xlim([0 1200]);
xlabel('Velocity (ft/s)');
ylabel('Altitude (ft)');
```

The high-speed part of the flight envelope shown in Figure 10.9 is not accurate due to the fact that the drag model does not include compressibility effects that are important in transonic and supersonic flight. The high altitude part of this flight envelope in Figure 10.9 is of uncertain practical value.

10.7 Steady Turning Flight Envelopes: Executive Jet Aircraft

Several flight envelopes that describe a wider set of steady climbing or descending and turning flight conditions can be defined for the executive jet aircraft. The subsequent analysis is based on an aircraft weight of 93,000 lbs.

One such flight envelope is the set of feasible values of the flight altitude, the airspeed, and the bank angle, for a fixed value of the flight path angle. A different flight envelope consists of the set of feasible values of the flight altitude, the airspeed, and the flight path angle, for a fixed value of the bank angle. These steady climbing or descending turning flight envelopes are defined in terms of three flight variables and hence each such feasible set must be visualized in three dimensions.

Steady Turning Flight Envelope for a Given Flight Path Angle

One steady flight envelope is developed by evaluating the Matlab function SCTFJ for various altitude and bank angle values, corresponding to the fixed flight path angle value of 3 degrees. Figure 10.10 shows the resulting surface that defines the boundary of this cross section of the steady flight envelope in three dimensions; this boundary surface is generated by the following Matlab m-file.

Matlab function 10.11 FigSCDTFJ.m

```
gamma=3*pi/180;
mu=linspace(0*pi/180,70*pi/180,40);
for i=1:size(mu,2)
    hmax(i)=fsolve(@(h) eqnFCCTJ(h,gamma,mu(i)),10000);
    h(i,:)=linspace(0,floor(hmax(i)),40);
    for j=1:size(h,2)
        [Vmax(i,j) VminTC(i,j) Vstall(i,j)]=SCTFJ(h(i,j),
        gamma,mu(i));
    end
end
Vmin=max(VminTC,Vstall);
surf(Vmax,mu*180/pi,h,'LineStyle','none');hold on;
surf(Vmin,mu*180/pi,h,'LineStyle','none');
xlabel('Velocity (ft/s)');
ylabel('Bank angle (deg)');
zlabel('Altitude (ft)');
```

Any triple of values for the flight altitude, airspeed, and bank angle that lies within the flight envelope illustrated in Figure 10.10 corresponds to a feasible steady flight condition with a fixed flight path angle of 3 degrees for which the engine constraints, the stall constraint, and the wing loading constraint are satisfied. For each such feasible triple of values, methodology has been provided to compute the pilot inputs, namely the elevator deflection and the throttle setting that give this steady climbing and turning flight condition. In addition, the methodology also allows computation of the associated angle of attack, values of the lift and drag coefficients, and values of the lift and drag.

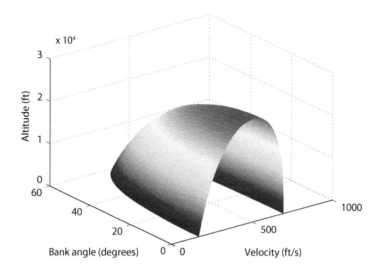

Figure 10.10. Steady turning flight envelope: executive jet aircraft ($\gamma = 3$ degrees).

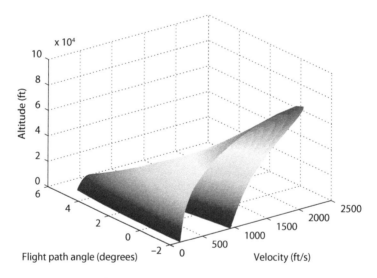

Figure 10.11. Steady turning flight envelope: executive jet aircraft ($\mu = 15$ degrees).

Steady Climbing and Turning Flight Envelope for Given Bank Angle

Another three-dimensional flight envelope is developed by evaluating the Matlab function SCTFJ for various altitude and flight path angle values, corresponding to the fixed bank angle value of 15 degrees. Figure 10.11 shows the resulting surface that defines the boundary of this flight envelope in three dimensions.

Any triple of values for the flight altitude, airspeed, and flight path angle that lies within the flight envelope illustrated in Figure 10.11 corresponds to a feasible steady flight condition with a fixed bank angle of 15 degrees for which the engine constraints, the stall constraint, and the wing loading constraint are satisfied. For each such feasible triple of values, methodology has been provided to compute the pilot inputs, namely the elevator deflection and the throttle setting. In addition, the methodology also allows computation of the associated angle of attack, values of the lift and drag coefficients, and values of the lift and drag cofficients, and values of the lift and drag.

Most General Steady Flight Envelope

The most general steady flight envelope for the executive jet aircraft is defined for helical flight of the aircraft, where the axis of the helix is vertical. This flight envelope is defined in terms of the flight altitude, the airspeed, the flight path angle, and the bank angle of the aircraft. This steady flight envelope is defined in terms of four flight variables and hence must be viewed as a feasible set in four dimensions. This generalizes the steady longitudinal flight envelope and the steady level turning flight envelope. In particular, the steady longitudinal flight envelope can be viewed as a three-dimensional cross section of the steady flight envelope corresponding to zero bank angle. The steady level turning flight envelope can be viewed as a three-dimensional cross section of the steady flight envelope corresponding to zero flight path angle.

Graphical representations of the most general steady flight envelope are necessarily challenging since they require representation of sets in four dimensions. The emphasis on computing various flight envelope cross sections is a natural way to, at least partially, represent the most general flight envelope data.

10.8 Steady Climbing and Turning Flight: General Aviation Aircraft

Assume that the propeller-driven general aviation aircraft, with a weight of 2,900 lbs, is in steady climbing and turning flight with a bank angle of 10 degrees and a rate of climb of 10 ft/s at an altitude of 10,000 ft. This bank angle corresponds to a load factor of 1.05, which is significantly below the maximum load factor of 2. Thus the wing loading constraint is not active.

Power Required

Suppose that the airspeed in this steady turning and climbing flight condition is 200 ft/s. The power required to maintain this turning and climbing flight condition is

$$P = W V_{climb} + \frac{1}{2}\rho V^3 S C_{D_0} + \frac{2K W^2}{\rho V S \cos^2 \mu} = 76,190 \text{ ft lbs/s} = 138.53 \text{ hp}.$$

The required power can be computed for various airspeeds. The power required curve, shown in Figure 10.12, is developed by the following Matlab m-file.

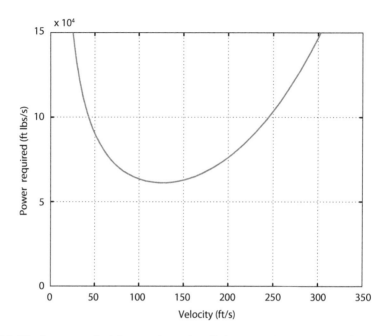

Figure 10.12. Power required for steady turning flight: general aviation aircraft (V_{climb}=10 ft/s, μ = 10 degrees).

Matlab function 10.12 PRSCTP.m
```
V=linspace(20,350,500);
W=2900; S=175; CD0=0.026; K=0.054; rho=1.7553e-3; mu=10*pi/180;
Vclimb=10;
P=W*Vclimb+1/2*rho*V.^3*S*CD0+2*K*W^2/rho./V/S/cos(mu)^2
plot(V,P);
xlabel('Velocity (ft/s)');
ylabel('Power required (ft lbs/s)');
grid on;
```

This power required curve for steady turning and climbing flight has the typical features: the power required for steady turning and climbing flight is large at low airspeeds and high airspeeds, and it has a minimum at an intermediate airspeed.

Minimum Required Power

As seen in Figure 10.12, there is a minimum required power to achieve a steady climbing turn with a bank angle of 10 degrees and a climb rate of 10 ft/s. This minimum required power can be determined by recognizing that the corresponding airspeed minimizes the power required to overcome the drag. The airspeed that achieves the minimum required power to overcome the drag for this steady turning and climbing flight condition is

$$V = \sqrt{\frac{2W}{\rho S \cos \mu} \sqrt{\frac{K}{3C_{D_0}}}} = 131.66 \text{ ft/s}.$$

At this steady climbing and turning flight condition, the lift coefficient and the pitch moment coefficient are

$$C_L = \frac{2W}{\rho V^2 S \cos \mu} = 1.2,$$

$$C_M = 0.$$

These expressions determine the aircraft angle of attack and the elevator deflection, which satisfy

$$0.02 + 0.12\alpha = 1.2,$$

$$0.12 - 0.08\alpha + 0.075\delta_e = 0.$$

The results are: the angle of attack is 9.85 degrees and the elevator deflection is 8.91 degrees. The value of the minimum power required for this steady turning and climbing flight condition is given by

$$P_{\min} = W V_{climb} + \frac{4}{3} \sqrt{\frac{2W^3}{\rho S \cos^3 \mu} \sqrt{3K^3 C_{D_0}}} = 61,185 \text{ ft lbs/s} = 111.24 \text{ hp}.$$

The corresponding throttle setting is

$$\delta_t = \frac{P_{\min}}{\eta P_{\max}^s} \left(\frac{\rho^s}{\rho}\right)^m = 0.57.$$

The turn radius corresponding to this minimum power steady turning and climbing flight condition is

$$r = \frac{V^2}{g \tan \mu} = 3,053 \text{ ft}.$$

In summary, the minimum power required for the general aviation aircraft to achieve a steady turn with a bank angle of 10 degrees and with a climb rate of 10 ft/s is 111.24 hp and occurs with the following pilot inputs: elevator deflection of 8.91 degrees and throttle setting of 0.57. The corresponding turn radius is 3,053 ft and the airspeed is 131.66 ft/s.

Maximum Airspeed

The maximum airspeed of the general aviation aircraft for steady climbing and turning flight with a bank angle of 10 degrees and a climb rate of 10 ft/s occurs at full throttle; that is, the engine power is a maximum. This maximum airspeed is found by solving the equation

$$W V_{climb} + \frac{1}{2}\rho V^3 S C_{D_0} + \frac{2K W^2}{\rho V S \cos^2 \mu} = \eta P^s_{max} \left(\frac{\rho}{\rho^s} \right)^m.$$

Multiplying the above equation by V and rearranging, the following fourth degree polynomial equation is obtained:

$$\frac{1}{2}\rho S C_{D_0} V^4 + \left(W V_{climb} - \eta P^s_{max} \left(\frac{\rho}{\rho^s} \right)^m \right) V + \frac{2K W^2}{\rho S \cos^2 \mu} = 0.$$

This equation can be solved numerically using the Matlab function MMSSCTP as follows:

Matlab function 10.13   MMSSCTP.m
```
function [Vmin,Vmax]=MMSSCTP
%Output: Minimum and maximum air speeds in a climbing turn
W=2900; S=175; CD0=0.026; K=0.054; Psmax=290*550; m=0.6; mu=25;
Vclimb=10;
[Ts ps rhos]=StdAtpUS(0);
[Th ph rhoh]=StdAtpUS(h);
[Vmin,Vmax]=sort(roots([1/2*rho*S*CD0 0 0 W*Vclimb-eta*Psmax*
(rho/rhos)^m 2*K*W^2/rho/S/cos(mu)^2]));
```

Matlab provides two positive solutions of 253.91 ft/s and 39.52 ft/s. These two solutions can also be obtained from the intersections of the power required curve and the curve that defines the maximum power that the propeller and engine can provide as shown in Figure 10.13.

The high-speed solution is the maximum airspeed of the aircraft in a steady climbing and turning flight with a bank angle of 10 degrees and a climb rate of 10 ft/s. At the given altitude, the speed of sound is 1,077.39 ft/s. Consequently, the aircraft Mach number for this steady turning and climbing flight condition is 0.236. At this maximum airspeed, the lift coefficient and the pitch moment coefficient are

$$C_L = \frac{2W}{\rho V^2 S \cos \mu} = 0.297,$$

$$C_M = 0.$$

These expressions allow determination that the aircraft angle of attack is 2.31 degrees and the elevator deflection is 0.87 degrees. The turn radius corresponding to this maximum speed steady turning and climbing flight condition is

$$r = \frac{V^2}{g \tan \mu} = 11,354 \text{ ft}.$$

In summary, the maximum airspeed of the general aviation aircraft in a steady turning and climbing flight condition at 10,000 ft with a bank angle of 10 degrees

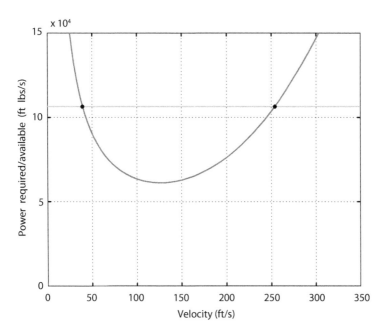

Figure 10.13. Maximum airspeed for steady turning flight: general aviation aircraft (V_{climb} = 10 ft/s, μ = 10 degrees).

and a climb rate of 10 ft/s is 253.9 ft/s and occurs with the following pilot inputs: elevator deflection of 0.87 degrees and with full throttle. The corresponding turn radius is 11,354 ft.

Minimum Airspeed

The minimum airspeed of the general aviation aircraft in a steady climbing turn at an altitude of 10,000 ft with a bank angle of 10 degrees and a climb rate of 10 ft/s must satisfy both the stall constraint and the engine power constraints. The stall speed for a steady climbing turn with a bank angle of 10 degrees and a climb rate of 10 ft/s is

$$V_{stall} = \sqrt{\frac{2W}{\rho S C_{L_{max}} \cos \mu}} = 89.38 \text{ ft/s}.$$

As seen previously, the minimum airspeed of the aircraft in a steady climbing turn due to the engine and propeller power constraint is 39.52 ft/s. In this case, the stall speed is larger than the minimum airspeed due to the power constraint, so the stall constraint is active and the engine and propeller power constraint is not active. Consequently, the minimum airspeed in this steady climbing turn is 89.38 ft/s.

At the minimum airspeed in this steady turning and climbing flight condition, the lift coefficient and the pitch moment coefficient are

$$C_L = \frac{2W}{\rho V^2 S \cos \mu} = 2.4,$$

$$C_M = 0.$$

These equations determine the aircraft angle of attack as 19.83 degrees and the elevator deflection as 19.55 degrees. The power required for this steady climbing turn is

$$P = W V_{climb} + \frac{1}{2} \rho V^3 S C_{D_0} + \frac{2K W^2}{\rho V S \cos^2 \mu} = 65,960 \,\text{ft lbs/s} = 119.9 \,\text{hp}.$$

The corresponding throttle setting is

$$\delta_t = \frac{P}{\eta P^s_{max}} \left(\frac{\rho^s}{\rho} \right)^m = 0.62.$$

The turn radius corresponding to this minimum speed steady turning and climbing flight condition is

$$r = \frac{V^2}{g \tan \mu} = 1,407 \,\text{ft}.$$

In summary, the minimum airspeed of the propeller-driven aircraft in a steady climbing turn at an altitude of 10,000 ft with a bank angle of 10 degrees and a climb rate of 10 ft/s is 89.38 ft/s and occurs with the following pilot inputs: elevator deflection of 19.55 degrees and throttle setting of 0.62. The corresponding turn radius is 1,407 ft.

Maximum Climb Rate in Turning Flight

The general aviation aircraft can maintain a steady turn with a bank angle of 10 degrees at many different climb rates. Due to the flight constraints that must be satisfied, especially the limit on the maximum power that the propeller and internal combustion engine can provide, there is a maximum climb rate. Flight conditions for a maximum climb rate in turning flight are analyzed in this section.

The climb rate is the ratio of the excess power and the weight, where the excess power is the difference between the engine power and the power to overcome the drag. To maximize the climb rate, it is clear that the power provided by the propeller and engine should be a maximum; that is, full throttle should be used. In addition, the airspeed should be selected to minimize the power to overcome the drag.

The airspeed for which the climb rate is maximum in a steady turn at an altitude of 10,000 ft is

$$V = \sqrt{\frac{2W}{\rho S \cos \mu} \sqrt{\frac{K}{3 C_{D_0}}}} = 126.3 \,\text{ft/s}.$$

The maximum climb rate in such a steady turn at an altitude of 10,000 ft is

$$(V_{climb})_{max} = \frac{\eta P^s_{max}}{W} \left(\frac{\rho}{\rho^s}\right)^m - \frac{4}{3} \sqrt{\frac{2W}{\rho S \cos^3 \mu}} \sqrt{3K^3 C_{D_0}} = 25.75 \text{ ft/s}.$$

At this flight condition for maximum climb rate, the lift coefficient and the pitch moment coefficient are

$$C_L = \frac{2W}{\rho V^2 S \cos \mu} = 1.184,$$
$$C_M = 0.$$

It follows that the aircraft angle of attack is 9.7 degrees and the elevator deflection is 8.74 degrees. The turn radius corresponding to this steady climbing turn is

$$r = \frac{V^2}{g \tan \mu} = 2,809.7 \text{ ft}.$$

In summary, the maximum climb rate of the general aviation aircraft turning with a bank angle of 10 degrees at 10,000 ft is 25.75 ft/s and occurs with the following pilot inputs: elevator deflection of 8.74 degrees and full throttle. The corresponding turn radius is 2,809.7 ft and the airspeed is 126.3 ft/s.

Minimum Radius Steady Climbing Turn

The general aviation aircraft is to execute a steady climbing turn at an altitude of 10,000 ft with a climb rate of 8 ft/s that has minimum turn radius. The analysis is based on a graphical two-dimensional representation of boundary curves that define the feasible set for the airspeed and bank angle values that satisfy all of the constraints.

Matlab m-files are presented that allow determination of the minimum turn radius. The maximum airspeed and the minimum airspeed due to the power constraint and the stall speed can be computed for various altitudes, climb rates, and bank angles. Define the Matlab function SCTFP that computes the maximum airspeed, the minimum airspeed due to the power constraint, and the stall speed for a specified altitude, climb rate, and bank angle. This Matlab function is useful for subsequent minimum radius turning flight and flight envelope computations for the general aviation aircraft.

Matlab function 10.14 SCTFP.m
```
function [Vmax VminPC Vstall]=SCTFP(h,Vclimb,mu);
%Input: altitude h (ft), rate of climb Vclimb (ft/s),
bank angle mu (rad)
%Output:  max air speed Vmax (ft/s), min air speed VminPC (ft/s),
stall speed Vstall (ft/s)
```

(continued)

(continued)

```
W=2900; S=175; CD0=0.026; K=0.054; Psmax=290*550; m=0.6;
CLmax=2.4; eta=0.8;
[Ts ps rhos]=StdAtpUS(0);
[T p rho]=StdAtpUS(h);
tmp=sort(roots([1/2*rho*S*CD0 0 0 W*Vclimb-eta*Psmax*
(rho/rhos)^m 2*K*W^2/rho/S/cos(mu)^2]));
Vmax=tmp(2);
VminPC=tmp(1);
Vstall=sqrt(2*W/rho/S/cos(mu)/CLmax);
```

This Matlab m-file can be used to compute and visualize the boundary curves that define the feasible set for the steady climbing turn at this altitude; by superimposing the curves of constant turn radius it is possible to identify the point in the feasible set that defines the minimum radius turn. The following Matlab m-file accomplishes this.

Matlab function 10.15 MinRCP.m
```
g=32.2; W=2900; S=175;
CD0=0.026; K=0.054; CL0=0.02; CLa=0.12; CM0=0.12; CMa=-0.08;
CMde=0.075; CLmax=2.4;
eta=0.8; Psmax=290*550; m=0.6;
rhos=2.3769e-3; h=10000;
[T p rho]=StdAtpUS(h);
nmax=2; Vclimb=8;
n=100;
mus=linspace(0*pi/180,58.5*pi/180,n);
Vs=zeros(n,n);
for k=1:n
    [Vmax0 VminTC0 Vstall0]=SCTFP(h,Vclimb,mus(k));
    Vmaxs(k)=Vmax0;
    VminTCs(k)=VminTC0;
    Vstalls(k)=Vstall0;
end
figure(1);
plot(VminTCs,mus*180/pi,'r',...
    Vmaxs,mus*180/pi,'r',...
    Vstalls,mus*180/pi,'b',...
    [0 300],acos(1/nmax)*180/pi*ones(2,1),'g');
ylim([0,70]);
xlim([0,300]);
xlabel('Velocity (ft/sec)');
ylabel('Bank angle (deg)');
figure(2);
plot(VminTCs,mus*180/pi,'r',...
    Vmaxs,mus*180/pi,'r',...
```

(continued)

(continued)

```
      Vstalls,mus*180/pi,'b',...
      [0 300],acos(1/nmax)*180/pi*ones(2,1),'g');hold on;
tmp=roots([1/2*rho*S*CLmax^2*K+1/2*rho*S*CD0 0 0
W*Vclimb-eta*Psmax*(rho/rhos)^m]);
Vopt=tmp(3);
muopt=acos(2*W/(Vopt^2*rho*S*CLmax));
Ropt=Vopt^2/g/tan(muopt);
Vss=linspace(0,400,n);
Rss=[0.2 0.5 1 2 3 4]*Ropt;
for j=1:size(Rss,2)
    for i=1:n
        muss(j,i)=atan(Vss(i)^2/g/Rss(j));
    end
    plot(Vss,muss(j,:)*180/pi,'k:');
end
plot(Vopt,muopt*180/pi,'ro');
ylim([40,70]);
xlim([50,200]);
xlabel('Velocity (ft/sec)');
ylabel('Bank angle (deg)');
```

The feasible set consists of the airspeed and the bank angle values that satisfy the three constraints; this feasible set is defined by the boundary curves shown in Figure 10.14(a), where the low-speed boundary curve indicates the stall limit, the inverted U boundary curve describes the maximum engine and propeller power limit, and the horizontal boundary curve indicates the wing loading limit. Several contour lines for fixed values of the turn radius are shown in Figure 10.14(b). It is seen that, at the minimum turn radius, the stall constraint and the power constraint are active, while the wing loading constraint is inactive.

The subsequent analysis of the minimum radius steady climbing turn for the general aviation aircraft is based on the fact that the stall constraint and the engine and propeller power constraint are active. These two conditions imply that the sum of the power required to climb and the power required to balance the drag at stall must equal the maximum power that the engine and propeller can produce, namely

$$W V_{\text{climb}} + \frac{1}{2} \rho V^3 S \left(C_{\text{DO}} + K C_{\text{L}_{\text{max}}}^2 \right) = \eta P_{\text{max}}^s \left(\frac{\rho}{\rho^s} \right)^m.$$

This is a cubic polynomial equation in the airspeed and can be solved to obtain the airspeed of 117.12 ft/s. Using this airspeed value, the bank angle can be obtained from the stall condition

$$\cos \mu = \frac{2W}{\rho V^2 S C_{\text{L}_{\text{max}}}} = 0.574.$$

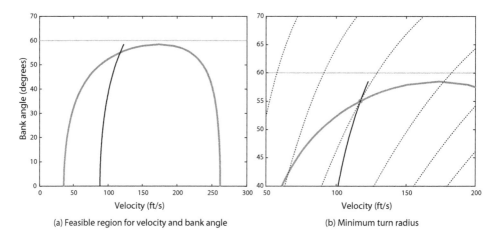

(a) Feasible region for velocity and bank angle   (b) Minimum turn radius

Figure 10.14. Minimum radius steady turn: general aviation aircraft ($V_{climb} = 8$ ft/s).

Thus the bank angle is 55 degrees. The turn radius, corresponding to full throttle, bank angle of 55 degrees, and climb rate of 8 ft/s at the airspeed of 117.12 ft/s, is

$$r = \frac{V^2}{g \tan \mu} = 298.25 \text{ ft}.$$

The expression for the lift coefficient and the fact that the pitch moment coefficient is zero determine the aircraft angle of attack as 19.83 degrees and the elevator deflection as 19.56 degrees.

These results are summarized as follows. The minimum radius steady climbing turn of the general aviation aircraft at an altitude of 10,000 ft with a climb rate of 8 ft/s occurs with the following pilot inputs: elevator deflection of 19.56 degrees, full throttle, and with a bank angle of 55 degrees. The minimum value of the turn radius is 298.25 ft and the airspeed is 117.12 ft/s.

Steady Climbing and Turning Flight Ceiling

The steady climbing and turning flight ceiling of the general aviation aircraft is computed, assuming the aircraft is in a steady banked turn with a bank angle of 10 degrees and a climb rate of 10 ft/s.

The flight ceiling for this steady climbing turn with a bank angle of 10 degrees and a climb rate of 10 ft/s is determined by solving the following equation:

$$W V_{climb} + \frac{4}{3}\sqrt{\frac{2W^3}{\rho S \cos^3 \mu}} \sqrt{3K^3 C_{D_0}} - \eta P_{max}^s \left(\frac{\rho}{\rho^s}\right)^m = 0,$$

where ρ is a function of altitude according to the standard atmospheric model. Define the Matlab function eqnFCCTP whose value is zero at the steady climbing turn flight ceiling.

Matlab function 10.16 eqnFCCTP.m
```
function error=eqnFCCTP(h,Vclimb,mu)
%Input: altitude h (ft), climb rate Vclimb (ft/s),
bank angle mu (rad)
%Output: error
W=2900; S=175; CD0=0.026; K=0.054; Psmax=290*550; m=0.6; eta=0.8;
[Ts ps rhos]=StdAtpUS(0);
[Th ph rhoh]=StdAtpUS(h);
error=4/3*sqrt(2*W^3/rhoh/S/cos(mu)^3*sqrt(3*K^3*CD0))
-eta*Psmax*(rhoh/rhos)^m;
```

The following Matlab command computes the steady climbing turn flight ceiling for which the value of the function defined in eqnFCCTP is zero.

```
hmax=fsolve(@(h) eqnFCCTP(h,10,10*pi/180),10000)
```

where 10,000 is an initial guess of the flight ceiling. After a few iterations, the flight ceiling for a steady climbing turn with a bank angle of 10 degrees and a climb rate of 10 ft/s is 28,549.58 ft.

In summary, the flight ceiling of the general aviation aircraft is 28,549.58 ft, assuming the aircraft is in a steady climbing turn with a bank angle of 10 degrees and a climb rate of 10 ft/s.

Steady Climbing and Turning Flight Envelope

A two-dimensional flight envelope of the general aviation aircraft is computed, assuming the aircraft is in a steady climbing turn with a bank angle of 10 degrees and a climb rate of 10 ft/s; this flight envelope is the set of feasible values of the flight altitude and the airspeed.

A two-dimensional graphical representation of this particular flight envelope, as shown in Figure 10.15, is generated by the following Matlab m-file. These computations use the Matlab m-file SCTFP that computes the maximum airspeed and minimum airspeed due to the maximum power constraint and the stall speed.

Matlab function 10.17 FigSCTFP.m
```
h=linspace(0,28549,500);
for k=1:size(h,2)
    [Vmax(k) VminTC(k) Vstall(k)]=SCTFP(h(k),10,10*pi/180);
end
Vmin=max(VminTC,Vstall);
area([Vmin Vmax(end:-1:1)],[h h(end:-1:1)],...
        'FaceColor',[0.8 1 1],'LineStyle','none');
```

(continued)

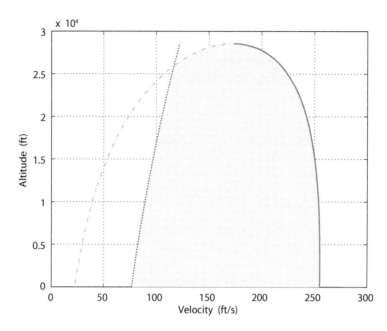

Figure 10.15. A steady turning flight envelope: general aviation aircraft ($V_{climb} = 10\,ft/s$, $\mu = 10$ degrees).

(continued)

```
hold on;
plot(Vmax,h,VminTC,h,Vstall,h);
grid on;
xlim([0 300]);
xlabel('Velocity (ft/s)');
ylabel('Altitude (ft)');
```

This flight envelope describes the set of all feasible altitude and airspeed values for which the general aviation aircraft can maintain a steady turn with a climb rate of 10 ft/s and a bank angle of 10 degrees.

10.9 Steady Descending and Turning Flight: General Aviation Aircraft

Assume that the propeller-driven general aviation aircraft, with a weight of 2,900 lbs, is in a steady descending turn with a bank angle of 10 degrees and a climb rate of −10 ft/s at an altitude of 10,000 ft. This bank angle corresponds to a load factor of 1.05, which is significantly below the maximum load factor of 2. Thus the wing loading constraint is not active.

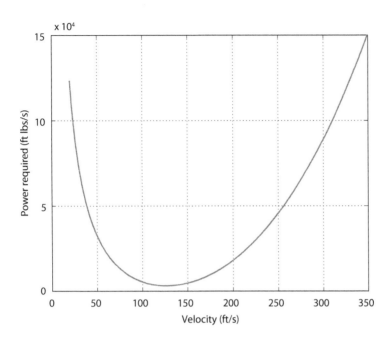

Figure 10.16. Power required for steady turning flight: general aviation aircraft ($V_{climb} = -10$ ft/s, $\mu = 10$ degrees).

Power Required

Suppose that the airspeed in this steady turning and descending flight condition is 200 ft/s. The power required to maintain this steady turn is

$$P = W V_{climb} + \frac{1}{2}\rho V^3 S C_{D_0} + \frac{2K W^2}{\rho V S \cos^2 \mu} = 18,190 \text{ ft lbs/s} = 33.07 \text{ hp.}$$

The required power can be computed for various airspeeds. The power required curve, shown in Figure 10.16, is developed by the following Matlab m-file.

Matlab function 10.18 PRSDTP.m
```
V=linspace(20,350,500);
W=2900; S=175; CD0=0.026; K=0.054; rho=1.7553e-3; mu=10*pi/180;
Vclimb=-10;
P=W*Vclimb+1/2*rho*V.^3*S*CD0+2*K*W^2/rho./V/S/cos(mu)^2
plot(V,P);
xlabel('Velocity (ft/s)');
ylabel('Power required (ft lbs/s)');
grid on;
```

This power required curve for steady turning and descending flight has the typical features: the power required for steady turning and descending flight is large at low airspeeds and high airspeeds, and it has a minimum at an intermediate airspeed.

Minimum Required Power

As seen in Figure 10.16, there is a minimum required power to achieve a steady descending turn a with bank angle of 10 degrees and a climb rate of $-10\,\text{ft/s}$. This minimum required power can be determined by recognizing that the corresponding airspeed minimizes the power required to overcome the drag. The airspeed that achieves the minimum required power to overcome the drag for this steady turning and descending flight condition is

$$V = \sqrt{\frac{2W}{\rho S \cos \mu}} \sqrt{\frac{K}{3C_{D_0}}} = 131.66\,\text{ft/s}.$$

At this steady flight condition, the lift coefficient and the pitch moment coefficient are

$$C_L = \frac{2W}{\rho V^2 S \cos \mu} = 1.2,$$

$$C_M = 0.$$

These expressions determine the aircraft angle of attack and the elevator deflection, which satisfy

$$0.02 + 0.12\alpha = 1.2,$$

$$0.12 - 0.08\alpha + 0.075\delta_e = 0.$$

The results are: the angle of attack is 9.85 degrees and the elevator deflection is 8.91 degrees. The value of the minimum power required for this steady turning and descending flight condition is given by

$$P_{\min} = W V_{climb} + \frac{4}{3}\sqrt{\frac{2W^3}{\rho S \cos^3 \mu}} \sqrt{3K^3 C_{D_0}} = 3,184.6\,\text{ft lbs/s} = 5.79\,\text{hp}.$$

The corresponding throttle setting is

$$\delta_t = \frac{P_{\min}}{\eta P^s_{\max}} \left(\frac{\rho^s}{\rho}\right)^m = 0.03.$$

The turn radius corresponding to this minimum power steady turning and descending flight condition is

$$r = \frac{V^2}{g \tan \mu} = 3,058.7\,\text{ft}.$$

In summary, the minimum power required for the general aviation aircraft to achieve a steady turn with a bank angle of 10 degrees and a climb rate of $-10\,\text{ft/s}$ is 5.79 hp

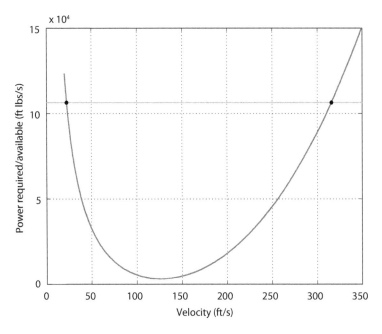

Figure 10.17. Maximum airspeed for steady turning flight: general aviation aircraft (V_{climb} = −10 ft/s, μ = 10 degrees).

and occurs with the following pilot inputs: elevator deflection of 8.91 degrees and a throttle setting of 0.03. The corresponding turn radius is 3,058.7 ft and the airspeed is 131.66 ft/s.

Maximum Airspeed

The maximum airspeed of the general aviation aircraft in a steady descending turn with a bank angle of 10 degrees and a climb rate of −10 ft/s can be obtained from the intersection of the power required curve and the curve that defines the maximum power that the propeller and engine can provide. This graphical approach is illustrated in Figure 10.17.

The maximum airspeed can also be determined by solving the equation

$$W V_{climb} + \frac{1}{2}\rho V^3 S C_{D_0} + \frac{2K W^2}{\rho V S \cos^2 \mu} = \eta P_{max}^s \left(\frac{\rho}{\rho^s}\right)^m.$$

Multiplying the above equation by V and rearranging, the following fourth degree polynomial equation is obtained:

$$\frac{1}{2}\rho S C_{D_0} V^4 + \left(W V_{climb} - \eta P_{max}^s \left(\frac{\rho}{\rho^s}\right)^m\right) V + \frac{2K W^2}{\rho S \cos^2 \mu} = 0.$$

This equation can be solved numerically using the Matlab function MMSSDTP as follows:

Matlab function 10.19 MMSSDTP.m
```
function [Vmin,Vmax]=MMSSDTP
%Output: Minimum and maximum air speeds in a descending turn
W=2900; S=175; CD0=0.026; K=0.054; Psmax=290*550; m=0.6; mu=25;
Vclimb=-10; eta=0.8;
[Ts ps rhos]=StdAtpUS(0);
[Th ph rhoh]=StdAtpUS(h);
[Vmin,Vmax]=sort(roots([1/2*rho*S*CD0 0 0 W*Vclimb-eta*Psmax*
(rho/rhos)^m 2*K*W^2/rho/S/cos(mu)^2]));
```

Matlab provides two positive solutions of 315.76 ft/s and 22.52 ft/s. The high-speed solution is the maximum airspeed of the aircraft in this steady descending turn. At the given altitude of 10,000 ft, the speed of sound is 1,077.39 ft/s. Consequently, the aircraft Mach number for this steady descending turn is 0.293. At this maximum airspeed, the lift coefficient and the pitch moment coefficient are

$$C_L = \frac{2W}{\rho V^2 S \cos \mu} = 0.192,$$
$$C_M = 0.$$

These expressions allow determination of the aircraft angle of attack as 1.44 degrees and the elevator deflection as -0.07 degrees. The turn radius corresponding to this maximum airspeed steady turning and descending flight condition is

$$r = \frac{V^2}{g \tan \mu} = 17,560.6 \, \text{ft.}$$

In summary, the maximum airspeed of the general aviation aircraft in a steady descending turn at 10,000 ft with a bank angle of 10 degrees and a climb rate of -10 ft/s is 315.76 ft/s and occurs with the following pilot inputs: elevator deflection of -0.07 degrees and with full throttle. The corresponding turn radius is 17,560.6 ft.

Minimum Airspeed

The minimum airspeed of the general aviation aircraft in a steady descending turn at an altitude of 10,000 ft with a bank angle of 10 degrees and a climb rate of -10 ft/s must satisfy both the stall and engine power constraints. The stall speed for a steady descending turn with a bank angle of 10 degrees and a climb rate of -10 ft/s is

$$V_{\text{stall}} = \sqrt{\frac{2W}{\rho S C_{L_{\max}} \cos \mu}} = 89.38 \, \text{ft/s.}$$

The minimum airspeed of the general aviation aircraft in a steady descending turn due to the engine power constraint is 22.52 ft/s. In this case, the stall speed is larger than the minimum airspeed due to the power constraint, so the stall constraint is active and the power constraint is not active. Consequently, the minimum airspeed in this steady descending turn is 89.38 ft/s. At the minimum airspeed in this steady descending turn, the lift coefficient $C_L = 2.4$ and the pitch moment coefficient $C_M = 0$. These determine the aircraft angle of attack as 19.83 degrees and the elevator deflection as 19.55 degrees. The value of the power required for this steady turning and descending flight condition is given by

$$P = W V_{climb} + \frac{1}{2} \rho V^3 S C_{D_0} + \frac{2 K W^2}{\rho V S \cos^2 \mu} = 77{,}619 \, \text{ft lbs/s} = 14.48 \, \text{hp}.$$

The corresponding throttle setting is

$$\delta_t = \frac{P}{\eta P_{\max}^s} \left(\frac{\rho^s}{\rho} \right)^m = 0.075.$$

The turn radius corresponding to this steady descending turn is

$$r = \frac{V^2}{g \tan \mu} = 1{,}407.02 \, \text{ft}.$$

In summary, the minimum airspeed of the propeller-driven aircraft in a steady descending turn at an altitude of 10,000 ft with a bank angle of 10 degrees and a climb rate of -10 ft/s occurs with the following pilot inputs: elevator deflection of 19.55 degrees and throttle setting of 0.075. The corresponding minimum airspeed is 89.38 ft/s and the turn radius is 1,407.02 ft.

Minimum Radius Steady Descending Turn

The general aviation aircraft is to execute a steady descending turn at an altitude of 10,000 ft with a climb rate of -10 ft/s that has minimum turn radius.

The Matlab m-file SCTFP can be used to compute and visualize the boundary curves that define the feasible set for the steady climbing turn at this altitude; by superimposing the curves of constant turn radius it is possible to identify the point in the feasible set that defines the minimum radius turn. The following Matlab m-file accomplishes this.

Matlab function 10.20 MinRDP.m

```
g=32.2; W=2900; S=175;
CD0=0.026; K=0.054; CL0=0.02; CLa=0.12; CM0=0.12; CMa=-0.08;
CMde=0.075;
CLmax=2.4;
eta=0.8; Psmax=290*550; m=0.6;
```

(continued)

(continued)

```
rhos=2.3769e-3; h=10000;
[T p rho]=StdAtpUS(h);
nmax=2; Vclimb=-10;
n=100;
mus=linspace(0*pi/180,68*pi/180,n);
Vs=zeros(n,n);
for k=1:n
    [Vmax0 VminTC0 Vstall0]=SCTFP(h,Vclimb,mus(k));
    Vmaxs(k)=Vmax0;
    VminTCs(k)=VminTC0;
    Vstalls(k)=Vstall0;
end
figure(1);
plot(VminTCs,mus*180/pi,'r',...
    Vmaxs,mus*180/pi,'r',...
    Vstalls,mus*180/pi,'b',...
    [0 350],acos(1/nmax)*180/pi*ones(2,1),'g');
ylim([0,70]);
xlim([0,350]);
xlabel('Velocity (ft/sec)');
ylabel('Bank angle (deg)');
figure(2);
plot(VminTCs,mus*180/pi,'r',...
    Vmaxs,mus*180/pi,'r',...
    Vstalls,mus*180/pi,'b',...
    [0 350],acos(1/nmax)*180/pi*ones(2,1),'g');hold on;
muopt=60*pi/180;
[Vmax0 Vtmp Vstall]=STFCP(h,Vclimb,muopt);
Vopt=Vstall;
Ropt=Vopt^2/g/tan(muopt);
Vss=linspace(0,400,n);
Rss=[0.2 0.5 1 2 3 4]*Ropt;
for j=1:size(Rss,2)
    for i=1:n
        muss(j,i)=atan(Vss(i)^2/g/Rss(j));
    end
    plot(Vss,muss(j,:)*180/pi,'k:');
end
plot(Vopt,muopt*180/pi,'ro');
ylim([0,70]);
xlim([0,350]);
xlabel('Velocity (ft/sec)');
ylabel('Bank angle (deg)');
```

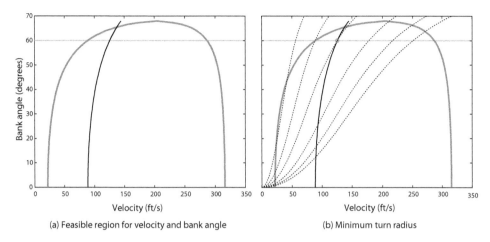

(a) Feasible region for velocity and bank angle

(b) Minimum turn radius

Figure 10.18. Minimum radius steady turn: general aviation aircraft ($V_{climb} = -10$ ft/s).

The feasible set consists of the airspeed and the bank angle values that satisfy the three constraints; this feasible set is defined by the boundary curves shown in figure 10.18(a), where the low-speed boundary curve indicates the stall limit, the inverted ∪ boundary curve characterizes the engine and propeller power limit, and the horizontal boundary curve indicates the wing loading limit. In Figure 10.18(b), several contour lines for fixed values of the turn radius are shown. It is seen that, at the minimum turn radius, the stall constraint and the wing loading constraint are active, while the engine and propeller power constraint is inactive.

Since the wing loading constraint is assumed to be active, the load factor is the maximum load factor of 2, which corresponds to a bank angle of 60 degrees. Since the stall constraint is assumed to be active, this implies that the lift coefficient is a maximum. This allows determination of the airspeed according to

$$V = \sqrt{\frac{2W}{\rho S C_{L_{max}} \cos \mu}} = 125.43 \text{ ft/s}.$$

The turn radius corresponding to a bank angle of 60 degrees and a climb rate of -10 ft/s at the airspeed of 125.43 ft/s is

$$r = \frac{V^2}{g \tan \mu} = 282.12 \text{ ft}.$$

The required power for this descending turn is

$$P = W V_{climb} + \frac{1}{2} \rho V^3 S C_{D_0} + \frac{2K W^2}{\rho V S \cos^2 \mu} = 73,171 \text{ ft lbs/s} = 133.03 \text{ hp}.$$

The corresponding throttle setting is

$$\delta_t = \frac{P}{\eta P_{\max}^s} \left(\frac{\rho^s}{\rho}\right)^m = 0.688.$$

Hence the engine and propeller power constraints, the wing loading constraint, and the stall constraint are satisfied.

The lift coefficient at stall is known and the fact that the pitch moment coefficient is zero allows determination that the aircraft angle of attack is 19.83 degrees and the elevator deflection is 19.56 degrees.

These results are summarized as follows. The minimum radius steady descending turn of the general aviation aircraft at an altitude of 10,000 ft with a climb rate of -10 ft/s occurs with the following pilot inputs: elevator deflection of 19.56 degrees, throttle setting of 0.688, and with a bank angle of 60 degrees. The minimum value of the turn radius is 282.12 ft.

Steady Descending and Turning Flight Ceiling

The steady descending and turning flight ceiling of the general aviation aircraft is computed, assuming the aircraft is in a steady banked turn with a bank angle of 10 degrees and a climb rate of -10 ft/s. The flight ceiling for this steady descending turn is determined by solving the following equation:

$$W V_{climb} + \frac{4}{3}\sqrt{\frac{2W^3}{\rho S \cos^3 \mu}} \sqrt{3K^3 C_{D_0}} - \eta P_{\max}^s \left(\frac{\rho}{\rho^s}\right)^m = 0,$$

where ρ is a function of the altitude according to the standard atmospheric model. The following Matlab command computes the steady descending turning flight ceiling for which the value of the function defined in eqnFCCTP is zero.

```
hmax=fsolve(@(h) eqnFCCTP(h,-10,10*pi/180),10000)
```

where 10,000 is an initial guess of the flight ceiling. After a few iterations, the flight ceiling for a steady descending turn with a bank angle of 10 degrees and a climb rate of -10 ft/s is 49,777.2 ft.

In summary, the flight ceiling of the general aviation aircraft is 49,777.2 ft, assuming the aircraft is in a steady descending turn with a bank angle of 10 degrees and a climb rate of -10 ft/s.

Steady Descending and Turning Flight Envelope

A two-dimensional flight envelope of the general aviation aircraft is computed, assuming the aircraft is in a steady descending turn with a bank angle of 10 degrees and a climb rate of -10 ft/s; this flight envelope is the set of feasible values of the flight altitude and the airspeed.

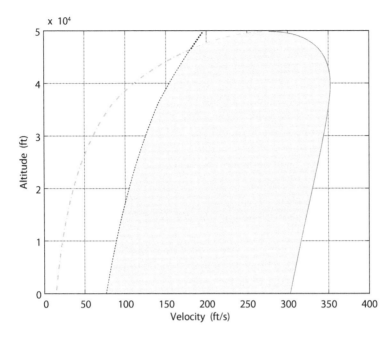

Figure 10.19. Steady turning flight envelope: general aviation aircraft ($V_{climb} = -10$ ft/s, $\mu = 10$ degrees).

   A graphical representation of this flight envelope of the general aviation aircraft at this fixed bank angle and fixed climb rate is developed using the previous analysis. The Matlab function SCTFP is used to compute the maximum airspeed and the minimum airspeed due to the engine power constraint, and the stall speed for various altitudes for the specified descent rate and bank angle. This particular flight envelope, as shown in Figure 10.19, is generated by the following Matlab m-file.

Matlab function 10.21 FigSDTFP.m
```
h=linspace(0,28549,500);
for k=1:size(h,2)
    [Vmax(k) VminTC(k) Vstall(k)]=SCTFP(h(k),-10,10*pi/180);
end
Vmin=max(VminTC,Vstall);
area([Vmin Vmax(end:-1:1)],[h h(end:-1:1)],...
    'FaceColor',[0.8 1 1],'LineStyle','none');
hold on;
plot(Vmax,h,VminTC,h,Vstall,h);
grid on;
xlim([0 300]);
xlabel('Velocity (ft/s)');
ylabel('Altitude (ft)');
```

10.10 Steady Turning Flight Envelopes: General Aviation Aircraft

Several flight envelopes that describe sets of steady climbing or descending and turning flight conditions can be defined for the propeller-driven general aviation aircraft. The subsequent analysis is based on an aircraft weight of 2,900 lbs.

One such flight envelope is the set of feasible values of the flight altitude, the airspeed, and the bank angle for a fixed value of the climb rate. A different flight envelope consists of the set of feasible values of the flight altitude, the airspeed, and the climb rate for a fixed value of the bank angle. These steady climbing or descending turning flight envelopes are defined in terms of three flight variables and hence must be visualized as feasible sets in three dimensions.

Steady Turning Flight Envelope for a Given Climb Rate

One particular steady flight envelope is developed by evaluating the Matlab SCTFP function for various altitude and bank angle values, corresponding to the fixed climb rate value of 10 ft/s. Figure 10.20 shows the resulting surface that defines the boundary of this cross section of the steady flight envelope; this boundary surface is generated by the following Matlab m-file.

Matlab function 10.22 FigSCDTFP.m
```
Vclimb=10;
mu=linspace(0,65*pi/180,40);
for i=1:size(mu,2)
    hmax(i)=fsolve(@(h) eqnFCCTP(h,Vclimb,mu(i)),10000);
    h(i,:)=linspace(0,floor(hmax(i)),40);
    for j=1:size(h,2)
        [Vmax(i,j) VminTC(i,j) Vstall(i,j)]=SCTFP(h(i,j),
        Vclimb,mu(i));
    end
end
Vmin=max(VminTC,Vstall);
surf(Vmax,mu*180/pi,h,'LineStyle','none');
hold on;
surf(Vmin,mu*180/pi,h,'LineStyle','none');
ylabel('\Bank angle (deg)');
zlabel('Altitude (ft)');
xlabel('Velocity (ft/s)');
```

Any triple of values for the flight altitude, airspeed, and bank angle that lies within the flight envelope illustrated in Figure 10.20 corresponds to a feasible steady flight condition with a fixed climb rate of 10 ft/s for which the engine constraints, the stall constraint, and the wing loading constraint are satisfied. For each such feasible triple of values, methodology has been provided to compute the pilot inputs, namely the elevator deflection and the throttle

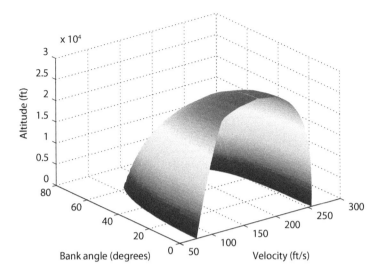

Figure 10.20. Steady turning flight envelope: general aviation aircraft ($V_{climb} = 10$ ft/s).

setting that give this steady climbing and turning flight condition. In addition, the values of the associated angle of attack, values of the lift and drag coefficients, and values of the lift and drag can be computed.

Steady Climbing and Turning Flight Envelope for a Given Bank Angle

Another three-dimensional flight envelope is developed in terms of the altitude, airspeed, and climb rate values, corresponding to the fixed bank angle value of 15 degrees. Figure 10.21 shows the resulting surface that defines the boundary of this flight envelope.

Any triple of values for the flight altitude, airspeed, and climb rate that lies within the flight envelope illustrated in Figure 10.21 corresponds to a feasible steady flight condition with a fixed bank angle of 15 degrees for which the propeller and engine constraints, the stall constraint, and the wing loading constraint are satisfied. For each such feasible triple of values, the pilot inputs, namely the elevator deflection and the throttle setting that give this steady climbing and turning flight condition, can be computed. In addition, the associated angle of attack, values of the lift and drag coefficients, and values of the lift and drag can be computed.

Most General Steady Flight Envelope

The most general steady flight envelope for the propeller-driven general aviation aircraft is defined for helical flight of the aircraft, where the axis of the helix is vertical. This flight envelope is defined in terms of the flight altitude, the airspeed, the climb rate, and the bank angle of the aircraft. This steady flight envelope is defined in terms of four flight variables and hence must be viewed as a feasible set in four dimensions. This generalizes the steady longitudinal flight envelope and the steady level turning flight envelope. In particular, the

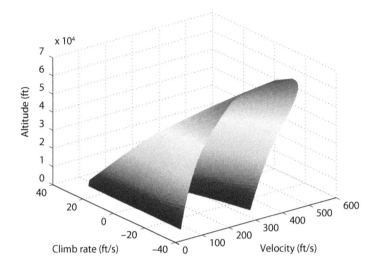

Figure 10.21. Steady turning flight envelope: general aviation aircraft ($\mu = 15$ degrees).

steady longitudinal flight envelope can be viewed as a three-dimensional cross section of the steady flight envelope corresponding to zero bank angle. The steady level turning flight envelope can be viewed as a three-dimensional cross section of the steady flight envelope corresponding to zero climb rate.

Graphical representations of the most general steady flight envelope are necessarily challenging since they require representation of sets in four dimensions. The emphasis on computing various flight envelope cross sections is a natural way to visualize the most general flight envelope data.

10.11 Conclusions

Steady climbing or descending turns have been studied in this chapter. Conditions for this most general form of steady flight have been analyzed, and several different performance measures have been characterized. Many different steady flight conditions have been developed for two specific categories of aircraft to illustrate the general results.

Since this type of flight is the most general steady flight, the results in this chapter characterize the most general steady flight conditions. That is, all of the steady flight results in the prior chapters can be viewed as special cases of steady climbing or descending turning flight.

10.12 Problems

10.1. Consider the executive jet aircraft in a steady banked turn, climbing with a flight path angle of 2 degrees, at an altitude of 33,000 ft. Assume the aircraft weight is a constant 73,000 lbs.

    (a) Assume the aircraft performs a banked climbing turn at an airspeed of 660 ft/s with a bank angle of 15 degrees. What is the glide angle?

What is the radius of the turn? What is the turn rate? What is the load factor? What are the required throttle and elevator deflection to maintain this steady climbing turn?

(b) Assume the aircraft performs a banked climbing turn at an airspeed of 660 ft/s with a bank angle of 45 degrees. What is the glide angle? What is the radius of the turn? What is the turn rate? What is the load factor? What are the required throttle and elevator deflection to maintain this steady climbing turn?

(c) Assume the aircraft performs a banked climbing turn with 100 percent throttle and a bank angle of 45 degrees. Determine all possible flight conditions for this coordinated turn. For each flight condition, determine the airspeed, the radius of the turn, the turn rate, and the load factor. What is the required elevator deflection to maintain this steady climbing turn?

10.2. Consider the executive jet aircraft in a steady banked turn, descending with a climb rate of -19 ft/s, at an altitude of 33,000 ft. Assume the aircraft weight is a constant 73,000 lbs.

(a) Assume the aircraft performs a descending banked turn at an airspeed of 660 ft/s with a bank angle of 15 degrees. What is the radius of the turn? What is the turn rate? What is the load factor? What are the required throttle and elevator deflection to maintain this steady descending turn?

(b) Assume the aircraft performs a descending banked turn at an airspeed of 660 ft/s with a bank angle of 45 degrees. What is the radius of the turn? What is the turn rate? What is the load factor? What are the required throttle and elevator deflection to maintain this steady descending turn?

(c) Assume the aircraft performs a descending banked turn with 100 percent throttle and a bank angle of 45 degrees. Determine all possible flight conditions for this coordinated turn. For each flight condition, determine the airspeed, the radius of the turn, the turn rate, and the load factor. What is the required elevator deflection to maintain this steady descending turn?

10.3. Consider the executive jet aircraft in a steady banked turn, with engine off, at an altitude of 33,000 ft. Assume the aircraft weight is a constant 73,000 lbs.

(a) Assume the aircraft performs a banked, gliding turn at an airspeed of 660 ft/s with a bank angle of 15 degrees. What is the glide angle? What is the radius of the turn? What is the turn rate? What is the load factor? What is the elevator deflection required to maintain this steady gliding turn?

(b) Assume the aircraft performs a banked, gliding turn at an airspeed of 660 ft/s with a bank angle of 45 degrees. What is the glide angle? What is the radius of the turn? What is the turn rate? What is the load factor? What are the required throttle and elevator deflection to maintain this steady gliding turn?

10.4. Consider the propeller-driven general aviation aircraft in a steady, banked turn, climbing with a climb rate of 15 ft/s, at an altitude of 18,000 ft. Assume the aircraft weight is a constant 2,900 lbs.

    (a) Assume the aircraft performs a climbing, banked turn at an airspeed of 260 ft/s with a bank angle of 22 degrees. What is the radius of the turn? What is the turn rate? What is the load factor? What are the required throttle and elevator deflection to maintain this steady climbing turn?

    (b) Assume the aircraft performs a climbing, banked turn at an airspeed of 260 ft/s with a bank angle of 40 degrees. What is the radius of the turn? What is the turn rate? What is the load factor? What are the required throttle and elevator deflection to maintain this steady climbing turn?

    (c) Assume the aircraft performs a climbing, banked turn with 100 percent throttle and a bank angle of 22 degrees. Determine all possible flight conditions for this coordinated turn. For each flight condition, determine the airspeed, the radius of the turn, the turn rate, and the load factor. What is the required elevator deflection to maintain this steady climbing turn?

10.5. Consider the propeller-driven general aviation aircraft in a steady banked turn, descending at a climb rate of −8 ft/s, at an altitude of 18,000 ft. Assume the aircraft weight is a constant 2,900 lbs.

    (a) Assume the aircraft performs a descending, banked turn at an airspeed of 260 ft/s with a bank angle of 22 degrees. What is the glide angle? What is the radius of the turn? What is the turn rate? What is the load factor? What are the required throttle and elevator deflection to maintain this steady gliding turn?

    (b) Assume the aircraft performs a descending, banked turn at an airspeed of 260 ft/s with a bank angle of 40 degrees. What is the glide angle? What is the radius of the turn? What is the turn rate? What is the load factor? What are the required throttle and elevator deflection to maintain this steady gliding turn?

    (c) Assume the aircraft performs a descending, banked turn with 100 percent throttle and a bank angle of 22 degrees. Determine all possible flight conditions for this coordinated turn. For each flight condition, determine the airspeed, the radius of the turn, the turn rate, and the load factor. What is the required elevator deflection to maintain this steady descending turn?

10.6. Consider the propeller-driven general aviation aircraft in a steady banked turn, with engine off, at an altitude of 18,000 ft. Assume the aircraft weight is a constant 2,900 lbs.

(a) Assume the aircraft performs a banked, gliding turn at an airspeed of 260 ft/s with a bank angle of 22 degrees. What is the glide angle? What is the radius of the turn? What is the turn rate? What is the load factor? What are the required throttle and elevator deflection to maintain this steady gliding turn?

(b) Assume the aircraft performs a banked, gliding turn at an airspeed of 260 ft/s with a bank angle of 40 degrees. What is the glide angle? What is the radius of the turn? What is the turn rate? What is the load factor? What are the required throttle and elevator deflection to maintain this steady gliding turn?

10.7. Consider the UAV in a steady banked turn, climbing with a flight path angle of 2 degrees, at an altitude of 2,000 ft. Assume the aircraft weight is a constant 45 lbs.

(a) Assume the aircraft performs a climbing, banked turn at an airspeed of 92 ft/s with a bank angle of 28 degrees. What is the radius of the turn? What is the turn rate? What is the load factor? What are the required throttle and elevator deflection to maintain this steady climbing turn?

(b) Assume the aircraft performs a climbing, banked turn at an airspeed of 92 ft/s with a bank angle of 50 degrees. What is the radius of the turn? What is the turn rate? What is the load factor? What are the required throttle and elevator deflection to maintain this steady climbing turn?

(c) Assume the aircraft performs a climbing, banked turn with 100 percent throttle and a bank angle of 28 degrees. Determine all possible flight conditions for this coordinated turn. For each flight condition, determine the airspeed, the radius of the turn, the turn rate, and the load factor. What is the required elevator deflection to maintain this steady level turn?

10.8. Consider the UAV in a steady banked turn, descending with a flight path angle of −2 degrees, at an altitude of 2,000 ft. Assume the aircraft weight is a constant 45 lbs.

(a) Assume the aircraft performs a descending, banked turn at an airspeed of 92 ft/s with a bank angle of 28 degrees. What is the glide angle? What is the radius of the turn? What is the turn rate? What is the load factor? What are the required throttle and elevator deflection to maintain this steady descending turn?

(b) Assume the aircraft performs a descending, banked turn at an airspeed of 92 ft/s with a bank angle of 50 degrees. What is the glide angle? What is the radius of the turn? What is the turn rate? What is the load factor? What are the required throttle and elevator deflection to maintain this steady descending turn?

(c) Assume the aircraft performs a descending, banked turn with 100 percent throttle and a bank angle of 28 degrees. Determine all possible flight conditions for this coordinated turn. For each flight condition, determine the airspeed, the radius of the turn, the turn rate, and the load factor. What is the required elevator deflection to maintain this steady descending turn?

10.9. Consider the UAV in a steady banked turn, with engine off, at an altitude of 2,000 ft. Assume the aircraft weight is a constant 45 lbs.

(a) Assume the aircraft performs a banked, gliding turn at an airspeed of 92 ft/s with a bank angle of 28 degrees. What is the glide angle? What is the radius of the turn? What is the turn rate? What is the load factor? What are the required throttle and elevator deflection to maintain this steady gliding turn?

(b) Assume the aircraft performs a banked, gliding turn at an airspeed of 92 ft/s with a bank angle of 50 degrees. What is the glide angle? What is the radius of the turn? What is the turn rate? What is the load factor? What are the required throttle and elevator deflection to maintain this steady gliding turn?

10.10. Consider the executive jet aircraft in a steady climbing or descending turn. Assume the aircraft weight is a constant 73,000 lbs.

(a) Suppose the aircraft is to maintain a maximum flight path angle while turning with a constant bank angle of 33 degrees. What is the maximum value of the flight path angle and what are the corresponding values of the airspeed and turn radius? What is the thrust required to maintain this steady climbing and turning flight condition? What are the corresponding throttle setting and elevator deflection?

(b) Suppose the aircraft is to maintain a minimum radius turn with a constant flight path angle of 2 degrees. What is the minimum value of the turn radius and what are the corresponding values of the airspeed and bank angle? What is the thrust required to maintain this steady climbing and turning flight condition? What are the corresponding throttle setting and elevator deflection?

10.11. Consider the propeller-driven general aviation aircraft in a steady climbing or descending turn. Assume the aircraft weight is a constant 2,900 lbs.

(a) Suppose the aircraft is to maintain a maximum climb rate while turning with a constant bank angle of 22 degrees. What is the maximum value of the climb rate and what are the corresponding values of the airspeed and turn radius? What is the power required to maintain this steady climbing and turning flight condition? What are the corresponding throttle setting and elevator deflection?

(b) Suppose the aircraft is to maintain a minimum radius turn with a constant climb rate of 3.3 ft/s. What is the minimum value of the turn radius and what are the corresponding values of the airspeed and bank angle? What is the power required to maintain this steady climbing and turning flight condition? What are the corresponding throttle setting and elevator deflection?

10.12. Consider the executive jet aircraft in a steady climbing or descending turn. Assume the aircraft weight is a constant 73,000 lbs.

(a) Suppose the aircraft is to maintain a steady climb and turn with a constant flight path angle of 2 degrees, a constant airspeed of 330 ft/s, and a constant bank angle of 33 degrees as it climbs from 21,000 ft to 32,000 ft. What is the thrust required to maintain this steady climbing and turning flight condition? What are the corresponding throttle setting and elevator deflection? Express these as functions of the air density. How long does it take the aircraft to climb from 21,000 ft to 32,000 ft? How many circular revolutions does it complete in this time period?

(b) Suppose the aircraft is to maintain a steady descent and turn with a constant flight path angle of −2 degrees, a constant airspeed of 330 ft/s, and a constant bank angle of 33 degrees as it descends from 21,000 ft to 5,000 ft. What is the thrust required to maintain this steady descending and turning flight condition? What are the corresponding throttle setting and elevator deflection? Express these as functions of the air density. How long does it take the aircraft to descend from 21,000 ft to 5,000 ft? How many circular revolutions does it complete in this time period?

10.13. Consider the propeller-driven general aviation aircraft in a steady climbing or descending turn. Assume the aircraft weight is a constant 2,900 lbs.

(a) Suppose the aircraft is to maintain a steady climb and turn with a constant climb rate of 3.3 ft/s, a constant airspeed of 230 ft/s, and a constant bank angle of 22 degrees as it climbs from 12,000 ft to 23,000 ft. What is the power required to maintain this steady climbing and turning flight condition? What are the corresponding throttle setting and elevator deflection? Express these as functions of the air density. How long does it take the aircraft to climb from 12,000 ft to 23,000 ft? How many circular revolutions does it complete in this time period?

(b) Suppose the aircraft is to maintain a steady descent and turn with a constant climb rate of −3.3 ft/s, a constant airspeed of 230 ft/s, and a constant bank angle of 22 degrees as it descends from 12,000 ft to 3,000 ft. What is the power required to maintain this steady descending and turning flight condition? What are the corresponding throttle setting and elevator deflection? Express these as functions of the air density. How long does it take the aircraft to descend from 12,000 ft to 3,000 ft? How many circular revolutions does it complete in this time period?

10.14. Consider the executive jet aircraft in steady turning flight. Is it possible that the aircraft can remain in steady turning flight with an airspeed of 260 ft/s, a flight path angle of 2 degrees, and a turn radius of 3,000 ft at an altitude of 12,000 ft? In other words, is this flight condition in the steady turning flight envelope?

If this flight condition is in the steady turning flight envelope, determine the throttle setting, the elevator deflection, and the bank angle required to maintain this flight condition.

If this flight condition is not in the steady turning flight envelope, which constraints are violated?

10.15. Consider the executive jet aircraft in steady turning flight. Is it possible that the aircraft can remain in steady turning flight with an airspeed of 500 ft/s, a flight path angle of −2 degrees, and a turn radius of 4,500 ft at an altitude of 9,000 ft? In other words, is this flight condition in the steady turning flight envelope?

If this flight condition is in the steady turning flight envelope, determine the throttle setting, the elevator deflection, and the bank angle required to maintain this flight condition.

If this flight condition is not in the steady turning flight envelope, which constraints are violated?

10.16. Consider the propeller-driven general aviation aircraft in steady turning flight. Is it possible that the aircraft can remain in steady level turning flight with an airspeed of 95 ft/s, a climb rate of 6 ft/s, and a turn radius of 1,400 ft at an altitude of 5,000 ft? In other words, is this flight condition in the steady turning flight envelope?

If this flight condition is in the steady turning flight envelope, determine the throttle setting, the elevator deflection, and the bank angle required to maintain this flight condition.

If this flight condition is not in the steady turning flight envelope, which constraints are violated?

10.17. Consider the propeller-driven general aviation aircraft in steady turning flight. Is it possible that the aircraft can remain in steady turning flight with an airspeed of 250 ft/s, a climb rate of −2 ft/s, and a turn radius of 2,000 ft at an altitude of 7,000 ft? In other words, is this flight condition in the steady turning flight envelope?

If this flight condition is in the steady turning flight envelope, determine the throttle setting, the elevator deflection, and the bank angle required to maintain this flight condition.

If this flight condition is not in the steady turning flight envelope, which constraints are violated?

11

Aircraft Range and Endurance in Steady Flight

In this chapter, aircraft range and aircraft endurance are studied assuming flight through a stationary atmosphere. The aircraft range is the distance that an aircraft travels in steady flight using a fixed amount of fuel. The aircraft endurance or the time of flight is the time that an aircraft remains in steady flight using a fixed amount of fuel. The maximum range and the maximum endurance in steady flight are important performance measures; they are determined as a function of important aircraft flight parameters.

Formulas for range and endurance in steady level flight are determined for aircraft powered by an ideal jet engine and for aircraft powered by a propeller driven by an ideal internal combustion engine. Steady level flight conditions are obtained that correspond to maximum range or maximum endurance. In addition, problems of range and endurance are determined for steady level turns, for steady longitudinal flight, and for steady climbing or descending turns, under the assumption that the change in flight altitude does not result in a large change in air density.

Range and endurance are important performance measures for many aircraft. The results developed in this chapter are important for flight analysis and for design of fixed-wing aircraft.

11.1 Fuel Consumption

Aircraft range and endurance are limited by the amount of fuel that an aircraft can carry, by the rate at which fuel is burned by the engine, by the efficiency of the propeller (if there is one), and by the aerodynamics of the aircraft. The rate at which fuel is burned depends on the type of engine. Here, range and endurance formulas are obtained in this chapter for two engine categories: an ideal jet engine and an ideal internal combustion engine that rotates a propeller, thereby producing thrust on the aircraft.

For an ideal jet engine, the rate at which fuel is burned is proportional to the thrust that the engine produces. Let W denote the total weight of the aircraft, including the fuel. Then the rate at which fuel is burned is given by

$$\frac{dW}{dt} = -cT, \tag{11.1}$$

where T denotes the thrust produced by the jet engine and c is the thrust specific fuel consumption rate. This consumption rate is an important engine parameter. It denotes the rate at which fuel is burned per unit of thrust produced by the engine. For an ideal jet engine this thrust specific fuel consumption rate is assumed to be a positive constant parameter independent of the flight conditions of the aircraft.

For an ideal internal combustion engine that drives a propeller, the rate at which fuel is burned is proportional to the power that the engine produces. As before, let W denote the total weight of the aircraft, including the fuel. Then the rate at which fuel is burned is given by

$$\frac{dW}{dt} = -cP, \tag{11.2}$$

where P denotes the power produced by the engine and c is the power specific fuel consumption rate. This consumption rate denotes the rate at which fuel is burned per unit of power produced by the engine. For an ideal internal combustion engine this power specific fuel consumption rate is assumed to be a positive constant parameter independent of the flight conditions of the aircraft.

11.2 Steady Flight Background

In the following sections, results are developed for aircraft range and aircraft endurance assuming steady flight. We follow standard convention to make the approximations that the thrust is much smaller than the lift, the angle of attack is a small angle in radian measure, and the flight path angle, or, equivalently, the climb rate, is small. The resulting equations for steady flight are exactly those used in the steady flight analyses of the prior chapters.

The main difference is that the weight of the aircraft, as a consequence of fuel consumption by the engine, is assumed to decrease slowly. This assumption that the aircraft weight including fuel slowly decreases does not invalidate the prior steady flight results. The fuel consumed to achieve a given range or endurance maneuver is also determined.

11.3 Range and Endurance of a Jet Aircraft in Steady Level Longitudinal Flight

The most important range and endurance results hold for steady level longitudinal flight, sometimes referred to as steady cruise. This section develops the most important results for an ideal jet aircraft.

Range of Jet Aircraft in Steady Level Longitudinal Flight

Since the engine burns fuel, the weight of the aircraft including fuel is not constant; it slowly decreases until the fuel is exhausted. However, the steady level flight results are valid as an approximation since the weight changes slowly.

We first write down the two fundamental differential equations that describe the airspeed and the rate of change of the aircraft weight. Let x denote the horizontal position of the aircraft, measured with respect to some reference, then steady level longitudinal flight implies that the airspeed is

$$\frac{dx}{dt} = V. \tag{11.3}$$

Further, for a jet engine the rate of change of the aircraft weight is

$$\frac{dW}{dt} = -cT. \tag{11.4}$$

Using the fact that the lift equals the weight in steady level flight and the expression for the lift, the airspeed can be written as

$$V = \sqrt{\frac{2W}{\rho S C_L}}. \tag{11.5}$$

Further, the required thrust in steady level flight can be written as

$$T = \frac{C_D}{C_L} W. \tag{11.6}$$

Consequently, we obtain

$$\frac{dx}{dt} = \sqrt{\frac{2W}{\rho S C_L}}, \tag{11.7}$$

$$\frac{dW}{dt} = -c \frac{C_D}{C_L} W. \tag{11.8}$$

Since the air density and the drag and lift coefficients are constant, equation (11.8) is a differential equation expressed in terms of the weight only; this differential equation can be integrated. Substituting the resulting time dependent weight expression into the differential equation (11.7), it can then be integrated to obtain the aircraft range.

However, we follow a slightly different approach that only makes use of elementary methods of calculus. Our objective is to obtain an expression for the aircraft range in terms of the fuel used during the flight duration. To this end, equation (11.1) can be rewritten as

$$dt = -\frac{dW}{cT}.$$

Multiply the above equation by the velocity of the aircraft to obtain

$$V dt = -\frac{V dW}{cT}.$$

Since $V = \frac{dx}{dt}$, it follows that

$$dx = -\frac{V \, dW}{cT}.$$

Therefore, the distance traveled by the aircraft, which is the range R, is obtained by integrating the above equation from the initial time when the weight of the aircraft including fuel is W_i to the terminal time when the weight of the aircraft and any remaining fuel is W_f. Clearly, $W_i > W_f$. The resulting expression for the range is

$$R = -\int_{W_i}^{W_f} \frac{V \, dW}{cT}. \tag{11.9}$$

Equation (11.9) is a general expression for the range of a jet aircraft. Now, a more explicit equation for range using the steady level flight assumptions is derived. Since $T = D$ from the steady level flight assumption,

$$R = -\int_{W_i}^{W_f} \frac{V \, dW}{cD}.$$

Multiply the above equation by $\frac{L}{W} = 1$ to obtain

$$R = -\int_{W_i}^{W_f} \frac{1}{c} \frac{L}{D} \frac{V \, dW}{W}.$$

Substitute equation (11.5) into the above equation to express the airspeed V in terms of C_L and W. Then

$$R = -\int_{W_i}^{W_f} \frac{1}{c} \sqrt{\frac{2}{\rho S} \frac{C_L^{\frac{1}{2}}}{C_D}} \frac{dW}{\sqrt{W}}.$$

Assuming that C_L, C_D, and ρ are constant throughout the flight, the range for a jet aircraft is given by

$$R = -\frac{1}{c} \sqrt{\frac{2}{\rho S} \frac{C_L^{\frac{1}{2}}}{C_D}} \int_{W_i}^{W_f} \frac{dW}{\sqrt{W}}.$$

Evaluating the integral leads to the main expression for the range of the jet aircraft:

$$R = \frac{2}{c} \sqrt{\frac{2}{\rho S}} \left(\frac{C_L^{\frac{1}{2}}}{C_D} \right) \left(\sqrt{W_i} - \sqrt{W_f} \right). \tag{11.10}$$

This formula gives the range for a jet-powered aircraft, assuming that the lift to drag ratio appearing in equation (11.10) and the flight altitude are maintained constant throughout the flight.

Equation (11.10) can also be solved to obtain the terminal weight of the aircraft for a given value of the range, assuming steady level flight with constant lift coefficient and drag coefficient. This assumes that sufficient fuel is available. The formula for the weight of fuel $W_{fuel} = W_i - W_f$ burned to achieve a given range is obtained as

$$W_{fuel} = W_i - \left(\sqrt{W_i} - \frac{cR}{2} \sqrt{\frac{\rho S}{2} \frac{C_D}{C_L^{\frac{1}{2}}}} \right)^2 . \qquad (11.11)$$

Consider the flight conditions to maximize the range of a jet aircraft with the fuel tank assumed to be full at the initial time and empty at the terminal time. According to equation (11.10), the range is maximized if the following conditions hold:

- minimum thrust specific fuel consumption rate c

- maximum value of the aerodynamics ratio $\frac{C_L^{\frac{1}{2}}}{C_D}$; it can easily be shown using calculus that the maximum value of this ratio, for the standard quadratic drag polar expression, occurs when the lift coefficient and the drag coefficient are

$$C_L = \sqrt{\frac{C_{D_0}}{3K}}, \qquad (11.12)$$

$$C_D = \frac{4}{3} C_{D_0} \qquad (11.13)$$

- minimum air density ρ

The first condition is clear. The range of a jet aircraft is maximized when the lift to drag ratio $\frac{C_L^{\frac{1}{2}}}{C_D}$ is maximized; that is, when the lift coefficient and the drag coefficient have the values indicated above. The range is also maximized by steady level flight at the highest possible altitude, namely the flight ceiling. It can be shown that the above conditions also minimize the fuel consumed for steady level flight of a fixed range.

Endurance of Jet Aircraft in Steady Level Longitudinal Flight

The aircraft endurance is the time that an aircraft remains in steady level flight on a fixed amount of fuel. Assume that the aircraft cruises in a steady level longitudinal flight condition as discussed earlier. The procedure to derive an endurance equation is similar to that used in the previous section.

The endurance of a jet aircraft can be obtained by solving the differential equations (11.7) and (11.8). We follow an alternative procedure to derive an endurance equation based on elementary methods of calculus. Equation (11.1) can be rewritten as

$$dt = -\frac{dW}{cT} .$$

The total time that the airplane remains in steady flight, which is the endurance E, is obtained by integrating the above equation from the initial time to the terminal time.

$$E = \int_0^E dt = -\int_{W_i}^{W_f} \frac{dW}{cT}, \tag{11.14}$$

where W_i is the initial weight of the aircraft including fuel, and W_f is the terminal weight of the aircraft and any remaining fuel. From the steady level flight assumption that $T = D$, the endurance is

$$E = -\int_{W_i}^{W_f} \frac{1}{cD} dW.$$

Multiply the above equation by $\frac{L}{W} = 1$ to obtain

$$E = -\int_{W_i}^{W_f} \frac{1}{c} \frac{L}{D} \frac{dW}{W}. \tag{11.15}$$

Assuming that the lift to drag ratio $\frac{L}{D} = \frac{C_L}{C_D}$ is constant throughout the flight, the endurance for an ideal jet aircraft is given by

$$E = -\frac{1}{c} \frac{C_L}{C_D} \int_{W_i}^{W_f} \frac{dW}{W}.$$

Evaluating the integral gives the main expression for the endurance of a jet aircraft

$$E = \frac{1}{c} \left(\frac{C_L}{C_D} \right) \ln \frac{W_i}{W_f}. \tag{11.16}$$

This endurance formula for a jet-powered aircraft is based on the assumption that the lift to drag ratio in equation (11.16) is maintained constant throughout the flight. Since the endurance formula for a jet aircraft does not depend on the air density, it follows that the endurance is independent of the flight altitude.

Equation (11.16) can also be solved to obtain the terminal weight of the aircraft for a given value of the flight duration, assuming steady level flight with the lift coefficient and the drag coefficient maintained constant. This assumes that sufficient fuel is available. The formula for the weight of fuel $W_{fuel} = W_i - W_f$ burned to achieve a given value of endurance is easily obtained from equation (11.16) as

$$W_{fuel} = W_i \left(1 - \exp \left(-c \frac{C_D}{C_L} E \right) \right). \tag{11.17}$$

Consider the flight conditions to maximize the endurance of a jet aircraft with a fuel tank assumed to be full at the initial time and empty at the terminal time. According to equation (11.16), the following conditions maximize the endurance of a jet aircraft:

- minimum power specific fuel consumption rate c

- maximum value of the aerodynamics lift to drag ratio $\frac{C_L}{C_D}$; it can easily be shown using calculus that the maximum value of this ratio, for the standard quadratic drag polar expression, occurs when the lift coefficient and the drag coefficient are:

$$C_L = \sqrt{\frac{C_{D_0}}{K}}, \tag{11.18}$$

$$C_D = 2C_{D_0}. \tag{11.19}$$

The first condition is essentially common sense. The second condition describes the aerodynamics condition for which the endurance of a jet aircraft is maximized; this occurs when the lift to drag ratio is maximized. This endurance according to equation (11.16) does not depend on the air density; consequently the endurance does not depend on the altitude at which the steady level flight occurs. It can be shown that the above conditions also minimize the fuel burned for steady level flight of a fixed endurance.

Explicit equations for range and endurance of a jet aircraft have been obtained. The expressions give simple formulas for range and endurance. The physical interpretations of the expressions provide insight into the properties of the aircraft range and endurance.

However, these equations should be used with caution, keeping in mind the assumptions that the aircraft is in steady level longitudinal flight with a small angle of attack. It is also assumed that C_L, C_D, ρ, and the fuel consumption rate c are constant throughout the flight. It is important to check that the flight condition is in the steady level longitudinal flight envelope and is compatible with the assumptions made to derive the equations. The range and endurance formulas are computed based on the assumption of a standard atmosphere that is stationary.

11.4 Range and Endurance of a General Aviation Aircraft in Steady Level Longitudinal Flight

This section develops important range and endurance results for ideal propeller-driven general aviation aircraft in steady level longitudinal cruise.

Range of Aircraft in Steady Level Longitudinal Flight

We first write down the two fundamental differential equations that describe the airspeed and the rate of change of the aircraft weight. Let x denote the horizontal position of the aircraft, measured with respect to an Earth-fixed frame. Steady level longitudinal flight implies that the airspeed is

$$\frac{dx}{dt} = V. \tag{11.20}$$

Further, for a propeller powered by an internal combustion engine the rate of change of the aircraft weight is

$$\frac{dW}{dt} = -cP = -\frac{c}{\eta}TV. \tag{11.21}$$

Since the lift equals the weight in steady level flight, the air speed can be written as

$$V = \sqrt{\frac{2W}{\rho S C_L}}. \tag{11.22}$$

Further, the required thrust in steady level flight can be written as

$$T = \frac{C_D}{C_L}W. \tag{11.23}$$

Consequently, we obtain

$$\frac{dx}{dt} = \sqrt{\frac{2W}{\rho S C_L}}, \tag{11.24}$$

$$\frac{dW}{dt} = -\frac{c}{\eta}\frac{C_D}{C_L^{\frac{3}{2}}}W^{\frac{3}{2}}\sqrt{\frac{2}{\rho S}}. \tag{11.25}$$

Since the air density and the drag and lift coefficients are constant, the differential equation (11.25) can be integrated and the resulting time dependent weight expression substituted into the differential equation (11.24). The resulting differential equation can then be integrated to obtain the the aircraft range.

We again follow a slightly different approach that is based on elementary methods of calculus. Our objective is to obtain an expression for the aircraft range in terms of the fuel burned during the flight duration. To this end, equation (11.2) can be rewritten as

$$dt = -\frac{dW}{cP}.$$

Multiply the above equation by the airspeed of the aircraft V to obtain

$$V dt = -\frac{V dW}{cP}.$$

Since $V = \frac{dx}{dt}$, the incremental distance traveled dx in time dt is

$$dx = -\frac{V dW}{cP}.$$

Therefore, the total distance traveled by the aircraft, which is the range R, is obtained by integrating the above equation from an initial time when the weight of the aircraft and fuel is W_i to a terminal time when the weight of the aircraft and remaining fuel is W_f. The result is

$$R = - \int_{W_i}^{W_f} \frac{V \, dW}{cP}. \tag{11.26}$$

Equation (11.26) is a general expression for the range of a propeller aircraft driven by an ideal internal combustion engine. A more explicit range equation can be derived using steady level flight assumptions. The required power is equal to the thrust multiplied by the velocity, $P = TV$. Since $T = D$ from the steady level flight assumption, the power provided by the engine is

$$P = \frac{1}{\eta} DV, \tag{11.27}$$

where η is the propeller efficiency. Substituting equation (11.27) into equation (11.26),

$$R = - \int_{W_i}^{W_f} \frac{\eta}{cD} \, dW.$$

Multiply the above equation by $\frac{L}{W} = 1$ to obtain

$$R = - \int_{W_i}^{W_f} \frac{\eta}{c} \frac{L}{D} \frac{dW}{W}.$$

Assume that the lift to drag ratio $\frac{L}{D} = \frac{C_L}{C_D}$ is constant throughout the flight. Then, the range for an ideal propeller-driven general aviation aircraft is given by

$$R = - \frac{\eta}{c} \frac{C_L}{C_D} \int_{W_i}^{W_f} \frac{dW}{W}.$$

Evaluation of the integral leads to the main expression for the range of a general aviation aircraft:

$$R = \frac{\eta}{c} \left(\frac{C_L}{C_D} \right) \ln \frac{W_i}{W_f}. \tag{11.28}$$

Equation (11.28) is known as the *Breguet range formula* for propeller-driven general aviation aircraft. It gives a simple expression for the range in steady level longitudinal flight, assuming that the lift to drag ratio in equation (11.28) is maintained constant throughout the flight.

Equation (11.28) can also be solved to obtain the terminal weight of the aircraft for a given value of the range, assuming steady level flight with the lift to drag ratio maintained constant. This assumes that sufficient fuel is available. The formula for the weight of fuel

$W_{fuel} = W_i - W_f$ burned to achieve a given value of range can be obtained from equation (11.28) as

$$W_{fuel} = W_i \left(1 - \exp \left(-\frac{cR}{\eta} \frac{C_D}{C_L} \right) \right). \tag{11.29}$$

Consider the flight conditions to maximize the range of an aircraft powered by an internal combustion engine and propeller with a fuel tank assumed to be full at the initial time and empty at the terminal time. According to equation (11.28), the following conditions are seen to maximize the aircraft range:

- maximum propeller efficiency η

- minimum power specific fuel consumption rate c

- maximum value of the aerodynamics lift to drag ratio $\frac{C_L}{C_D}$; it can easily be shown using calculus that the maximum lift to drag ratio, assuming a standard quadratic drag expression, occurs when the lift coefficient and the drag coefficient are

$$C_L = \sqrt{\frac{C_{D_0}}{K}}, \tag{11.30}$$

$$C_D = 2C_{D_0}. \tag{11.31}$$

The first two conditions are common sense. The last condition describes the aerodynamics condition for which the range of a propeller aircraft driven by an internal combustion engine is maximized; this occurs when the lift to drag ratio is maximized. This range according to equation (11.28) does not depend on the air density; consequently the range does not depend on the altitude at which the steady level longitudinal flight occurs. It can be shown that the above conditions also minimize the fuel burned for steady level longitudinal flight of a fixed range.

Endurance of Propeller Aircraft in Steady Level Longitudinal Flight

The aircraft endurance is the time that an aircraft remains in steady level flight using a fixed amount of fuel. The endurance of a propeller aircraft can be obtained by solving the differential equations (11.24) and (11.25).

We follow an alternative procedure to derive an endurance equation that uses elementary methods of calculus. Thus equation (11.2) can be rewritten as

$$dt = -\frac{dW}{cP}.$$

The total time that the airplane remains in steady flight, which is the endurance E, is obtained by integrating the above equation from the initial time to the terminal time.

This gives

$$E = -\int_{W_i}^{W_f} \frac{dW}{cP},\qquad(11.32)$$

where W_i is the initial weight of the aircraft including fuel and W_f is the terminal weight of the aircraft. Substituting equation (11.27) into equation (11.32),

$$E = -\int_{W_i}^{W_f} \frac{\eta}{cDV}\,dW.$$

Multiply the above equation by $\frac{L}{W} = 1$ to obtain

$$E = -\int_{W_i}^{W_f} \frac{\eta}{c}\frac{L}{D}\frac{dW}{VW},$$

$$E = -\int_{W_i}^{W_f} \frac{\eta}{c}\frac{C_L}{C_D}\frac{dW}{VW}.\qquad(11.33)$$

In steady level flight

$$W = L = \frac{1}{2}\rho V^2 S C_L.$$

The airspeed V is expressed in terms of C_L and W as

$$V = \sqrt{\frac{2W}{\rho C_L S}}.\qquad(11.34)$$

Substituting equation (11.34) into equation (11.33), we obtain

$$E = -\int_{W_i}^{W_f} \frac{\eta}{c}\frac{C_L^{\frac{3}{2}}}{C_D}\sqrt{\frac{\rho S}{2}}\frac{dW}{W^{\frac{3}{2}}}.\qquad(11.35)$$

Assume that C_L, C_D, ρ, and η are constant during the flight. Then, the endurance for an ideal aircraft powered by a propeller is

$$E = -\frac{\eta}{c}\frac{C_L^{\frac{3}{2}}}{C_D}\sqrt{\frac{\rho S}{2}}\int_{W_i}^{W_f} \frac{dW}{W^{\frac{3}{2}}}.$$

Evaluation of the integral leads to the main expression for the endurance of a propeller-driven general aviation aircraft:

$$E = \frac{\eta}{c} \left(\frac{C_L^{\frac{3}{2}}}{C_D} \right) \sqrt{2\rho S} \left(\frac{1}{\sqrt{W_f}} - \frac{1}{\sqrt{W_i}} \right). \qquad (11.36)$$

This endurance formula, for a propeller-driven general aviation aircraft, is valid so long as the lift to drag ratio appearing in equation (11.36) and the flight altitude are maintained constant throughout the flight.

Equation (11.36) can also be solved to obtain the terminal weight of the aircraft for a given value of the flight duration, assuming steady level flight with the lift coefficient and the drag coefficient maintained constant. This assumes that sufficient fuel is available. The formula for the weight of fuel burned to achieve a given value of endurance is easily obtained from equation (11.36) as

$$W_{fuel} = W_i - \left(\frac{1}{\sqrt{W_i}} + \frac{cE}{\eta} \frac{C_D}{C_L^{\frac{3}{2}}} \right)^{-2}. \qquad (11.37)$$

Consider the flight conditions to maximize the endurance of an aircraft powered by a propeller-driven internal combustion engine with a fuel tank assumed to be full at the initial time and empty at the terminal time. According to equation (11.36), the following conditions are seen to maximize the aircraft endurance:

- maximum propeller efficiency η

- minimum power specific fuel consumption rate c

- maximum value of the aerodynamics ratio $\frac{C_L^{\frac{3}{2}}}{C_D}$; it can easily be shown using calculus that the maximum value of this ratio, assuming the standard quadratic drag polar expression, occurs when the lift coefficient and the drag coefficient are:

$$C_L = \sqrt{\frac{3C_{D_0}}{K}}, \qquad (11.38)$$

$$C_D = 4C_{D_0}. \qquad (11.39)$$

- maximum density of air ρ

The first two conditions are identical with the maximum range conditions. The endurance of an aircraft driven by an internal combustion engine and propeller is maximized when the aerodynamics lift and drag coefficients are as given in equations (11.38) and (11.39). The endurance depends on the air density. If an aircraft flies at a lower altitude, in which the air density is higher, then its endurance is increased. It can be shown that the above conditions also minimize the fuel required for steady level flight with a fixed endurance.

11.5 Range and Endurance of a Jet Aircraft in a Steady Level Turn

We present range and endurance expressions for a jet aircraft in a steady level turn. Since the engine burns fuel, the weight of the aircraft including fuel is not constant; it slowly decreases. The steady level turning flight results are valid since the aircraft weight changes slowly. Since the turn occurs in a horizontal plane, the air density is constant.

Following the development in section 11.3 modified for a steady level turn, the velocity along the circular path, denoted by the time rate of change of the circular arc length s, and the time rate of change of the aircraft weight, can be shown to be given by

$$\frac{ds}{dt} = \sqrt{\frac{2W}{\rho S C_L \cos \mu}}, \tag{11.40}$$

$$\frac{dW}{dt} = -c \frac{C_D}{C_L} \frac{1}{\cos \mu} W. \tag{11.41}$$

The bank angle, the drag coefficient, and the lift coefficient are assumed constant throughout the steady level turn.

The distance traveled by the aircraft along the circular arc defined by the steady level turn is the steady level turning range R. Following the development in section 11.3, the steady level turning range, assuming that the lift coefficient, the drag coefficient, and the flight altitude are maintained constant throughout the flight, is

$$R = \frac{2}{c} \sqrt{\frac{2}{\rho S}} \left(\frac{(C_L \cos \mu)^{\frac{1}{2}}}{C_D} \right) \left(\sqrt{W_i} - \sqrt{W_f} \right). \tag{11.42}$$

This equation gives the steady level turning range in terms of the initial weight and the terminal weight of the aircraft including fuel and the bank angle. The range decreases as the bank angle increases. This equation also shows that the aerodynamics flight condition that maximizes the steady level turning range is that the lift coefficient and the drag coefficient be chosen to maximize the aerodynamics ratio $\frac{C_L^{\frac{1}{2}}}{C_D}$.

The steady level turning flight endurance, assuming that the lift coefficient and the drag coefficient are maintained constant, can be shown to be given by

$$E = \frac{1}{c} \left(\frac{C_L \cos \mu}{C_D} \right) \ln \frac{W_i}{W_f}. \tag{11.43}$$

This equation gives the steady level turn endurance in terms of the initial weight and the terminal weight of the aircraft including fuel and the bank angle. The endurance decreases as the bank angle increases. This equation also shows that the aerodynamics flight condition that maximizes the steady level turning endurance is that the lift coefficient and the drag coefficient be chosen to maximize the aerodynamics ratio $\frac{C_L}{C_D}$.

11.6 Range and Endurance of a General Aviation Aircraft in a Steady Level Turn

We present range and endurance expressions for a propeller aircraft in a steady level turn. Since the internal combustion engine burns fuel, the weight of the aircraft including fuel is not constant; it slowly decreases. However, the steady level turning flight results remain valid since the aircraft weight changes slowly. Since the turn occurs in a horizontal plane, the air density is constant.

Following the development in section 11.4, modified for a steady level turn, the velocity along the circular path, denoted by the time rate of change of the circular arc length s, and the time rate of change of the aircraft weight, can be shown to be given by

$$\frac{ds}{dt} = \sqrt{\frac{2W}{\rho S C_L \cos \mu}}, \tag{11.44}$$

$$\frac{dW}{dt} = -\frac{c}{\eta} \frac{C_D}{C_L} \frac{1}{\cos \mu} \sqrt{\frac{2W}{\rho S C_L \cos \mu}} W. \tag{11.45}$$

The bank angle, the drag coefficient, and the lift coefficient are assumed constant throughout the steady level turn.

The range of the aircraft along the circular path defined by the steady level turn is the steady level turning range R. Following the development in section 11.4, the steady level turning range, assuming that the lift coefficient, the drag coefficient, and the flight altitude are maintained constant throughout the flight, is

$$R = \frac{\eta}{c} \left(\frac{C_L \cos \mu}{C_D} \right) \ln \frac{W_i}{W_f}. \tag{11.46}$$

This equation gives the steady level turn range in terms of the initial weight and the terminal weight of the aircraft including fuel and the bank angle. The steady level turn range decreases as the bank angle increases. This equation also shows that the aerodynamics flight condition that maximizes the steady level turn range is that the lift coefficient and the drag coefficient be chosen to maximize the aerodynamics ratio $\frac{C_L}{C_D}$.

The steady level turning flight endurance, assuming that the lift coefficient, the drag coefficient, and the bank angle are maintained constant, can be shown to be given by

$$E = \frac{\eta}{c} \left(\frac{(C_L \cos \mu)^{\frac{3}{2}}}{C_D} \right) \sqrt{2 \rho S} \left(\frac{1}{\sqrt{W_f}} - \frac{1}{\sqrt{W_i}} \right). \tag{11.47}$$

This equation gives the steady level turn endurance in terms of the initial weight and the terminal weight of the aircraft including fuel and the bank angle. The steady level turn endurance decreases as the bank angle increases. This equation also shows that the aerodynamics flight condition that maximizes the steady level turning endurance is that the

lift coefficient and the drag coefficient be chosen to maximize the aerodynamics ratio $\frac{C_L^{\frac{3}{2}}}{C_D}$.

11.7 Range and Endurance of a Jet Aircraft in a Steady Turn

We now present range and endurance expressions for a jet aircraft in a steady climbing or descending turn. Since the engine burns fuel, the weight of the aircraft including fuel is not constant; it slowly decreases. The steady turning flight results are valid since the aircraft weight changes slowly. Since the turn does not occur in a horizontal plane, the air density is not constant. The results in this section are restricted to the case that the altitude change is small so that the air density is suitably approximated by a constant value.

Following the development in section 11.3 modified for a steady climbing or descending turn, the velocity along the helical path, denoted by the time rate of change of the arc length s, and the time rate of change of the aircraft weight can be shown to be given by

$$\frac{ds}{dt} = \sqrt{\frac{2W}{\rho S C_L \cos \mu}}, \tag{11.48}$$

$$\frac{dW}{dt} = -c \left(\gamma + \frac{C_D}{C_L \cos \mu} \right) W. \tag{11.49}$$

The bank angle, the flight path angle, and the drag and lift coefficients are assumed constant throughout the steady climbing or descending turn.

The distance traveled by the aircraft along the helical path defined by the steady climbing or descending turn is the steady turning range R. Following the development in section 11.3, the steady turning range can be shown to be

$$R = \frac{2}{c} \sqrt{\frac{2}{\rho S}} \left(\frac{(C_L \cos \mu)^{\frac{1}{2}}}{\gamma C_L \cos \mu + C_D} \right) \left(\sqrt{W_i} - \sqrt{W_f} \right). \tag{11.50}$$

This equation gives the steady turning range in terms of the initial weight and the terminal weight of the aircraft including fuel and the bank angle and the flight path angle. This equation shows that the aerodynamics flight condition that maximizes the climbing or descending turning range is that the lift coefficient and the drag coefficient be chosen to maximize the ratio

$$\left(\frac{(C_L \cos \mu)^{\frac{1}{2}}}{\gamma C_L \cos \mu + C_D} \right).$$

This result shows that the lift coefficient and the drag coefficient that maximize the range depend on both the bank angle and the flight path angle.

The steady climbing or descending turn endurance, assuming that the bank angle, the flight path angle, the lift coefficient, and the drag coefficient are maintained constant, can be shown to be given by

$$E = \frac{1}{c} \left(\frac{C_L \cos \mu}{\gamma C_L \cos \mu + C_D} \right) \ln \frac{W_i}{W_f}. \tag{11.51}$$

This equation gives the steady turn endurance in terms of the initial weight and the terminal weight of the aircraft including fuel and the bank angle and flight path angle. The steady turn endurance decreases as the bank angle and the flight path angle increase. This equation also shows that the aerodynamics flight condition that maximizes the climbing or descending turn endurance is that the lift coefficient and the drag coefficient be chosen to maximize the ratio

$$\left(\frac{C_L \cos \mu}{\gamma C_L \cos \mu + C_D} \right).$$

This result shows that the lift coefficient and the drag coefficient to maximize the endurance do depend on both the bank angle and the flight path angle.

11.8 Range and Endurance of a General Aviation Aircraft in a Steady Turn

We present range and endurance expressions for a propeller-driven general aviation aircraft in a steady climbing or descending turn. Since the engine burns fuel, the weight of the aircraft including fuel is not constant; it slowly decreases. The steady turning flight results remain valid since the aircraft weight changes slowly. Since the turn does not occur in a horizontal plane, the air density is not constant. The results in this section are restricted to the case that the altitude change is small so that the air density is suitably approximated by a constant value.

Following the development in section 11.4 modified for a steady climbing or descending turn, the velocity along the helical path, denoted by the time rate of change of the helical arc length s, and the time rate of change of the aircraft weight can be shown to be given by

$$\frac{ds}{dt} = \sqrt{\frac{2W}{\rho S C_L \cos \mu}}, \tag{11.52}$$

$$\frac{dW}{dt} = -\frac{c}{\eta} \left(\gamma + \frac{C_D}{C_L \cos \mu} \right) \sqrt{\frac{2W}{\rho S C_L \cos \mu}} \, W. \tag{11.53}$$

The bank angle, the flight path angle, and the drag and lift coefficients are assumed constant throughout the steady turn.

The range of the aircraft along the helical path defined by the steady climbing or descending turn is the steady turning range R. Following the development in section 11.4 for a steady climbing or descending turn, the steady turning range can be shown to be given by

$$R = \frac{\eta}{c} \left(\frac{C_L \cos \mu}{\gamma C_L \cos \mu + C_D} \right) \ln \frac{W_i}{W_f}. \tag{11.54}$$

This equation gives the steady turning range in terms of the initial weight and the terminal weight of the aircraft including fuel and the bank angle and the flight path angle. The steady turning range decreases as the bank angle and the flight path angle increase. This equation also shows that the aerodynamics flight condition that maximizes the climbing or descending turn range is that the lift coefficient and the drag coefficient be chosen to maximize the ratio

$$\left(\frac{C_L \cos \mu}{\gamma C_L \cos \mu + C_D}\right).$$

This result shows that the lift coefficient and the drag coefficient that maximize the range depend on both the bank angle and the flight path angle.

The endurance, assuming the general aviation aircraft is in a steady climbing or descending turn, can be shown to be given by

$$E = \frac{\eta}{c}\left(\frac{(C_L \cos \mu)^{\frac{3}{2}}}{\gamma C_L \cos \mu + C_D}\right)\sqrt{2\rho S}\left(\frac{1}{\sqrt{W_f}} - \frac{1}{\sqrt{W_i}}\right). \qquad (11.55)$$

This equation gives the steady turn endurance in terms of the initial weight and the terminal weight of the aircraft including fuel and the bank angle and the flight path angle. The steady turn endurance decreases as the bank angle and the flight path angle increase. This equation also shows that the aerodynamics flight condition that maximizes the climbing or descending turn endurance is that the lift coefficient and the drag coefficient be chosen to maximize the ratio

$$\left(\frac{(C_L \cos \mu)^{\frac{3}{2}}}{\gamma C_L \cos \mu + C_D}\right).$$

This result shows that the lift coefficient and the drag coefficient that maximize the endurance depend on both the bank angle and the flight path angle.

11.9 Maximum Range and Maximum Endurance: Executive Jet Aircraft

Maximum range and maximum endurance for the executive jet aircraft are analyzed. Steady flight conditions are determined that correspond to maximum range and to maximum endurance.

Maximum Range in Steady Level Longitudinal Flight

Suppose that the executive jet aircraft is flying at an altitude of 35,000 ft to achieve maximum range. From the results of section 11.3, the range of a jet aircraft is maximized when the aerodynamics ratio $\frac{C_L^{\frac{1}{2}}}{C_D}$ is maximized. The corresponding values of C_L, C_D, and $\frac{C_L^{\frac{1}{2}}}{C_D}$ are

$$C_L = \sqrt{\frac{C_{D_0}}{3K}} = 0.316,$$

$$C_D = \frac{4}{3}C_{D_0} = 0.02,$$

$$\left(\frac{C_L^{\frac{1}{2}}}{C_D}\right) = \frac{3}{4}\left(\frac{1}{3KC_{D_0}^3}\right)^{\frac{1}{4}} = 28.12.$$

These results describe the aerodynamics that characterize the maximum range of the executive jet aircraft in flight at an altitude of 35,000 ft. The corresponding airspeed, for this value of lift coefficient, can be written as a function of the aircraft weight as

$$V = \sqrt{\frac{2W}{\rho S}\sqrt{\frac{3K}{C_{D_0}}}}.$$

This analysis shows that the airspeed of the executive jet aircraft to achieve maximum range depends on the aircraft weight: as fuel is burned and the weight of the aircraft slowly decreases, the airspeed of the aircraft must be slowly reduced according to this relationship. The required thrust to maximize the range is

$$T = 4W\sqrt{\frac{KC_{D_0}}{3}}.$$

The required throttle setting to maximize the range is given by

$$\delta_t = 4W\sqrt{\frac{KC_{D_0}}{3}}\frac{1}{T_{\max}^s}\left(\frac{\rho^s}{\rho}\right)^m.$$

This analysis shows that the throttle setting to maximize the executive jet aircraft range depends on the aircraft weight: as fuel is burned and the weight of the aircraft slowly decreases, the throttle setting must be slowly reduced according to this relationship.

Graphical plots showing the variations of airspeed and throttle with aircraft weight are obtained from the following Matlab m-file.

Matlab function 11.1 MRSLFJ.m
```
S=950; CD0=0.015; K=0.05; Tsmax=12500; m=0.6; h=35000;
[Ts ps rhos]=StdAtpUS(0);
[T p rho]=StdAtpUS(h);
Wi=73000; Wf=Wi-28000;
W=linspace(Wi,Wf,200);
V=sqrt(2*W/rho/S*sqrt(3*K/CD0));
```

(continued)

(continued)

```
sigma=4*W*sqrt(K*CD0/3)/(Tsmax*(rho/rhos)^m);
figure;
plot(W,V);
xlim([Wf Wi]);
xlabel('Weight (lbs)');
ylabel('Velocity (ft/s)');
figure;
plot(W,sigma);
xlim([Wf Wi]);
xlabel('Weight (lbs)');
ylabel('Throttle setting');
```

The results of these computations are shown in Figure 11.1.

Assuming these flight conditions that maximize the aircraft range at an altitude of 35,000 ft are satisfied, the value of the maximum range is now computed. The thrust specific fuel consumption rate is $c = 0.69/\text{hr} = 1.917 \times 10^{-4}/\text{s}$, the initial weight of the aircraft, including a full fuel tank, is $W_i = 73,000$ lbs, the terminal weight of the aircraft, with an empty fuel tank, is $W_f = 45,000$ lbs, and the optimal aerodynamics ratio for maximum range flight of the executive jet aircraft is $\frac{C_L^{\frac{1}{2}}}{C_D} = 28.12$. The air density is $7.3654 \times 10^{-4}\text{slug}/\text{ft}^3$ at an altitude of 35,000 ft.

The maximum range of the executive jet aircraft, in steady level cruise, is

$$R = \frac{2}{c}\sqrt{\frac{2}{\rho S}\frac{C_L^{\frac{1}{2}}}{C_D}}\left[\sqrt{W_i} - \sqrt{W_f}\right] = 2.8796 \times 10^7 \text{ ft} = 5,453.8 \text{ miles}.$$

This analysis has developed the flight conditions at an altitude of 35,000 ft that achieve maximum range of 5,453.8 miles for the executive jet aircraft. The optimal flight conditions have been determined and the maximum range has been computed. The corresponding time of flight is computed as

$$E = \frac{1}{c}\frac{C_L}{C_D}\ln\frac{W_i}{W_f} = 39,874 \text{ s} = 11 \text{ hr, } 4 \text{ min, } 34 \text{ s}.$$

Maximum Endurance in Steady Level Longitudinal Flight

Suppose that the executive jet aircraft is flying at an altitude of 35,000 ft to achieve maximum endurance. From the results of section 11.3, the endurance of a jet aircraft is maximized when the aerodynamics ratio $\frac{C_L}{C_D}$ is maximized. The corresponding optimal

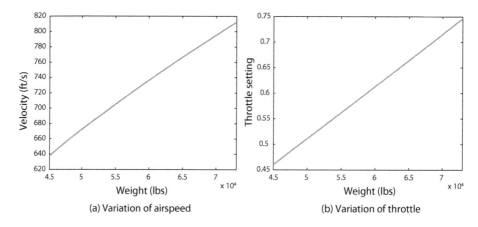

(a) Variation of airspeed   (b) Variation of throttle

Figure 11.1. *Maximum range steady level longitudinal flight: executive jet aircraft.*

values of C_L, C_D, and $\frac{C_L}{C_D}$ are given by

$$C_L = \sqrt{\frac{C_{D_0}}{K}} = 0.548,$$

$$C_D = 2C_{D_0} = 0.03,$$

$$\left(\frac{C_L}{C_D}\right) = \frac{1}{2}\sqrt{\frac{1}{KC_{D_0}}} = 18.26.$$

The airspeed that maximizes the aircraft endurance can be written as a function of the aircraft weight as

$$V = \sqrt{\frac{2W}{\rho S}\sqrt{\frac{K}{C_{D_0}}}}.$$

This analysis shows that the airspeed of the aircraft that maximizes the endurance depends on the aircraft weight: as fuel is burned and the weight of the aircraft slowly decreases, the airspeed of the aircraft must be slowly reduced according to this relationship. The required thrust that maximizes the endurance is

$$T = 2W\sqrt{KC_{D_0}}.$$

The throttle setting is given by

$$\delta_t = 2W \frac{\sqrt{KC_{D_0}}}{T^s_{\max}} \left(\frac{\rho^s}{\rho} \right)^m.$$

This analysis shows that the throttle setting that maximizes the endurance depends on the aircraft weight: as fuel is burned and the weight of the aircraft slowly decreases, the airspeed of the aircraft must be slowly reduced according to this relationship.

Graphical plots of the variations of the airspeed and throttle setting that maximize the endurance as the weight of the aircraft changes are obtained using the following Matlab commands. The computed results are shown in Figure 11.2.

Matlab function 11.2 MESLFJ.m

```
S=950; CD0=0.015; K=0.05; Tsmax=12500; m=0.6; h=35000;
[Ts ps rhos]=StdAtpUS(0);
[T p rho]=StdAtpUS(h);
Wi=73000; Wf=Wi-28000;
W=linspace(Wi,Wf,200);
V=sqrt(2*W/rho/S*sqrt(K/CD0));
sigma=2*W*sqrt(K*CD0)/(Tsmax*(rho/rhos)^m);
figure;
plot(W,V);
xlim([Wf Wi]);
xlabel('Weight (lbs)');
ylabel('Velocity (ft/s)');
figure;
plot(W,sigma);
xlim([Wf Wi]);
xlabel('Weight (lbs)');
ylabel('Throttle setting');
```

Assuming these flight conditions to maximize the aircraft endurance at an altitude of 35,000 ft are satisfied, the value of the maximum endurance is now computed. This computation is based on values of the relevant constant parameters: the thrust specific fuel consumption rate $c = 0.69/\text{hr} = 1.917 \times 10^{-4}/\text{s}$, the initial weight of the aircraft, including a full fuel tank, $W_i = 73{,}000$ lbs, the terminal weight of the aircraft, with an empty fuel tank, $W_f = 45{,}000$ lbs, and the optimal value of the aerodynamics ratio $\frac{C_L}{C_D} = 18.26$.

The maximum endurance of the executive jet aircraft, in steady level cruise, is

$$E = \frac{1}{c} \frac{C_L}{C_D} \ln \frac{W_i}{W_f} = 4.6085 \times 10^4 \, \text{s} = 12 \, \text{hr}, 48 \, \text{min}, 4.6 \, \text{s}.$$

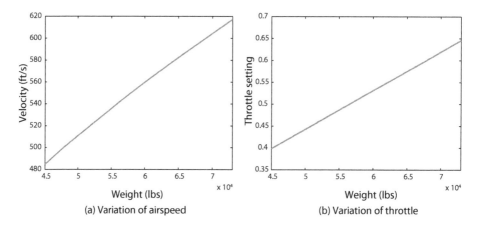

Figure 11.2. Maximum endurance steady level longitudinal flight: executive jet aircraft.

The corresponding aircraft range is computed as

$$R = \frac{2}{c}\sqrt{\frac{2}{\rho S}\frac{C_L^{\frac{1}{2}}}{C_D}}\left[\sqrt{W_i} - \sqrt{W_f}\right] = 2.5272 \times 10^7 \text{ ft} = 4{,}786.3 \text{ miles}.$$

Maximum Range in Steady Level Turning Flight

Suppose that the executive jet aircraft is flying at an altitude of 10,000 ft in a steady turn with a bank angle of 30 degrees. It is desired to maximize the steady level turning range for a steady flight segment, assuming the initial aircraft weight is 52,000 lbs and the terminal aircraft weight is 51,600 lbs; that is, 400 lbs of fuel are consumed.

The steady level turn range is maximum if the lift coefficient and the drag coefficient maximize $\frac{C_L^{\frac{1}{2}}}{C_D}$, which gives $C_L = 0.316$ and $C_D = 0.02$. The flight altitude is maintained constant throughout.

The maximum range for this flight segment of the executive jet aircraft is

$$R = \frac{2}{c}\sqrt{\frac{2}{\rho S}\frac{C_L^{\frac{1}{2}}}{C_D}}\left(\cos \mu\right)^{\frac{1}{2}}\left(\sqrt{W_i} - \sqrt{W_f}\right) = 2.628 \times 10^5 \text{ ft} = 49.7 \text{ miles},$$

and the turn radius, based on the initial weight of the aircraft and fuel, is

$$r = \frac{2}{\rho g C_L}\frac{1}{\sin \mu}\frac{W_i}{S} = 21{,}2580 \text{ ft}.$$

This corresponds to almost two complete circular revolutions in the horizontal plane.

Maximum Endurance in Steady Level Turning Flight

Suppose that the executive jet aircraft is flying at an altitude of 10,000 ft in a steady turn with a bank angle of 30 degrees. It is desired to maximize the steady level turn endurance for a steady flight segment, assuming the initial aircraft weight is 52,000 lbs and the terminal aircraft weight is 51,600 lbs; that is, 400 lbs of fuel are consumed.

The steady level turn endurance is maximum if the lift coefficient and the drag coefficient maximize $\frac{C_L}{C_D}$, which gives $C_L = 0.548$ and $C_D = 0.03$. The flight altitude is maintained constant throughout the flight.

The maximum endurance for this flight segment of the executive jet aircraft is

$$E = \frac{1}{c}\frac{C_L}{C_D}\ln\frac{W_i}{W_f} = 735.8\,\text{s} = 12\,\text{min},\ 15.8\,\text{s},$$

and the turn radius, based on the initial weight of the aircraft and fuel, is

$$r = \frac{2}{\rho g C_L}\frac{1}{\sin\mu}\frac{W_i}{S} = 12{,}240\,\text{ft}.$$

11.10 Maximum Range and Maximum Endurance: General Aviation Aircraft

Maximum range and maximum endurance for the propeller-driven general aviation aircraft are analyzed. Steady flight conditions are determined that correspond to maximum range and to maximum endurance.

Maximum Range in Steady Level Longitudinal Flight

Suppose that the propeller-driven general aviation aircraft is flying at an altitude of 20,000 ft to achieve maximum range. From the results of section 11.4, the range of the aircraft is maximized when the aerodynamic ratio $\frac{C_L}{C_D}$ is maximized. The corresponding values of C_L, C_D, and $\frac{C_L}{C_D}$ are

$$C_L = \sqrt{\frac{C_{D_0}}{K}} = 0.694,$$

$$C_D = 2C_{D_0} = 0.052,$$

$$\left(\frac{C_L}{C_D}\right) = \frac{1}{2}\sqrt{\frac{1}{KC_{D_0}}} = 13.344.$$

These results describe the aerodynamics that characterize the maximum range of the general aviation aircraft in flight at an altitude of 20,000 ft. The corresponding airspeed, for this value of the lift coefficient, can be written as a function of the aircraft weight as

$$V = \sqrt{\frac{2W}{\rho S} \sqrt{\frac{K}{C_{D_0}}}}.$$

This analysis shows that the airspeed of the aircraft to achieve maximum range depends on the aircraft weight: as fuel is burned and the weight of the aircraft slowly decreases, the airspeed of the aircraft must be slowly reduced according to this relationship.

The required power to maximize the range is

$$P = 4\sqrt{\frac{2W^3}{3\rho S} \sqrt{K^3 C_{D_0}}},$$

so that the required throttle setting to maximize the range is given by

$$\delta_t = 4\sqrt{\frac{2W^3}{3\rho S} \sqrt{K^3 C_{D_0}}} \frac{1}{\eta P_{\max}^s} \left(\frac{\rho^s}{\rho}\right)^m.$$

This analysis shows that the throttle setting to maximize the general aviation aircraft range depends on the aircraft weight: as fuel is burned and the weight of the aircraft slowly decreases, the throttle setting must be slowly reduced according to this relationship.

Graphical plots showing the variations of airspeed and throttle with aircraft weight are obtained from the following Matlab m-file.

Matlab function 11.3 MRSLFP.m
```
S=175; CD0=0.026; K=0.054; eta=0.8; Psmax=290*550; m=0.6; h=20000;
[Ts ps rhos]=StdAtpUS(0);
[T p rho]=StdAtpUS(h);
Wi=2900; Wf=Wi-370;
W=linspace(Wi,Wf,200);
V=sqrt(2*W/rho/S*sqrt(K/CD0));
P=4*sqrt(2*W.^3/3/rho/S*sqrt(K^3*CD0));
sigma=P/(eta*Psmax*(rho/rhos)^m);
figure;
plot(W,V);
xlim([Wf Wi]);
xlabel('Weight (lbs)');
ylabel('Velocity (ft/s)');
figure;
plot(W,sigma);
xlim([Wf Wi]);
xlabel('Weight (lbs)');
ylabel('Throttle setting');
```

The results are shown in Figure 11.3.

Assuming these flight conditions that maximize the aircraft range at an altitude of 20,000 ft are satisfied, the value of the maximum range is now computed. The power specific fuel consumption rate is $c = 0.45 \frac{\text{lbs}}{\text{hr hp}} = 2.273 \times 10^{-7}$ 1/ft, the initial weight

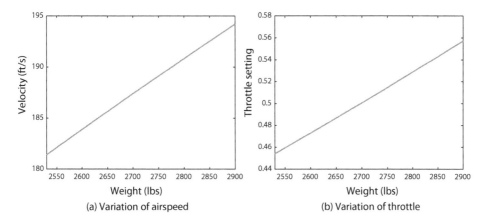

Figure 11.3. *Maximum range steady level longitudinal flight: general aviation aircraft.*

of the aircraft, including a full fuel tank, is $W_i = 2{,}900$ lbs, the terminal weight of the aircraft, with an empty fuel tank, is $W_f = 2{,}530$ lbs, and the optimal aerodynamics ratio for maximum range is $\frac{C_L}{C_D} = 13.3440$. The air density is 1.2664×10^{-3} slug/ft^3 at an altitude of 20,000 ft.

The maximum aircraft range of the general aviation aircraft is

$$R = \frac{\eta}{c}\frac{C_L}{C_D}\ln\frac{W_i}{W_f} = 6.411\times10^6 \text{ ft} = 1{,}214.2 \text{ miles.}$$

This analysis has developed the flight conditions at an altitude of 20,000 ft that achieve maximum range of 1,214.2 miles for the general aviation aircraft. The optimal flight conditions have been determined. The corresponding time of flight of the aircraft in this flight condition is

$$E = \frac{\eta}{c}\frac{C_L^{\frac{3}{2}}}{C_D}\sqrt{2\rho S}\left(\frac{1}{\sqrt{W_f}} - \frac{1}{\sqrt{W_i}}\right) = 22{,}413 \text{ s} = 6 \text{ hr, } 13 \text{ min, } 33 \text{ s.}$$

Maximum Endurance in Steady Level Longitudinal Flight

Suppose that the propeller-driven general aviation aircraft is flying at an altitude of 20,000 ft to achieve maximum endurance. From the results in section 11.4, the endurance of the aircraft is maximized when the aerodynamics ratio $\frac{C_L^{\frac{3}{2}}}{C_D}$ is maximized.

The corresponding values of C_L, C_D, and $\frac{C_L^{\frac{3}{2}}}{C_D}$ are given by

$$C_L = \sqrt{\frac{3C_{D_0}}{K}} = 1.2,$$

$$C_D = 4C_{D_0} = 0.104,$$

$$\left(\frac{C_L^{\frac{3}{2}}}{C_D}\right) = \frac{1}{4}\left(\frac{3}{KC_{D_0}^{\frac{1}{3}}}\right)^{\frac{3}{4}} = 12.67.$$

The airspeed that maximizes the aircraft endurance can be written as a function of the aircraft weight as

$$V = \sqrt{\frac{2W}{\rho S}\sqrt{\frac{K}{3C_{D_0}}}}.$$

This analysis shows that the airspeed of the aircraft that maximizes the endurance depends on the aircraft weight: as fuel is burned and the weight of the aircraft slowly decreases, the airspeed of the aircraft must be slowly reduced according to this relationship.

The required power that maximizes the endurance is

$$P = 4\sqrt{\frac{2W^3}{3\rho S}\sqrt{\frac{K^3 C_{D_0}}{3}}},$$

so that the throttle setting is given by

$$\delta_t = 4\sqrt{\frac{2W^3}{3\rho S}\sqrt{\frac{K^3 C_{D_0}}{3}}}\frac{1}{\eta P_{\max}^s}\left(\frac{\rho^s}{\rho}\right)^m.$$

This analysis shows that the throttle setting that maximizes the endurance depends on the aircraft weight: as fuel is burned and the weight of the aircraft slowly decreases, the throttle setting of the aircraft must be slowly reduced according to this relationship.

Graphical plots of the variations of the airspeed and throttle setting that maximize the endurance as the weight of the aircraft changes are obtained using the following Matlab m-file.

Matlab function 11.4 MESLFP.m
```
S=175; CD0-0.026; K = 0.054; eta=0.8; Psmax=290*550; m=0.6; h=20000;
[Ts ps rhos]=StdAtpUS(0);
[T p rho]=StdAtpUS(h);
Wi=2900; Wf=Wi-370;
W=linspace(Wi,Wf,200);
V=sqrt(2*W/rho/S*sqrt(K/3/CD0));
P=4*sqrt(2*W.^3/3/rho/S*sqrt(K^3*CD0/3));
ts=P/(eta*Psmax*(rho/rhos)^m);
```

(continued)

(continued)

```
figure;
plot(W,V);
xlim([Wf Wi]);
xlabel('Weight (lbs)');
ylabel('Velocity (ft/s)');
figure;
plot(W,ts);
xlim([Wf Wi]);
xlabel('Weight (lbs)');
ylabel('Throttle setting');
```

The results are shown in Figure 11.4.

Assuming the flight conditions to maximize the endurance at an altitude of 20,000 ft are satisfied, the value of the maximum endurance is now computed. This computation is based on values of the relevant constant parameters: the power specific fuel consumption rate $c = 0.45 \frac{\text{lbs}}{\text{hr hp}} = 2.273 \times 10^{-7} \, 1/\text{ft}$, the propeller efficiency $\eta = 0.8$, the initial weight of the aircraft, including a full fuel tank, $W_i = 2{,}900 \, \text{lbs}$, the terminal weight of the aircraft, with an empty fuel tank, $W_f = 2{,}530 \, \text{lbs}$, and the optimal value of the aerodynamics ratio $\frac{C_L^{\frac{3}{2}}}{C_D} = 12.67$.

The maximum endurance of the general aviation aircraft is

$$E = \frac{\eta}{c} \frac{C_L^{\frac{3}{2}}}{C_D} \sqrt{2\rho S} \left(\frac{1}{\sqrt{W_f}} - \frac{1}{\sqrt{W_i}} \right) = 25{,}395 \, \text{s} = 7 \, \text{hr}, \ 3 \, \text{min}, \ 15.3 \, \text{s}.$$

This analysis has developed the flight conditions at an altitude of 20,000 ft that achieve maximum endurance of 7 hr, 3 min, 15.3 s. The optimal flight conditions have been determined. The corresponding range of the aircraft in this flight condition is computed as

$$R = \frac{\eta}{c} \frac{C_L}{C_D} \ln \frac{W_i}{W_f} = 5.543 \times 10^6 \, \text{ft} = 1{,}049.8 \, \text{miles}.$$

Maximum Range in Steady Level Turning Flight

Suppose that the propeller-driven general aviation aircraft is flying at an altitude of 10,000 ft in a steady level turn with a bank angle of 30 degrees. It is desired to maximize the steady level turn range of a flight segment, assuming the initial aircraft weight is 2,650 lbs and the terminal aircraft weight is 2,640 lbs; that is, 10 lbs of fuel are consumed.

The steady level turn range is maximum if the lift coefficient and the drag coefficient maximize $\frac{C_L}{C_D}$, which gives $C_L = 0.694$ and $C_D = 0.052$. The flight altitude is maintained constant throughout the flight. The maximum range of this flight segment of the general aviation aircraft is

$$R = \frac{\eta}{c} \frac{C_L}{C_D} \cos \mu \ln \frac{W_i}{W_f} = 1.5419 \times 10^5 \, \text{ft} = 29.2 \, \text{miles}$$

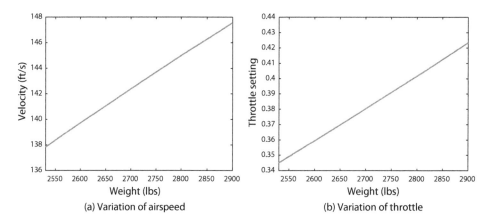

Figure 11.4. *Maximum endurance steady level longitudinal flight: general aviation aircraft.*

and the turn radius, based on the initial weight of aircraft and fuel, is

$$r = \frac{2}{\rho g C_L} \frac{1}{\sin \mu} \frac{W_i}{S} = 2{,}675 \text{ ft},$$

so that slightly more than nine complete revolutions on the level circular path can be completed.

Maximum Endurance in Steady Level Turning Flight

Suppose that the propeller-driven general aviation aircraft is flying at an altitude of 10,000 ft in a steady level turn with a bank angle of 30 degrees. It is desired to maximize the steady level turn endurance for a flight segment, assuming the initial aircraft weight is 2,650 lbs and the terminal aircraft weight is 2,640 lbs; that is, 10 lbs of fuel are consumed. The steady level turn endurance is maximum if the lift coefficient and the drag coefficient maximize $\frac{C_L^{\frac{3}{2}}}{C_D}$, which gives $C_L = 1.2$ and $C_D = 0.104$. The flight altitude is maintained constant throughout the flight. The maximum endurance for this flight segment of the general aviation aircraft is

$$E = \frac{\eta}{c} \frac{(C_L \cos \mu)^{\frac{3}{2}}}{C_D} \sqrt{2\rho S} \left(\frac{1}{\sqrt{W_f}} - \frac{1}{\sqrt{W_i}} \right) = 1{,}282 \text{ s} = 21 \text{ min}, \ 21.6 \text{ s}.$$

The turn radius, based on the initial weight of aircraft and fuel, is

$$r = \frac{2}{\rho g C_L} \frac{1}{\sin \mu} \frac{W_i}{S} = 1{,}547 \text{ ft}.$$

11.11 Conclusions

This chapter has developed formulas for range and endurance for aircraft in steady flight, thereby obtaining conditions for maximization of the range and conditions for maximization of endurance. These are extremely important performance measures for many aircraft, especially for steady level longitudinal flight, and they are often used as a basis for preliminary aircraft design. The development has been applied to determine the range and endurance properties for the executive jet aircraft and for the propeller-driven general aviation aircraft.

11.12 Problems

11.1. Consider the executive jet aircraft in steady level longitudinal flight with a constant lift coefficient of 0.42.

   (a) What is the range of the aircraft in flight at an altitude of 5,000 ft? What are the initial airspeed, elevator deflection, and throttle setting with a full fuel tank? What are the terminal airspeed, elevator deflection, and throttle setting when the fuel tank is empty?

   (b) What is the range of the aircraft in flight at an altitude of 31,000 ft? What are the initial airspeed, elevator deflection, and throttle setting with a full fuel tank? What are the terminal airspeed, elevator deflection, and throttle setting when the fuel tank is empty?

   (c) What is the endurance of the aircraft in flight at an altitude of 5,000 ft? What are the initial airspeed, elevator deflection, and throttle setting with a full fuel tank? What are the terminal airspeed, elevator deflection, and throttle setting when the fuel tank is empty?

   (d) What is the endurance of the aircraft in flight at an altitude of 31,000 ft? What are the initial airspeed, elevator deflection, and throttle setting with a full fuel tank? What are the terminal airspeed, elevator deflection, and throttle setting when the fuel tank is empty?

11.2. Consider the propeller-driven general aviation aircraft in steady level longitudinal flight with a constant lift coefficient of 0.95.

   (a) What is the range of the aircraft in flight at an altitude of 2,000 ft? What are the initial airspeed, elevator deflection, and throttle setting with a full fuel tank? What are the terminal airspeed, elevator deflection, and throttle setting when the fuel tank is empty?

   (b) What is the range of the aircraft in flight at an altitude of 20,000 ft? What are the initial airspeed, elevator deflection, and throttle setting with a full fuel tank? What are the terminal airspeed, elevator deflection, and throttle setting when the fuel tank is empty?

   (c) What is the endurance of the aircraft in flight at an altitude of 2,000 ft? What are the initial airspeed, elevator deflection, and throttle setting with a full fuel tank? What are the terminal airspeed, elevator deflection, and throttle setting when the fuel tank is empty?

(d) What is the endurance of the aircraft in flight at an altitude of 20,000 ft? What are the initial airspeed, elevator deflection, and throttle setting with a full fuel tank? What are the terminal airspeed, elevator deflection, and throttle setting when the fuel tank is empty?

11.3. Consider the UAV in steady level longitudinal flight with a constant lift coefficient of 0.7.

(a) What is the range of the aircraft in flight at an altitude of 2,000 ft? What are the initial airspeed, elevator deflection, and throttle setting with a full fuel tank? What are the terminal airspeed, elevator deflection, and throttle setting when the fuel tank is empty?

(b) What is the range of the aircraft in flight at an altitude of 10,000 ft? What are the initial airspeed, elevator deflection, and throttle setting with a full fuel tank? What are the terminal airspeed, elevator deflection, and throttle setting when the fuel tank is empty?

(c) What is the endurance of the aircraft in flight at an altitude of 2,000 ft? What are the initial airspeed, elevator deflection, and throttle setting with a full fuel tank? What are the terminal airspeed, elevator deflection, and throttle setting when the fuel tank is empty?

(d) What is the endurance of the aircraft in flight at an altitude of 10,000 ft? What are the initial airspeed, elevator deflection, and throttle setting with a full fuel tank? What are the terminal airspeed, elevator deflection, and throttle setting when the fuel tank is empty?

11.4. Consider the executive jet aircraft in steady level longitudinal flight.

(a) What is the maximum range of the aircraft in flight at an altitude of 5,000 ft? What are the initial airspeed, elevator deflection, and throttle setting with a full fuel tank? What are the terminal airspeed, elevator deflection, and throttle setting when the fuel tank is empty? What is the flight time required to complete this maximum range flight?

(b) What is the maximum range of the aircraft in flight at an altitude of 31,000 ft? What are the initial airspeed, elevator deflection, and throttle setting with a full fuel tank? What are the terminal airspeed, elevator deflection, and throttle setting when the fuel tank is empty? What is the flight time required to complete this maximum range flight?

(c) What is the maximum endurance of the aircraft in flight at an altitude of 5,000 ft? What are the initial airspeed, elevator deflection, and throttle setting with a full fuel tank? What are the terminal airspeed, elevator deflection, and throttle setting when the fuel tank is empty? What is the range covered by this maximum endurance flight?

(d) What is the maximum endurance of the aircraft in flight at an altitude of 31,000 ft? What are the initial airspeed, elevator deflection, and throttle setting with a full fuel tank? What are the terminal airspeed, elevator deflection, and throttle setting when the fuel tank is empty? What is the range covered during this maximum endurance flight?

11.5. Consider the propeller-driven general aviation aircraft in steady level longitudinal flight.

    (a) What is the maximum range of the aircraft in flight at an altitude of 2,000 ft? What are the initial airspeed, elevator deflection, and throttle setting with a full fuel tank? What are the terminal airspeed, elevator deflection, and throttle setting when the fuel tank is empty? What is the flight time required to complete this maximum range flight?

    (b) What is the maximum range of the aircraft in flight at an altitude of 20,000 ft? What are the initial airspeed, elevator deflection, and throttle setting with a full fuel tank? What are the terminal airspeed, elevator deflection, and throttle setting when the fuel tank is empty? What is the flight time required to complete this maximum range flight?

    (c) What is the maximum endurance of the aircraft in flight at an altitude of 2,000 ft? What are the initial airspeed, elevator deflection, and throttle setting with a full fuel tank? What are the terminal airspeed, elevator deflection, and throttle setting when the fuel tank is empty? What is the range covered during this maximum endurance flight?

    (d) What is the maximum endurance of the aircraft in flight at an altitude of 20,000 ft? What are the initial airspeed, elevator deflection, and throttle setting with a full fuel tank? What are the terminal airspeed, elevator deflection, and throttle setting when the fuel tank is empty? What is the range covered during this maximum endurance flight?

11.6. Consider the UAV in steady level longitudinal flight.

    (a) What is the maximum range of the aircraft in flight at an altitude of 1,000 ft? What are the initial airspeed, elevator deflection, and throttle setting with a full fuel tank? What are the terminal airspeed, elevator deflection, and throttle setting when the fuel tank is empty? What is the flight time required to complete this maximum range flight?

    (b) What is the maximum range of the aircraft in flight at an altitude of 10,000 ft? What are the initial airspeed, elevator deflection, and throttle setting with a full fuel tank? What are the terminal airspeed, elevator deflection, and throttle setting when the fuel tank is empty? What is the flight time required to complete this maximum range flight?

    (c) What is the maximum endurance of the aircraft in flight at an altitude of 1,000 ft? What are the initial airspeed, elevator deflection, and throttle setting with a full fuel tank? What are the terminal airspeed, elevator deflection, and throttle setting when the fuel tank is empty? What is the range covered during this maximum endurance flight?

    (d) What is the maximum endurance of the aircraft in flight at an altitude of 10,000 ft? What are the initial airspeed, elevator deflection, and throttle setting with a full fuel tank? What are the terminal airspeed, elevator deflection, and throttle setting when the fuel tank is empty? What is the range covered during this maximum endurance flight?

11.7. Consider the executive jet aircraft in steady level longitudinal flight at a constant altitude of 35,000 ft, with constant but not optimal values of the lift coefficient and the drag coefficient.

    (a) Suppose that the range in this flight condition is 4,900 miles. What are the values of the lift coefficient and the drag coefficient? What is the corresponding time of flight?

    (b) Suppose that the endurance in this flight condition is 43,100 s. What are the values the lift coefficient and the drag coefficient? What is the corresponding range?

11.8. Consider the general aviation aircraft in steady level longitudinal flight at a constant altitude of 20,000 ft, with constant but not optimal values of the lift coefficient and the drag coefficient.

    (a) Suppose that the range in this flight condition is 1,050 miles. What are the values of the lift coefficient and the drag coefficient? What is the corresponding time of flight?

    (b) Suppose that the endurance in this flight condition is 23,200 s. What are the values of the lift coefficient and the drag coefficient? What is the corresponding range?

11.9. Consider the executive jet aircraft in steady level turning flight maneuver with a bank angle of 45 degrees. Assume the initial weight at the beginning of the turning maneuver is 65,000 lbs and the terminal weight at the end of the turning maneuver is 64,700 lbs.

    (a) What is the maximum distance traveled by the aircraft in turning flight at an altitude of 5,000 ft? What are the airspeed, elevator deflection, and throttle setting during this maneuver? What is the flight time required to complete this turning flight maneuver?

    (b) What is the maximum distance traveled by the aircraft in turning flight at an altitude of 31,000 ft? What are the airspeed, elevator deflection, and throttle setting during this maneuver? What is the flight time required to complete this turning flight maneuver?

    (c) What is the maximum time of flight of the aircraft in turning flight at an altitude of 5,000 ft? What are the airspeed, elevator deflection, and throttle setting during this maneuver? What is the distance traveled during this turning flight maneuver?

    (d) What is the maximum time of flight of the aircraft in turning flight at an altitude of 31,000 ft? What are the airspeed, elevator deflection, and throttle setting during this maneuver? What is the distance traveled during this turning flight maneuver?

11.10. Consider the propeller-driven general aviation aircraft in steady level turning flight maneuver with a bank angle of 45 degrees. Assume the initial weight at the beginning of the turning maneuver is 2,700 lbs and the terminal weight at the end of the turning maneuver is 2,675 lbs.

(a) What is the maximum distance traveled by the aircraft in turning flight at an altitude of 2,000 ft? What are the airspeed, elevator deflection, and throttle setting during this maneuver? What is the flight time required to complete this turning flight maneuver?

(b) What is the maximum distance traveled by the aircraft in turning flight at an altitude of 20,000 ft? What are the airspeed, elevator deflection, and throttle setting during this maneuver? What is the flight time required to complete this turning flight maneuver?

(c) What is the maximum time of flight of the aircraft in turning flight at an altitude of 2,000 ft? What are the airspeed, elevator deflection, and throttle setting during this maneuver? What is the distance traveled during this turning flight maneuver?

(d) What is the maximum time of flight of the aircraft in turning flight at an altitude of 20,000 ft? What are the airspeed, elevator deflection, and throttle setting during this maneuver? What is the distance traveled during this turning flight maneuver?

11.11. Consider the UAV in steady level turning flight maneuver with a bank angle of 45 degrees. Assume the initial weight at the beginning of the turning maneuver is 40 lbs and the terminal weight at the end of the turning maneuver is 39 lbs.

(a) What is the maximum distance traveled by the aircraft in turning flight at an altitude of 1,000 ft? What are the airspeed, elevator deflection, and throttle setting during this maneuver? What is the flight time required to complete this turning flight maneuver?

(b) What is the maximum distance traveled by the aircraft in turning flight at an altitude of 10,000 ft? What are the airspeed, elevator deflection, and throttle setting during this maneuver? What is the flight time required to complete this turning flight maneuver?

(c) What is the maximum time of flight of the aircraft in turning flight at an altitude of 1,000 ft? What are the airspeed, elevator deflection, and throttle setting during this maneuver? What is the distance traveled during this turning flight maneuver?

(d) What is the maximum time of flight of the aircraft in turning flight at an altitude of 10,000 ft? What are the airspeed, elevator deflection, and throttle setting during this maneuver? What is the distance traveled during this turning flight maneuver?

11.12. Consider the executive jet aircraft in a steady climbing turning flight maneuver with a bank angle of 45 degrees and a known constant value of the flight path angle.

(a) Plot the value of the lift coefficient as a function of the flight path angle, assuming that for each flight path angle the lift coefficient is selected to maximize the steady climbing and turning range of the aircraft. Also, plot the value of the aerodynamic factor that appears in the steady climbing and turning range formula as a function of the flight path angle. The flight path angle in your plots should vary from -20 degrees to $+20$ degrees.

(b) Plot the value of the lift coefficient as a function of the flight path angle, assuming that for each flight path angle the lift coefficient is selected to maximize the steady climbing and turning endurance of the aircraft. Also, plot the value of the aerodynamic factor that appears in the steady climbing and turning endurance formula as a function of the flight path angle. The flight path angle in your plots should vary from -20 degrees to $+20$ degrees.

11.13. Consider the propeller-driven general aviation aircraft in a steady climbing turning flight maneuver with a bank angle of 40 degrees and a known constant value of the flight path angle.

(a) Plot the value of the lift coefficient as a function of the flight path angle, assuming that for each flight path angle the lift coefficient is selected to maximize the steady climbing and turning range of the aircraft. Also, plot the value of the aerodynamic factor that appears in the steady climbing and turning range formula as a function of the flight path angle. The flight path angle in your plots should vary from -20 degrees to $+20$ degrees.

(b) Plot the value of the lift coefficient as a function of the flight path angle, assuming that for each flight path angle the lift coefficient is selected to maximize the steady climbing and turning endurance of the aircraft. Also, plot the value of the aerodynamic factor that appears in the steady climbing and turning endurance formula as a function of the flight path angle. The flight path angle in your plots should vary from -20 degrees to $+20$ degrees.

12

Aircraft Maneuvers and Flight Planning

The steady flight results developed in the previous chapters provide the fundamental theoretical basis for analysis of steady flight and for related flight performance measures. In this sense, those results are presented from the perspective of an aerospace engineer who is interested in aircraft design and aircraft performance.

The steady flight results developed in the previous chapters are now examined from the perspective of flight operations, for example, from the perspective of a pilot who wants to maneuver the aircraft from one steady flight condition to another or, more generally, to plan a predefined sequence of steady flight maneuvers. The results in the prior chapters are now expressed in a form that allows determination of how pilot inputs, namely the elevator, ailerons, rudder, and throttle setting, are determined to achieve a desired flight maneuver or a sequence of flight maneuvers.

Background on static flight stability is given in section 12.1. Several relatively simple maneuvers are studied in section 12.2: a flight maneuver defined by a change in the elevator, with no other changes in the pilot inputs; a flight maneuver defined by a change in the throttle, with no other changes in the pilot inputs; a flight maneuver defined by a change in the bank angle, with no other changes in the pilot inputs; a flight maneuver defined by simultaneous changes in the elevator, the throttle, and the bank angle. A change in pilot inputs means that the inputs are changed from values associated with one steady flight condition to values associated with another steady flight condition and then the inputs are maintained constant at these values. The detailed flight dynamics associated with the flight maneuver from one steady flight condition to another steady flight condition are ignored in this analysis. In section 12.4 these methods are applied to flight planning problems that guarantee that an aircraft follows a sequence of waypoints.

12.1 Static Flight Stability

Before describing flight maneuvers, it is important to introduce a flight stability assumption. The analysis of the prior chapters has been concerned with describing steady flight. In addition to being able to characterize the conditions for steady flight, it is important to know if this steady flight condition has the following stability property: small perturbations from the steady flight condition do not cause the aircraft flight to diverge from this steady flight condition. If this property is valid, this flight condition is said to be stable and small initial perturbations from steady flight result in subsequent perturbations from steady flight that remain small; otherwise the flight condition is unstable. It is assumed that the pilot inputs are not changed throughout.

Flight stability is a subject of great practical importance and a careful study requires the introduction of more advanced concepts. Consequently, we introduce here the simplified notions of static flight stability, namely static pitch stability and static directional stability, that characterize the effects of longitudinal perturbations from a steady flight condition and the effects of lateral perturbations from a steady flight condition, respectively.

The most important factors in flight stability are captured by the expressions for the pitch moment and for the yaw moment. The aircraft design details of the wing, the horizontal tail, and the vertical tail have a strong influence on the aircraft pitch moment and the aircraft yaw moment. These influences are described in the references. Here only the main results are summarized.

Recall that the aircraft pitch moment depends on the aircraft pitch moment coefficient, and the pitch moment coefficient depends on the aircraft angle of attack and the elevator deflection according to equation (3.62):

$$C_M = C_{M_0} + C_{M_\alpha} \alpha + C_{M_{\delta_e}} \delta_e. \tag{12.1}$$

The conditions for static pitch stability can be shown to be given by the inequalities

$$C_{M_0} > 0, \tag{12.2}$$

$$C_{M_\alpha} < 0. \tag{12.3}$$

These pitch stability conditions guarantee that there is an angle of attack for which the pitch moment is zero, and the aerodynamic pitch moment naturally corrects for small longitudinal perturbations in the angle of attack. In particular, the negative value for C_{M_α} guarantees that if the angle of attack is slightly increased (decreased) then the aerodynamic pitch moment, according to equation (12.1), is slightly decreased (increased). According to the usual convention in defining the angle of attack and the pitch moment, the aerodynamic pitch moment naturally compensates for any angle of attack perturbation. A positive value for C_{M_0} guarantees that if there is no elevator deflection steady flight occurs for a positive angle of attack. In general, the elevator deflection is adjusted to achieve zero aerodynamic pitch moment in steady flight; the value of $C_{M_{\delta_e}}$ does not affect the static pitch stability of the aircraft.

This pitch stability property is a characteristic of the aircraft design features, most importantly the geometry of the wing and the location of the center of mass of the aircraft. The dominant factors that determine the sign and magnitude of the aerodynamic pitch moment coefficient derivative C_{M_α} are the lift coefficient derivative C_{L_α} and the distance h from the center of mass to the neutral point (a fictitious point for which the aerodynamic pitch moment is exactly zero) along the aircraft-fixed x_A-axis; that is,

$$C_{M_\alpha} = C_{L_\alpha} h. \tag{12.4}$$

Note that the neutral point defines the point on the aircraft-fixed x_A-axis for which the pitch moment coefficient is zero and also independent of the angle of attack. Since typically $C_{L_\alpha} > 0$, it follows that the aircraft has static pitch stability if the center of mass of the aircraft is forward of this neutral point. The aircraft is statically unstable in pitch if the center of mass of the aircraft is aft of this neutral point.

Also, recall that the aircraft yaw moment depends on the aircraft yaw moment coefficient. The yaw moment coefficient depends on the side-slip angle, the ailerons deflection, and the rudder deflection according to equation (3.65):

$$C_N = C_{N_\beta}\beta + C_{N_{\delta_a}}\delta_a + C_{N_{\delta_r}}\delta_r. \tag{12.5}$$

The condition for static directional stability is given by the inequality

$$C_{N_\beta} > 0. \tag{12.6}$$

This directional stability condition guarantees that the aerodynamic yaw moment naturally corrects for small lateral perturbations in the side-slip angle. In particular, a positive value for C_{N_β} guarantees that if the side-slip angle is positive (negative) then the aerodynamic yaw moment is positive (negative). According to the usual convention in defining the yaw axis and the side-slip angle as described in chapter 3, the aerodynamic yaw moment naturally compensates for any side-slip angle perturbation. The values of $C_{N_{\delta_a}}$ and $C_{N_{\delta_r}}$ do not affect the static directional stability of the aircraft. This directional flight stability property is a characteristic of the aircraft design features, most importantly the vertical tail.

Practical aircraft flight, for most aircraft and for most types of steady flight conditions, requires flight stability. Hence, it is common to design aircraft so that they have static pitch stability and static directional stability. Otherwise, it may be necessary to add flight control systems that provide stability augmentation.

12.2 Flight Maneuvers

The pilot controls the elevator, the rudder, the ailerons, and the throttle. The primary use of the rudder is to guarantee coordinated flight, that is, flight for which the side-slip angle is zero or, equivalently, the aircraft velocity vector lies in the plane of mass symmetry of the aircraft; the rudder is especially important when the aircraft flies through crosswinds. The primary purpose of the ailerons is to control the bank angle of the aircraft; the details fall within the domain of flight dynamics. In the sequel, the bank angle is viewed as a pilot input in spite of the fact that the pilot actually controls the bank angle only indirectly through the ailerons. To keep the description simple, the following discussion views the effective pilot inputs as the elevator deflection, the throttle, and the bank angle.

Flight Maneuvers Defined by a Change in the Elevator Deflection

In this section, longitudinal flight maneuvers defined by a change in the elevator deflection are studied, assuming that initially the aircraft is in a steady longitudinal flight condition and, after the elevator is deflected, the aircraft reaches a terminal steady longitudinal flight condition. There is no change in the throttle or the bank angle.

The following steps describe a procedure to determine the resulting steady longitudinal flight conditions of an aircraft after the flight maneuver is terminated:

- determine the value of the aircraft angle of attack using the condition that the pitch moment is zero, taking into account the known value of the elevator deflection

- determine the value of the lift coefficient using the model for the lift coefficient as a function of the angle of attack

- determine the value of the aircraft airspeed using the fact that the lift must equal the weight

- determine the flight path angle from the thrust or power equation, taking into account the known value of the thrust or power and the computed value of the airspeed

In summary, a change in the elevator deflection causes a change in the angle of attack, the flight path angle, and the airspeed of the aircraft, as well as a change in the lift and drag on the aircraft; the values of all of these quantities can be computed using the methods presented previously. Since the aircraft is initially in steady longitudinal flight, the bank angle remains zero and the aircraft remains flying in the fixed vertical plane throughout the maneuver.

This approach to determine the steady longitudinal flight conditions that arise from a change in the elevator deflection uses the equations derived previously for steady longitudinal flight. Once the terminal flight condition that arises from the maneuver is determined, it must be checked that this flight condition lies in the flight envelope for steady longitudinal flight, that is, the resulting lift coefficient is less than the lift coefficient at stall; if this condition is not satisfied then there is no steady flight and the aircraft might be in a dynamic stall. Since there is no change in the throttle, the engine is guaranteed to remain within its operating range.

Flight Maneuvers Defined by a Change in the Throttle Setting

In this section, longitudinal flight maneuvers defined by a change in the throttle setting are studied, assuming that initially the aircraft is in steady longitudinal flight and, after the throttle setting is changed, the aircraft reaches a terminal steady longitudinal flight condition. There is no change in the elevator deflection or the bank angle.

The following steps describe a procedure to determine the resulting steady longitudinal flight conditions of an aircraft after the flight maneuver is terminated:

- Since there is no change in the elevator there is no change in the angle of attack and hence there is no change in the lift coefficient.

- Since the lift equals the weight and there is no change in the lift coefficient, there is no change in the aircraft airspeed.

- determine the flight path angle from the thrust or power equation, taking into account the value of the thrust or power corresponding to the terminal throttle setting and the airspeed.

In summary, a change in the throttle does not cause a change in the angle of attack, the airspeed, or the lift or drag on the aircraft, but it does cause a change the flight path angle of the aircraft; the value of the flight path angle can be computed. Since the aircraft is initially in steady longitudinal flight, the bank angle remains zero and the aircraft remains flying in the fixed vertical plane throughout the maneuver.

This approach to determine the steady longitudinal flight condition that arises from a change in the throttle uses the equations derived previously for steady longitudinal flight. Once the terminal flight condition that arises from the maneuver is determined, it must be checked that this flight condition lies in the flight envelope for steady longitudinal flight, that is, the resulting lift coefficient is less than the lift coefficient at stall; if this condition is not satisfied then there is no steady flight and the aircraft might be in a dynamic stall condition. If the maneuver is defined by a step change in the throttle where the final throttle setting is between 0 and 1, then the engine is guaranteed to remain within its operating range.

Flight Maneuvers Defined by a Change in Bank Angle

The aircraft is initially in steady longitudinal flight. The ailerons are differentially deflected as follows: they are moved from their initial zero deflection to some non-zero deflection for a short period of time, and finally they are brought back to zero deflection and maintained at zero deflection thereafter. The bank angle of the aircraft is assumed to reach a terminal constant value, which is known. There is no change in the elevator deflection or in the throttle.

The following steps describe a procedure to determine the resulting steady turning flight conditions of an aircraft after the flight maneuver is terminated:

- Since the aircraft is initially in steady longitudinal flight, this maneuver causes a change in the bank angle of the aircraft from zero to some terminal constant value, which is assumed to be known.

- Since there is no change in the elevator deflection there is no change in the aircraft angle of attack.

- Since there is no change in the angle of attack, there is no change in the lift coefficient.

- Since the vertical component of the lift, which depends on the bank angle, equals the weight, the terminal airspeed in the turn can be determined.

- Determine the flight path angle from the thrust or power equation, taking into account the value of the thrust or power produced by the engines, the airspeed and the bank angle.

- The turn radius can be determined from the airspeed and the bank angle.

In summary, a change in the bank angle causes the aircraft to turn and causes a change in the flight path angle and the airspeed of the aircraft, as well as a change in the lift and drag on the aircraft; the values of the quantities that define this steady turning flight condition can be computed.

This approach to determine the steady turning flight condition that arises from a change in the bank angle uses the equations derived previously for steady turning flight. Once the terminal flight condition that arises from the maneuver is determined, it must be checked that this flight condition lies in the flight envelope for steady turning flight, that is, the resulting lift coefficient is less than the lift coefficient at stall; if this condition is not satisfied then there is no steady flight and the aircraft would be in a dynamic stall condition. The bank angle should also satisfy the load factor constraint for steady turning flight. Since there is no change in the throttle setting, the engine is guaranteed to remain within its operating limits.

Flight Maneuvers Defined by Simultaneous Changes in Elevator, Throttle, and Bank Angle

The above descriptions illustrate the approach for determining the terminal steady flight condition that arises from a flight maneuver for which only a single pilot input is changed. If two or more pilot inputs are changed simultaneously, the equations for steady longitudinal flight or steady turning flight can be used to determine the terminal steady flight condition, namely the angle of attack, the flight path angle, the airspeed, and the turn radius, as well as the lift and drag. As indicated previously, the maneuver results in a terminal steady flight condition only if the new steady flight condition lies in the steady flight envelope.

12.3 Pilot Inputs That Achieve a Desired Flight Condition

The results for steady longitudinal flight and for steady turning flight can also be used to determine the pilot inputs, namely the elevator, throttle, and bank angle, that achieve a desired flight condition. Typical examples of desired flight conditions are:

- steady longitudinal flight with a desired flight path angle and desired airspeed;

- steady longitudinal flight with a desired power or thrust level and a desired airspeed;

- steady turning flight with a desired turn radius, a desired power or thrust level, and a specified airspeed;

- steady turning flight with desired turn radius, a desired flight path angle, and a desired airspeed.

These various steady flight maneuver problems can be solved using the steady flight development of the previous chapters. In fact, this flexibility demonstrates the utility of mathematical models for steady aircraft flight. These mathematical models provide the basis for examining many different steady flight problems that aerospace engineers must often solve.

12.4 Flight Plans Defined by a Sequence of Waypoints

Considering straight line paths and circular paths as special cases of helical paths, coordinated flight along helical paths represents the most general type of steady flight. Most aircraft fly along a concatenation of helical flight segments, that is, helical flight segments appropriately patched together. Flight paths that characterize the route of an aircraft, for example, in takeoff, climb, turn, cruise, turn, descent, and landing, are most often defined by a concatenation of helical paths. Equivalently, most typical aircraft routes, planned by the pilot, the flight management computer, or air traffic control, are defined by a concatenation of helical flight paths. Along each such helical path segment, the aircraft is in steady coordinated flight with constant airspeed, constant flight path angle or climb rate, and constant bank angle. The waypoints define the position vectors in the ground frame at which the different helical segments are joined. As the aircraft passes a waypoint, it transitions from the prior steady flight condition to the next steady flight condition; during these transition periods the flight is not steady and flight dynamics effects are important.

The waypoints are typically defined by specifying a fixed length sequence of desired position vectors in three dimensions: two position variables that describe each waypoint location in a horizontal plane according to a ground-based origin and east and west directions, and one position variable for the altitude of the waypoint. The first waypoint is typically the takeoff location of the aircraft; the last waypoint in the sequence is typically the landing location of the aircraft.

3D Flight Planning

A 3D flight plan describes an aircraft path that satisfies a sequence of specified waypoints. A 3D flight plan consists of specifications for the steady flight conditions that define each steady flight segment between successive waypoints. In particular, the steady flight conditions for each flight segment are defined by constant airspeed, constant flight path angle or climb rate, and constant bank angle. The 3D flight plan may also specify the angle of attack, elevator deflection, and throttle setting required to achieve steady flight during each segment between successive waypoints.

3D flight planning is accomplished by first constructing a path between each pair of successive waypoints. A path between waypoints may be a straight line path segment (during takeoff, cruise, or landing), a circular arc (during a horizontal turn), or, more generally, a helical path. Typically, there are many possible concatenations of flight segments that meet the waypoint specifications. For flight maneuvers in a horizontal plane, paths are often selected so that the slope of the path leading to a waypoint equals the slope of the path leaving the waypoint; such a flight plan leads to a smooth flight transition as the aircraft passes the waypoint. Matching of the slopes of the path leading to a waypoint and the path leaving the waypoint is not possible if either flight segment corresponds to climbing or descending flight.

Once the straight line or helical paths are selected between waypoints, specific flight conditions on each flight segment can be determined using the steady flight results presented in the previous chapters. One way to select specific flight conditions on each flight segment is to achieve some flight optimality property. Two common optimality objectives follow:

- The flight plan might be selected to minimize the fuel burned from takeoff to landing. This can be accomplished by selecting the steady flight conditions for each flight segment to minimize the fuel burned during that flight segment. For example, for jet aircraft the flight conditions should be selected to minimize the required thrust during that flight segment; for propeller-driven general aviation aircraft the flight conditions should be selected to minimize the required power during that flight segment. During periods of steady level cruising flight, the flight conditions should be selected for maximum range, which is equivalent to minimum fuel consumption for a fixed range, according to the engine characteristics.

- The flight plan might be selected to minimize the total elapsed time from takeoff to landing. This can be accomplished by selecting the steady flight conditions for each flight segment that maximize the airspeed, subject to all flight constraints, during that flight segment.

During short duration steady flight segments, the aircraft weight and the air density can be assumed to be constant; this enables relatively straightforward analysis of the steady flight conditions.

For longer duration steady climbing or descending flight segments for which there are significant changes in the aircraft weight and the air density, flight planning is more complicated and requires the use of models that characterize the flight dynamics. In these cases, results for range, endurance, and fuel requirements are generally not available, and computational methods must be used to solve the relevant differential equations; see chapter 13. The pilot inputs, elevator deflection, and throttle setting must be slowly adjusted to satisfy the steady flight conditions as the aircraft weight and the air density change slowly.

4D Flight Planning

In addition to specification of the position vector that defines a waypoint, a time instant may also be specified for when the aircraft should be located at that waypoint. In this way, a flight plan that satisfies a sequence of waypoints and timing specifications is referred to as a 4D flight plan. A 4D flight plan consists of specifications for the steady flight conditions that define each steady flight segment between successive waypoints. The steady flight conditions for each flight segment are defined by constant airspeed, constant flight path angle or climb rate, and constant bank angle. The flight plan may also specify the angle of attack, elevator deflection, and throttle setting required to achieve steady flight during each segment between successive waypoints.

4D flight planning is accomplished by constructing a path between each pair of successive waypoints. A path between waypoints may be a straight line path segment (during takeoff, cruise, or landing), a circular arc (during a horizontal turn), or, more generally, a helical path. Typically, there are many possible concatenations of flight segments that meet the waypoint specifications.

Once the flight path segments are selected, the constant airspeed along each steady flight segment should be selected to satisfy the specified timing constraints associated with the waypoints. Of course, on each steady flight segment the airspeed and the other flight conditions must be chosen to be consistent with steady flight, making sure that the flight constraints are all satisfied.

During short duration steady flight segments, the aircraft weight can be assumed to be constant; this enables relatively straightforward analysis of the required steady flight conditions.

12.5 A Flight Planning Problem: Executive Jet Aircraft

In this section, a flight planning problem for the executive jet aircraft is formulated and analyzed based on coordinated flight maneuvers defined by a specified sequence of waypoints.

The specified waypoints are defined with respect to a ground-fixed frame: the x and y axes are in a horizontal plane located at an altitude of 5,000 ft with the x-axis pointing due east and the y-axis pointing due north; the z-axis defines the altitude above this horizontal plane.

Waypoint	Name	Coordinates (x, y, z) in ft
1	climb	$(0, 0, 0)$
2	cruise	$(2.1 \times 10^5, 0, 1.0 \times 10^4)$
3	descent	$(1.47 \times 10^6, 0, 1.0 \times 10^4)$
4		$(1.88 \times 10^6, 0, 0)$

The aircraft is assumed to begin at waypoint 1, just after takeoff, with a full fuel tank and is intended to end at waypoint 4, just before landing. The flight plan should be fuel efficient; that is, a near minimum amount of fuel should be required to complete the flight.

Development of a Flight Plan

Since the waypoints imply that the complete flight occurs in a ground-fixed vertical plane defined by $y = 0$, the flight plan is developed to guarantee that the aircraft always flies within this plane. The executive jet aircraft should fly according to the following specifications:

- steady climbing flight between waypoints 1 and 2 following a straight line path between these two waypoints with a constant flight path angle of 2.73 degrees,

- steady level cruising flight between waypoints 2 and 3 following a straight line constant altitude path between these two waypoints,

- steady descending flight between waypoints 3 and 4 following a straight line path between these two waypoints with a constant flight path angle of -1.365 degrees.

The overall flight consists of a concatenation of these three straight line flight segments. Each of these three flight segments is analyzed in detail. A summary of results is given.

Flight Conditions between Waypoint 1 and Waypoint 2

A fuel efficient climb occurs when the thrust is minimum. We compute the airspeed for the minimum thrust climb and the value of this minimum thrust at waypoint 1. The airspeed at waypoint 1 is

$$V_1 = \sqrt{\frac{2W_1}{\rho S}} \sqrt{\frac{K}{C_{D_0}}} = 370.1 \text{ ft/s}$$

and the value of the minimum thrust at waypoint 1 is

$$T_1 = W_1 \gamma + 2W_1 \sqrt{K C_{D_0}} = 7,473.2 \text{ lbs.}$$

During the steady climbing flight segment, the flight conditions are adjusted to maintain the flight path angle constant at 2.73 degrees and the airspeed constant at 370.1 ft/s. This should result in a fuel efficient climbing flight segment. The time of flight during the steady climbing flight segment is easily computed; since the climb rate is 17.63 ft/s and the required altitude change is 10,000 ft, the steady climbing flight time during this flight segment is 567.2 s.

Although the thrust required for steady climbing flight during this flight segment changes slowly as the air density and the aircraft weight change slowly, we approximate the rate at which the aircraft weight changes due to fuel consumption using the value of the thrust at waypoint 1. This approximation is sufficiently accurate for the present analysis. This leads to the rate of change of the aircraft weight given by

$$\frac{dW}{dt} = -cT_1 = 1.433 \text{ lbs/s.}$$

This differential equation can be integrated directly to obtain the weight of fuel burned during this flight segment as 812.8 lbs; hence the aircraft weight at waypoint 2 is approximately 72,187.2 lbs.

The elevator deflection and the throttle setting must be adjusted to compensate for the slow change in the aircraft density and the aircraft weight as fuel is burned by the jet engine. Expressions for the elevator deflection and the throttle setting as functions of the air density aircraft and the aircraft weight in a steady climb have been developed previously. It can be shown that all flight constraints are satisfied for the steady climbing flight segment.

Flight Conditions between Waypoint 2 and Waypoint 3

For this jet aircraft, a fuel efficient cruise is obtained by using the results that minimize the total fuel required to achieve the specified cruising range of 1.26×10^6 ft at an altitude of 15,000 ft. This implies that the aerodynamic coefficients should be selected to maximize the aerodynamics ratio $\frac{C_L^{\frac{1}{2}}}{C_D}$, which gives $C_L = 0.316$ and $C_D = 0.02$. These should be maintained constant throughout the cruising flight segment.

During the steady cruising flight segment, the fuel burned can be determined by using the range equation for a jet aircraft; in particular the range equation

$$\frac{2}{c} \sqrt{\frac{2}{\rho S} \frac{C_L^{\frac{1}{2}}}{C_D}} \left[\sqrt{W_2} - \sqrt{W_3} \right] = 1.2 \times 10^6 \, \text{ft}$$

can be solved for the aircraft weight W_3 at waypoint 3 based on the specified range of 1.26×10^6 ft and the aircraft weight 72,187.2 lbs at waypoint 2. This computation gives the aircraft weight at waypoint 3 as 70,336.6 lbs so that the weight of fuel burned during the steady cruise flight segment is 1,850.6 lbs. The time of flight during the steady cruising flight segment can be determined from the endurance formula

$$E = \frac{1}{c} \frac{C_L}{C_D} \ln \frac{W_2}{W_3} = 2,140.4 \, \text{s},$$

using the steady level flight conditions $C_L = 0.316$ and $C_D = 0.02$ and the aircraft weight values at waypoints 2 and 3.

The elevator deflection and the throttle setting must be adjusted to compensate for the slow change in the aircraft weight as fuel is burned by the jet engine. Expressions for the elevator deflection and the throttle setting as functions of the aircraft weight in a steady cruise have been developed previously. It can be shown that all flight constraints are satisfied for the steady cruising flight segment.

Flight Conditions between Waypoint 3 and Waypoint 4

A fuel efficient descent occurs when the thrust is minimum. We compute the airspeed for the minimum thrust and the value of this minimum thrust at waypoint 3. The value of airspeed at waypoint 3 on this flight segment is

$$V_3 = \sqrt{\frac{2W_3}{\rho S} \sqrt{\frac{K}{C_{D_0}}}} = 363.3 \, \text{ft/s},$$

and the value of the minimum thrust at waypoint 3 is

$$T_3 = W_3 \gamma + 2W_3 \sqrt{K C_{D_0}} = 2,176.7 \, \text{lbs}.$$

During this steady descending flight segment, the flight conditions are adjusted to maintain the flight path angle constant at -1.365 degrees and the airspeed constant at 363.3 ft/s. This should result in a fuel efficient descending flight segment. The time of flight during the steady descending flight segment is easily computed; since the descent rate is 8.82 ft/s and the required altitude change is 10,000 ft, the steady descending flight time during this flight segment is 1,134.4 sec.

Although the thrust required for steady descending flight during this flight segment changes slowly as the air density and the aircraft weight change slowly, we approximate the rate at which the aircraft weight changes due to fuel consumption using the value of the thrust at waypoint 3. This approximation leads to the rate of change of the aircraft weight given by

$$\frac{dW}{dt} = -cT_3 = 0.417 \, \text{lbs/s}.$$

This differential equation can be integrated directly to obtain the weight of fuel burned during this flight segment as 473.3 lbs; hence the aircraft weight at waypoint 4 is approximately 69,863.2 lbs.

The elevator deflection and the throttle setting must be adjusted to compensate for the slow change in the aircraft density and the aircraft weight as fuel is burned by the jet engine. Expressions for the elevator deflection and the throttle setting as functions of the air density aircraft and the aircraft weight in a steady descent have been developed previously. It can be shown that all flight constraints are satisfied for the steady descending flight segment.

Summary of the Flight Plan

The following table describes the cumulative fuel burned by the jet engine as the aircraft flies from waypoint 1 to waypoint 4.

Waypoints	Name	Cumulative fuel burned in lbs
1–2	climb	812.8
2–3	cruise	2663.4
3–4	descent	3136.7

The following table describes the cumulative time of flight from waypoint 1 to waypoint 4.

Waypoints	Name	Cumulative flight time in sec
1–2	climb	567.2
2–3	cruise	2707.7
3–4	descent	3842.1

Hence, this flight plan describes the three steady flight segments of the jet aircraft as it flies from waypoint 1 to waypoint 4. The total required fuel is 3,136.7 lbs and the required time to complete this flight is 3,842.1 s = 1 hr, 4 min, 2.1 s. The total distance flown is 1,890,000 ft = 357.95 miles, so that the average speed is 335.5 mph.

12.6 A Flight Planning Problem: General Aviation Aircraft

In this section, a flight planning problem for the propeller-driven general aviation aircraft is formulated and analyzed based on coordinated flight maneuvers defined by a specified sequence of waypoints.

The specified waypoints are defined with respect to a ground-fixed frame: the x and y axes are in a horizontal plane located at an altitude of 1,000 ft with the x-axis pointing due east and the y-axis pointing due north; the z-axis defines the altitude above this horizontal plane.

Waypoint	Name	Coordinates (x, y, z) in ft
1	climb	$(0, 0, 0)$
2	level turn	$(125000, 0, 5000)$
3	cruise	$(125000, 20000, 5000)$
4	level turn	$(-125000, 20000, 5000)$
5	descent	$(-125000, 0, 5000)$
6		$(0, 0, 0)$

The aircraft is assumed to begin at waypoint 1, just after takeoff, with a full fuel tank and is intended to end at waypoint 6, just before landing. The flight plan should be fuel efficient; that is, a near minimum amount of fuel should be required to complete the flight.

Development of a Flight Plan

Since the aircraft is required to satisfy the waypoint specifications, the aircraft should fly according to the following specifications:

- steady climbing flight between waypoints 1 and 2 following a straight line path between these two waypoints with a constant flight path angle of 2.29 degrees

- steady level left turn between waypoints 2 and 3 following a horizontal circular path between these two waypoints with a constant turn radius of 10,000 ft; the flight direction changes by 180 degrees during this flight segment

- steady level cruising flight between waypoints 3 and 4 following a straight line constant altitude path between these two waypoints

- steady level left turn between waypoints 4 and 5 following a horizontal circular path between these two waypoints with a constant turn radius of 10,000 ft; the flight direction changes by 180 degrees during this flight segment

- steady descending flight between waypoints 5 and 6 following a straight line path between these two waypoints with a constant flight path angle of -2.29 degrees

The overall flight consists of a concatenation of these five flight segments. Each of these five flight segments is analyzed in detail. Then a summary of results is given.

Flight Conditions between Waypoint 1 and Waypoint 2

A fuel efficient climb occurs when the power is minimum. We compute the airspeed for the minimum power climb and the value of this minimum power at waypoint 1. This gives the value of airspeed at waypoint 1 as

$$V_1 = \sqrt{\frac{2W_1}{\rho S} \sqrt{\frac{K}{3C_{D_0}}}} = 109.5 \, \text{ft/s},$$

and the value of the minimum power at waypoint 1 as

$$P_1 = W_1 V_1 \gamma + \frac{4}{3}\sqrt{\frac{2W_1^3}{\rho S} \sqrt{3K^3 C_{D_0}}} = 36{,}496.4 \, \text{ft lbs/s}.$$

During the steady climbing flight segment, the flight conditions are adjusted to maintain the flight path angle constant at 2.29 degrees and the airspeed constant at 109.5 ft/s. This should result in a fuel efficient climbing flight segment. The time of flight during the steady climbing flight segment is easily computed; since the climb rate is 4.38 ft/s and the required altitude change is 5,000 ft, the steady climbing flight time during this flight segment is 11, 41.6 s.

Although the power required for steady climbing flight during this flight segment changes slowly as the air density and the aircraft weight change slowly, we approximate the rate at which the aircraft weight changes by

$$\frac{dW}{dt} = -cP_1 = 0.01037 \, \text{lbs/s}.$$

This differential equation can be integrated directly to obtain the weight of fuel burned during this flight segment as 11.8 lbs; hence the aircraft weight at waypoint 2 is approximately 2,888.2 lbs.

In principle, the elevator deflection and the throttle setting must be adjusted to compensate for the slow change in the aircraft weight as fuel is burned by the internal combustion engine; in this case the change in weight of the aircraft is small so that the elevator deflection and throttle setting can be maintained constant. It can be shown that all flight constraints are satisfied for the steady climbing flight segment.

Flight Conditions between Waypoint 2 and Waypoint 3

A minimum fuel steady level turn for a propeller aircraft implies that the aerodynamic coefficients should be selected to maximize the ratio $\frac{C_L}{C_D}$, thereby giving the values $C_L = 0.694$ and $C_D = 0.052$. Furthermore, the bank angle should be chosen to satisfy the turn radius specification at waypoint 2, namely

$$r = \frac{2}{\rho g C_L} \frac{1}{\sin\mu} \frac{W_2}{S} = 10{,}000 \, \text{ft}.$$

Solving this equation gives the required bank angle $\mu = -4.3$ degrees, where the negative sign is used provide a left turn. These values characterize a fuel efficient steady level turn. To achieve the specified steady level turning range of 31,415.9 ft at an altitude of 6,000 ft requires satisfaction of the turning range equation

$$\frac{\eta}{c} \frac{C_L}{C_D} \cos \mu \ln \left(\frac{W_2}{W_3}\right) = 31,415.9 \,\text{ft}.$$

This equation can be solved for the aircraft weight W_3 at waypoint 3 based on the specified steady level turning range of 31,416 ft, a bank angle of -4.3 degrees, and the aircraft weight of 2,888.2 lbs at waypoint 2. This computation gives the aircraft weight at waypoint 3 as 2,882.8 lbs so that the weight of fuel burned during this flight segment is 5.4 lbs. The time of flight during the steady turning flight segment can be determined from the endurance formula

$$E = \frac{\eta}{c} \frac{(C_L \cos \mu)^{\frac{3}{2}}}{C_D} \sqrt{2\rho S} \left(\frac{1}{\sqrt{W_2}} - \frac{1}{\sqrt{W_3}}\right) = 202.8 \,\text{s},$$

using the steady level flight conditions $C_L = 0.694$ and $C_D = 0.052$, bank angle $\mu = -4.3$ degrees, and the aircraft weight values at waypoints 2 and 3.

In principle, the elevator deflection and the throttle setting must be adjusted to compensate for the slow change in the aircraft weight as fuel is burned by the internal combustion engine; in this case the change in weight is small so that the elevator deflection and throttle setting can be maintained constant. It can be shown that all flight constraints are satisfied for the steady level turning flight segment.

Flight Conditions between Waypoint 3 and Waypoint 4

For this propeller aircraft, a fuel efficient cruise is obtained by using the results that minimize the total fuel required to achieve the specified cruising range of 25,0000 ft at an altitude of 6,000 ft. This implies that the aerodynamic coefficients $C_L = 0.694$ and $C_D = 0.052$ should be maintained constant throughout the cruising flight segment. During the steady cruising flight segment, the fuel burned can be determined by using the range equation for a jet aircraft; in particular the range equation

$$\frac{\eta}{c} \frac{C_L}{C_D} \ln \left(\frac{W_3}{W_4}\right) = 250,000 \,\text{ft}$$

can be solved for the aircraft weight W_4 at waypoint 4 based on the specified range of 250,000 ft and the aircraft weight of 2,882.8 lbs at waypoint 3. This computation gives the aircraft weight at waypoint 4 as 2,862.5 lbs so that the weight of fuel burned during the steady cruise flight segment is 15.3 lbs. The time of flight during the steady cruising flight segment can be determined from the endurance formula

$$E = \frac{\eta}{c} \frac{C_L^{\frac{3}{2}}}{C_D} \sqrt{2\rho S} \left(\frac{1}{\sqrt{W_4}} - \frac{1}{\sqrt{W_3}} \right) = 1{,}617.1 \text{ s},$$

using the steady level flight conditions $C_L = 0.694$ and $C_D = 0.052$ and the aircraft weight values at waypoints 3 and 4.

The elevator deflection and the throttle setting must be adjusted to compensate for the slow change in the aircraft weight as fuel is burned by the internal combustion engine; in this case the change in weight is small so that the elevator deflection and throttle setting can be maintained constant. It can be shown that all flight constraints are satisfied for the steady level turning flight segment.

Flight Conditions between Waypoint 4 and Waypoint 5

The only difference in this fourth steady level turning flight segment and the second steady level turning flight segment is that the weight at the beginning of this flight segment, waypoint 4, is 2,867.5 lbs. The results are easily obtained following the procedure indicated for the second flight segment. During this turning flight segment, the lift coefficient and the drag coefficient $C_L = 0.694$ and $C_D = 0.052$ and the bank angle $\mu = -4.3$ degrees. These values characterize a fuel efficient steady level turn. To achieve the specified steady level turning range of 31,415.9 ft at an altitude of 6,000 ft requires combustion of 5.4 lbs of fuel. Hence, the aircraft weight at waypoint 5 is 2,862.1 lbs. The time of flight during this steady level turning flight segment is 202.8 s.

The elevator deflection and the throttle setting must be adjusted to compensate for the slow change in the aircraft weight as fuel is burned by the internal combustion engine; in this case the change in weight is small so that the elevator deflection and throttle setting can be maintained constant. It can be shown that all flight constraints are satisfied for the steady level turning flight segment.

Flight Conditions between Waypoint 5 and Waypoint 6

A fuel efficient descent occurs when the power is minimum. We compute the airspeed for the minimum power descent and the value of this minimum power at waypoint 5. This gives the value of airspeed at waypoint 5 on this flight segment as

$$V_5 = \sqrt{\frac{2W_5}{\rho S} \sqrt{\frac{K}{3C_{D_0}}}} = 141.2 \text{ ft/s}$$

and the value of the minimum power at waypoint 5 is

$$P_5 = W_5 V_5 \gamma + \frac{4}{3} \sqrt{\frac{2W^3}{\rho S}} \sqrt{3K^3 C_{D_0}} = 45{,}103.8 \text{ ft lbs/s}.$$

During the steady descending flight segment, the flight conditions are adjusted to maintain the flight path angle constant at -2.29 degrees and the airspeed constant at 141.2 ft/s. This should result in a fuel efficient descending flight segment. The time of flight during the steady descending flight segment is easily computed: since the descent rate is 5.64 ft/s and the required altitude change is 5,000 ft, the steady descending flight time during this flight segment is 886.5 s. Although the power required for steady descending flight during this flight segment changes slowly as the air density and the aircraft weight change slowly, we approximate the rate at which the aircraft weight changes due to fuel consumption by

$$\frac{dW}{dt} = -cP_5 = 0.0128 \text{ lbs/s}.$$

This differential equation is integrated directly to obtain the weight of fuel burned during this flight segment as 11.3 lbs; hence the aircraft weight at waypoint 6 is approximately 2,850.8 lbs. The elevator deflection and the throttle setting must be adjusted to compensate for the slow change in the aircraft weight as fuel is burned by the internal combustion engine; in this case the change in weight is small so that the elevator deflection and throttle setting can be maintained constant. It can be shown that all flight constraints are satisfied for the steady climbing flight segment.

Summary of the Flight Plan

The following table describes the cumulative fuel burned by the jet engine from takeoff to landing.

Waypoints	Name	Cumulative fuel burned in lbs
1–2	climb	11.8
2–3	level turn	17.2
3–4	cruise	32.5
4–5	level turn	37.9
5–6	descent	49.2

The following table describes the cumulative time of flight from takeoff to landing.

Waypoints	Name	Cumulative flight time in sec
1–2	climb	1141.6
2–3	level turn	1344.4
3–4	cruise	2961.5
4–5	level turn	3164.3
5–6	descent	4050.8

Hence, this flight plan describes the five steady flight segments of the propeller aircraft as it flies from waypoint 1 to waypoint 6. The total required fuel is 49.2 lbs and the required time to complete this flight is 4,050.8 s $=$ 1 hr, 7 min, 30 s. The total distance flown is 531,416 ft $=$ 100.6 miles, so that the average speed is 89.5 mph.

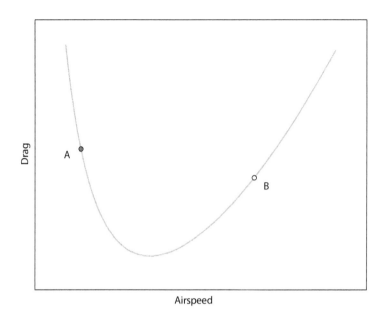

Figure 12.1. Drag versus airspeed curve for steady level flight.

12.7 Conclusions

The purpose of this chapter has been to describe, in a conceptual way, how the methods for steady flight analysis can be used to describe aircraft maneuvers between steady flight conditions, including steady flight planning defined by a sequence of waypoints. Flight planning problems constitute a major aircraft performance topic. Many complex flight planning problems can be formulated and studied using the methods described.

12.8 Problems

12.1. An aircraft is initially flying in steady level flight. Assume the aircraft is stable and assume the drag versus airspeed curve for steady level flight of the aircraft is given in figure 12.1. Flight conditions A and B are indicated on this curve.

Assume the pilot makes a constant change in the elevator that causes the nose of the aircraft to pitch up; no change is made to the throttle, rudder, or ailerons.

(a) If the aircraft is initially in steady level flight at point A, describe the resulting change in the steady flight conditions. How do the angle of attack and the airspeed change? Does the aircraft return to a steady level flight condition? Does it climb at a constant rate? Does it descend at a constant rate? Describe your reasoning.

(b) If the aircraft is initially in steady level flight at point B, describe the resulting changes in the steady flight conditions. How do the angle of attack and the airspeed change? Does the aircraft return to a steady level flight condition? Does it climb at a constant rate? Does it descend at a constant rate? Describe your reasoning.

Assume the pilot makes a constant increase in the throttle setting; no change is made to, the elevator, rudder, or ailerons.

(c) If the aircraft is initially in steady level flight at point A, describe the resulting change in the steady flight conditions. How do the angle of attack and the airspeed change? Does the aircraft return to a steady level flight condition? Does it climb at a constant rate? Does it descend at a constant rate? Describe your reasoning.

(d) If the aircraft is initially in steady level flight at point B, describe the resulting change in the steady flight conditions. How do the angle of attack and the airspeed change? Does the aircraft return to a steady level flight condition? Does it climb at a constant rate? Does it descend at a constant rate? Describe your reasoning.

All subsequent problems in this section are based on the aircraft data provided in appendix B.

12.2. Consider the executive jet aircraft carrying out a sequence of longitudinal coordinated flight maneuvers defined by a sequence of waypoints described below. The total time of flight is sufficiently short that the weight of the executive jet aircraft can be assumed to be a constant 73,000 lbs and the air density during the flight can be assumed to be constant.

The specified waypoints are defined with respect to a ground-fixed frame: the x and y axes are in a horizontal plane located at an altitude of 5,000 ft with the x-axis pointing due east and the y-axis pointing due north; the z-axis defines the altitude above this horizontal plane.

Waypoint	Name	Coordinates (x, y, z) in ft	Time Instant in sec
1	climb	$(0, 0, 0)$	0
2	cruise	$(20000, 0, 2000)$	60
3	descend	$(26666, 0, 2000)$	80
4		$(46666, 0, 0)$	140

The aircraft should be in steady climbing flight between waypoints 1 and 2; it should be in steady level flight between waypoints 2 and 3, and it should be in steady descending flight between waypoints 3 and 4.

(a) Develop a flight plan that specifies a constant airspeed and constant climb rate for each of the three steady flight segments that satisfy the waypoint specifications.

    (b) For each of the steady flight segments constructed in part (a), determine the elevator deflection and the throttle setting that guarantee the desired steady flight conditions along that flight segment. Check that all flight constraints are satisfied for all steady flight segments.

    (c) On the basis of your flight plan in part (a), give plot(s) that describe the 3D path of the aircraft.

12.3. Consider the propeller-driven general aviation aircraft carrying out a sequence of longitudinal coordinated flight maneuvers defined by a sequence of waypoints described below. The total time of flight is sufficiently short that the weight of the general aviation aircraft can be assumed to be a constant 2,900 lbs and the air density during the flight can be assumed to be constant.

The specified waypoints are defined with respect to a ground-fixed frame: the x and y axes are in a horizontal plane located at an altitude of 4,000 ft with the x-axis pointing due east and the y-axis pointing due north; the z-axis defines the altitude above this horizontal plane.

Waypoint	Name	Coordinates (x, y, z) in ft	Time Instant in sec
1	climb	(0, 0, 0)	0
2	cruise	(18000, 0, 2000)	100
3	descend	(25200, 0, 2000)	140
4		(43200, 0, 0)	240

The aircraft should be in steady climbing flight between waypoints 1 and 2; it should be in steady level flight between waypoints 2 and 3, and it should be in steady descending flight between waypoints 3 and 4.

    (a) Develop a flight plan that specifies a constant airspeed and constant climb rate for each of the three steady flight segments that satisfy the waypoint specifications.

    (b) For each of the steady flight segments constructed in part (a), determine the elevator deflection and the throttle setting that guarantee the desired steady flight conditions along that flight segment. Check that all flight constraints are satisfied for all steady flight segments.

    (c) On the basis of your flight plan in part (a), give plot(s) that describe the 3D path of the aircraft.

12.4. Consider the UAV carrying out a sequence of longitudinal coordinated flight maneuvers defined by a sequence of waypoints described below. The total time of flight is sufficiently short that the weight of the UAV can be assumed to be a constant 45 lbs and the air density during the flight can be assumed to be constant.

The specified waypoints are defined with respect to a ground-fixed frame: the x and y axes are in a horizontal plane located at an altitude of 1,000 ft with the x-axis pointing due east and the y-axis pointing due north; the z-axis defines the altitude above this horizontal plane.

Waypoint	Name	Coordinates $(x,\ y,\ z)$ in ft	Time Instant in sec
1	climb	$(0, 0, 0)$	0
2	cruise	$(1000, 0, 100)$	20
3	descend	$(3000, 0, 100)$	50
4		$(4000, 0, 0)$	70

The aircraft should be in steady climbing flight between waypoints 1 and 2; it should be in steady level flight between waypoints 2 and 3, and it should be in steady descending flight between waypoints 3 and 4.

    (a) Develop a flight plan that specifies a constant airspeed and constant climb rate for each of the three steady flight segments that satisfy the waypoint specifications.

    (b) For each of the steady flight segments constructed in part (a), determine the elevator deflection and the throttle setting that guarantee the desired steady flight conditions along that flight segment. Check that all flight constraints are satisfied for all steady flight segments.

    (c) On the basis of your flight plan in part (a), give plot(s) that describe the 3D path of the aircraft.

12.5. Consider the executive jet aircraft carrying out a sequence of longitudinal and turning coordinated flight maneuvers defined by a sequence of waypoints described below. The total time of flight is sufficiently short that the weight of the executive jet aircraft can be assumed to be a constant 73,000 lbs and the air density during the flight can be assumed to be constant.

The specified waypoints are defined with respect to a ground-fixed frame: the x and y axes are in a horizontal plane located at an altitude of 5,000 ft with the x-axis pointing due east and the y-axis pointing due north; the z-axis defines the altitude above this horizontal plane.

Waypoint	Name	Coordinates $(x,\ y,\ z)$ in ft
1	climb	$(0, 0, 0)$
2	left turn	$(20000, 0, 2000)$
3	descend	$(20000, -10000, 2000)$
4		$(0, -10000, 0)$

The aircraft should be in steady climbing flight between waypoints 1 and 2; it should be in steady level turn between waypoints 2 and 3, and it should be in steady descending flight between waypoints 3 and 4.

    (a) Select a flight plan for which the elapsed time from waypoint 1 to waypoint 4 is near minimum. Your flight plan should specify the constant airspeed, the constant climb rate, and the bank angle for each of the three steady flight segments that satisfy the waypoint specifications.

(b) For each of the steady flight segments constructed in part (a), determine the elevator deflection and the throttle setting that guarantee the desired steady flight conditions along that flight segment. Check that all flight constraints are satisfied for all steady flight segments.

(c) What is the elapsed flight time from waypoint 1 to waypoint 4?

(d) On the basis of your flight plan in part (a), give plot(s) that describe the 3D path of the aircraft.

12.6. Consider the propeller-driven general aviation aircraft carrying out a sequence of longitudinal and turning coordinated flight maneuvers defined by a sequence of waypoints described below. The total time of flight is sufficiently short that the weight of the general aviation aircraft can be assumed to be a constant 2,900 lbs and the air density during the flight can be assumed to be constant.

The specified waypoints are defined with respect to a ground-fixed frame: the x and y axes are in a horizontal plane located at an altitude of 4,000 ft with the x-axis pointing due east and the y-axis pointing due north; the z-axis defines the altitude above this horizontal plane.

Waypoint	Name	Coordinates (x, y, z) in ft
1	climb	(0, 0, 0)
2	right turn	(18000, 0, 2000)
3	descend	(21600, 3600, 2000)
4		(21600, 21600, 0)

The aircraft should be in steady climbing flight between waypoints 1 and 2; it should be in steady level turn between waypoints 2 and 3, and it should be in steady descending flight between waypoints 3 and 4.

(a) Select a flight plan for which the elapsed time from waypoint 1 to waypoint 4 is near minimum. Your flight plan should specify the constant airspeed, the constant climb rate, and the bank angle for each of the three steady flight segments that satisfy the waypoint specifications.

(b) For each of the steady flight segments constructed in part (a), determine the elevator deflection and the throttle setting that guarantee the desired steady flight conditions along that flight segment. Check that all flight constraints are satisfied for all steady flight segments.

(c) What is the elapsed flight time from waypoint 1 to waypoint 4.

(d) On the basis of your flight plan in part (a), give plot(s) that describe the 3D path of the aircraft.

12.7. Consider the UAV carrying out a sequence of longitudinal and turning coordinated flight maneuvers defined by a sequence of waypoints described below. The total time of flight is sufficiently short that the weight of the UAV can be assumed to be a constant 45 lbs and the air density during the flight can be assumed to be constant.

The specified waypoints are defined with respect to a ground-fixed frame: the x and y axes are in a horizontal plane located at an altitude of 1,000 ft with the x-axis pointing due east and the y-axis pointing due north; the z-axis defines the altitude above this horizontal plane.

Waypoint	Name	Coordinates (x, y, z) in ft
1	climb	(0, 0, 0)
2	right turn	(1000, 0, 100)
3	descend	(1000, 0, 100)
4		(2000, 0, 0)

The aircraft should be in steady climbing flight between waypoints 1 and 2; it should be in steady level turn between waypoints 2 and 3, and it should be in steady descending flight between waypoints 3 and 4.

(a) Select a flight plan for which the elapsed time from waypoint 1 to waypoint 4 is near minimum. Your flight plan should specify the constant airspeed, the constant climb rate, and the bank angle for each of the three steady flight segments that satisfy the waypoint specifications.

(b) For each of the steady flight segments constructed in part (a), determine the elevator deflection and the throttle setting that guarantee the desired steady flight conditions along that flight segment. Check that all flight constraints are satisfied for all steady flight segments.

(c) What is the elapsed flight time from waypoint 1 to waypoint 4.

(d) On the basis of your flight plan in part (a), give plot(s) that describe the 3D path of the aircraft.

12.8. Consider the executive jet aircraft carrying out a sequence of longitudinal coordinated flight maneuvers defined by a sequence of waypoints described below.

The specified waypoints are defined with respect to a ground-fixed frame: the x and y axes are in a horizontal plane located at an altitude of 5,000 ft with the x-axis pointing due east and the y-axis pointing due north; the z-axis defines the altitude above this horizontal plane.

Waypoint	Name	Coordinates (x, y, z) in ft
1	climb	(0, 0, 0)
2	cruise	(210000, 0, 21000)
3	descend	(1470000, 0, 21000)
4		(1680000, 0, 0)

The executive jet aircraft should be in steady climbing flight between waypoints 1 and 2; it should be in steady level flight between waypoints

2 and 3, and it should be in steady descending flight between waypoints 3 and 4.

Assume the aircraft takes off and has a full fuel tank at waypoint 1; ignore the mass of fuel burned during the climb segment and during the descent segment but not during the cruise segment.

(a) Develop a flight plan that specifies the airspeed and the flight path angle for each of the three steady flight segments that satisfy the waypoint specifications. Select a flight plan for which the elapsed flight time is minimum.

(b) For the climbing flight segment and the descending flight segment constructed in part (a), determine the elevator deflection and the throttle setting that guarantee the desired steady flight conditions along that flight segment; these expressions should be in terms of the air density. For the cruising flight segment determine the elevator deflection and the throttle as functions of the aircraft weight. Check that all flight constraints are satisfied for all steady flight segments.

(c) What amount of fuel is required to complete the flight? What is the elapsed flight time from waypoint 1 to waypoint 4.

(d) On the basis of your flight plan in part (a), give plot(s) that describe the 3D path of the executive jet aircraft.

12.9. Consider the general aviation aircraft carrying out a sequence of longitudinal coordinated flight maneuvers defined by a sequence of waypoints described below.

The specified waypoints are defined with respect to a ground-fixed frame: the x and y axes are in a horizontal plane located at an altitude of 1,000 ft with the x-axis pointing due east and the y-axis pointing due north; the z-axis defines the altitude above this horizontal plane.

Waypoint	Name	Coordinates $(x, \ y, \ z)$ in ft
1	climb	$(0, 0, 0)$
2	cruise	$(120000, 0, 12000)$
3	descend	$(2280000, 0, 12000)$
4		$(2400000, 0, 0)$

The general aviation aircraft should be in steady climbing flight between waypoints 1 and 2; it should be in steady level flight between waypoints 2 and 3, and it should be in steady descending flight between waypoints 3 and 4.

Assume the aircraft takes off with a full fuel tank at waypoint 1; ignore the mass of fuel burned during the climbing flight segment and during the descending flight segment but not during the cruising flight segment.

(a) Develop a flight plan that specifies the airspeed and the flight path angle for each of the three steady flight segments that satisfy the waypoint specifications. Select a flight plan that is fuel efficient.

(b) For the climbing flight segment and the descending flight segment constructed in part (a), determine the elevator deflection and the throttle setting that guarantee the desired steady flight conditions along that flight segment; these expressions should be in terms of the air density. For the cruising flight segment determine the the elevator deflection and the throttle as functions of the aircraft weight. Check that all flight constraints are satisfied for all steady flight segments.

(c) What amount of fuel is required to complete the flight? What is the elapsed flight time from waypoint 1 to waypoint 4.

(d) On the basis of your flight plan in part (a), give plot(s) that describe the 3D path of the general aviation aircraft.

From Steady Flight to Flight Dynamics

This chapter provides a mathematical description for the translational dynamics of a fixed-wing aircraft in flight. This description is referred to as an aircraft translational flight dynamics model, and it is described in terms of sets of ordinary differential equations.

The assumptions that lead to the aircraft translational dynamics model are first stated; the differential equations of motion are then derived, presented, and interpreted. The differential equations of motion characterize the three-dimensional translational dynamics of the aircraft center of mass, under the influence of aerodynamics, gravity, and propulsion forces. Consequently, these are three degrees of freedom flight descriptions. These differential equations *do not* explicitly capture the rotational dynamics of the aircraft. Ignoring the rotational dynamics of the aircraft is often justified in analysis of open-loop flight maneuvers, flight planning, and evaluation of dynamic flight performance measures; these flight properties are primarily dependent on the translational dynamics of the aircraft.

The three-dimensional flight equations for the translational dynamics of the aircraft presented here have been developed and studied in some detail in Vinh (1995); similar equations have been presented in the appendix of Eshelby (2000). Differential equations specifically for longitudinal translational flight have been given for a paper airplane model in Stengel (2009). These differential equations for aircraft translational flight have, perhaps, not received the attention they deserve. Their main utility is that they provide a natural and clear connection with the algebraic equations used for steady flight analysis and they are the simplest model that captures dynamic aircraft performance features.

These differential equations can be contrasted with commonly used differential equations for rigid body flight that explicitly describe both the translational and rotational dynamics of the aircraft. The former differential equations describe the three degrees of freedom for translational dynamics of the aircraft, ignoring the rotational dynamics of the aircraft. The latter differential equations describe the six degrees of freedom for both translational and rotational dynamics of the aircraft. Both sets of differential equations describe the aircraft dynamics in three spatial dimensions. The rigid body differential equations are primarily used for dynamic stability analysis and for design and analysis of feedback control systems. There is a large literature on the rigid body differential equations of motion. (For rigid body flight dynamics, see Etkin [1972], Pamadi [1998], and Stengel [2004].)

13.1 Flight Dynamics Assumptions

The aircraft translational dynamics equations are based on Newton's laws for translational motion of a rigid body, viewing the aircraft as a rigid body with a plane of mass symmetry. The aircraft is assumed to have its mass concentrated at its center of mass. The aircraft is acted on by aerodynamics, gravitational, and propulsive forces.

The following assumptions are made in the subsequent development:

- The aircraft is an ideal rigid body with a plane of mass symmetry.

- The mass of the aircraft is constant.

- Flat Earth is assumed.

- Gravity is constant and uniform.

- The atmosphere is stationary.

- The propulsion system provides a thrust force on the aircraft that acts along the body-fixed x_A-axis of the aircraft; the net thrust provided by the engine (in the case of an ideal jet engine) or the power provided by the engine and propeller (in the case of an ideal internal combustion engine) is proportional to the throttle setting; no additional engine dynamics are included here.

- The velocity vector always lies in the plane of mass symmetry of the aircraft; that is, there is no side-slipping of the aircraft.

- The aerodynamics forces consist of the lift force and the drag force defined according to standard aerodynamics principles; the drag force acts opposite to the velocity vector and the lift force acts normal to the velocity vector in the aircraft plane of mass symmetry; the angle between the lift force and the vertical is the bank angle about the velocity vector.

- An inertial or ground-fixed frame is defined whose x_E and y_E axes lie in a horizontal plane and whose z_E-axis is vertical, with positive axis up.

- A velocity frame is defined with origin located at the aircraft center of mass, with positive x_V-axis along the velocity vector of the aircraft in the direction of motion, positive z_V-axis normal to the velocity vector and lying in the vertical plane, and the positive y_V-axis in the horizontal plane completing a right-hand frame; this frame is referred to as the velocity frame; the velocity frame is neither an inertial frame nor a body-fixed frame.

The following dynamic flight variables are introduced; the notation is consistent with the notation used in the prior chapters in studying steady flight. In contrast with the situation for steady flight, these flight variables are not necessarily constant but vary with time according to the governing differential equations that are subsequently described:

- (x, y, z) are components of the position vector of the aircraft with respect to the ground frame.

- (V, σ, γ) are the magnitude of the velocity vector of the aircraft referred to as the airspeed, the aircraft heading angle, and the flight path angle defined by the direction of the velocity vector in the ground frame.

- L, D are lift and drag forces defined according to standard aerodynamic principles.

- C_L, C_D are corresponding lift coefficient and drag coefficient.

- α is the aircraft angle of attack.

- μ is the bank angle of the aircraft about its velocity vector.

- T is the magnitude of the thrust vector.

Some important aircraft parameters for flight dynamics follow:

- W is the constant aircraft weight.

- g is constant acceleration of gravity.

- ρ is the air density; it is a function of the altitude according to the standard atmospheric model.

- S is the constant wing surface area.

- Aerodynamics data are given by constants $C_{L0}, C_{L\alpha}, C_{D0}, K$.

13.2 Differential Equations for the Translational Flight Dynamics

The translational differential equations of motion are expressed in terms of the airspeed of the aircraft, the heading angle, and the flight path angle; these flight variables define the direction of the velocity vector with respect to the ground frame.

Under the stated assumptions, the translational dynamics equations follow from Newton's laws that the time rate of change of the linear (translational) momentum of the aircraft is the net external force on the aircraft. Expressing this in the velocity frame, which is a rotating frame, Newton's law is

$$m \left(\begin{bmatrix} \frac{dV}{dt} \\ 0 \\ 0 \end{bmatrix} + \omega \times \begin{bmatrix} V \\ 0 \\ 0 \end{bmatrix} \right) = F. \tag{13.1}$$

The left side of the vector equation describes the components of the time rate of change of the momentum vector of the aircraft in the velocity frame; the right-hand side is the components of the net force vector on the aircraft in the velocity frame. In this expression, ω denotes the angular velocity vector of the velocity frame, expressed in the velocity frame, and F denotes the net force vector acting on the aircraft in the velocity frame. The cross product term in the acceleration expression reflects the fact that the velocity frame rotates and is not inertial.

The angular velocity vector of the velocity frame can be expressed in terms of the time derivative of the heading angle, the time derivative of the flight path angle, and the flight path angle according to

$$\omega = \begin{bmatrix} -\frac{d\sigma}{dt} \sin \gamma \\ \frac{d\gamma}{dt} \\ \frac{d\sigma}{dt} \cos \gamma \end{bmatrix}. \tag{13.2}$$

Thus the acceleration vector of the aircraft in the velocity frame is

$$m \left(\begin{bmatrix} \frac{dV}{dt} \\ 0 \\ 0 \end{bmatrix} + \omega \times \begin{bmatrix} V \\ 0 \\ 0 \end{bmatrix} \right) = \begin{bmatrix} \frac{dV}{dt} \\ V \cos \gamma \frac{d\sigma}{dt} \\ -V \frac{d\gamma}{dt} \end{bmatrix}. \tag{13.3}$$

The net force vector on the aircraft is the vector sum of the force of gravity, the lift and drag aerodynamic forces, and the propulsive force. The net force vector, expressed in the velocity frame, is

$$F = W \begin{bmatrix} -\sin \gamma \\ 0 \\ \cos \gamma \end{bmatrix} + \begin{bmatrix} -D \\ L \sin \mu \\ -L \cos \mu \end{bmatrix} + T \begin{bmatrix} \cos \alpha \\ \sin \alpha \sin \mu \\ -\sin \alpha \cos \mu \end{bmatrix}. \tag{13.4}$$

Expressions (13.3) and (13.4) can be substituted into the above form of Newton's law given by equation (13.1) to obtain the differential equations of motion for aircraft flight dynamics:

$$\frac{W}{g} \frac{dV}{dt} = -W \sin \gamma - D + T \cos \alpha, \tag{13.5}$$

$$\frac{W}{g} V \cos \gamma \frac{d\sigma}{dt} = L \sin \mu + T \sin \alpha \sin \mu, \tag{13.6}$$

$$\frac{W}{g} V \frac{d\gamma}{dt} = -W \cos \gamma + L \cos \mu + T \sin \alpha \cos \mu. \tag{13.7}$$

The lift and drag forces in the translational dynamics equations are given by

$$L = \frac{1}{2}\rho V^2 S C_L, \tag{13.8}$$

$$D = \frac{1}{2}\rho V^2 S C_D. \tag{13.9}$$

The lift coefficient and drag coefficient are functions of the angle of attack according to

$$C_L = C_{L_0} + C_{L_\alpha}\alpha, \tag{13.10}$$

and the quadratic drag polar expression is

$$C_D = C_{D_0} + K C_L^2. \tag{13.11}$$

These expressions for the lift and drag are essentially the same as those used for steady flight analysis. Theoretical aerodynamics suggests that it may be important to use more complicated expressions. For example, the lift coefficient and the drag coefficient may have a significant dependence on the time derivative of the angle of attack and on the pitch rate, which is the time derivative of the sum of the angle of attack and the flight path angle.

The translational kinematics differential equations express the components of the velocity vector of the aircraft in the ground frame

$$\frac{dx}{dt} = V \cos\sigma \cos\gamma, \tag{13.12}$$

$$\frac{dy}{dt} = V \sin\sigma \cos\gamma, \tag{13.13}$$

$$\frac{dz}{dt} = V \sin\gamma. \tag{13.14}$$

The above equations of motion, namely equations (13.5), (13.6), (13.7), (13.12), (13.13), and (13.14), for the three-dimensional translational dynamics of an aircraft are expressed as six first order differential equations for the three aircraft translational degrees of freedom. The following observations can be made:

- There are six state variables for the above equations: the three translational position components of the aircraft in the ground frame, namely the aircraft airspeed, the heading angle, and the flight path angle that describe the magnitude and direction of the velocity vector.

- There are three aircraft input variables, namely the thrust magnitude, the angle of attack, and the bank angle about the velocity vector.

- If the lift coefficient is assumed to depend on the time rate of change of the angle of attack, it is natural to view the angle of attack as an additional state variable and to view the time rate of change of the angle of attack as an aircraft input variable.

- Since the atmosphere is stationary, the time rate of change of the airspeed, the time rate of change of the heading angle, and the time rate of change of the flight path angle do not depend on the horizontal position variables x and y and depend on the altitude variable z only through the air density term. This implies that the four differential equations, namely equations (13.5), (13.6), (13.7), and (13.14), for the airspeed, the heading angle, the flight path angle, and the altitude can be considered alone. If one makes the additional assumption that the altitude change is small so that the air density is essentially constant, then the three differential equations, namely equations (13.5), (13.6), and (13.7), for the aircraft airspeed, the heading angle, and the flight path angle may be considered alone.

- Important constraints are imposed as inequality constraints:
  - propulsion constraints on the thrust or power magnitude that arise due to engine limitations;
  - stall constraint that reflects the aerodynamics of the aircraft;
  - load constraint that reflects the structural limitations of the aircraft wing.

The above six differential equations and the associated three constraints characterize the aircraft translational dynamics in three spatial dimensions. These are the simplest such differential equations that can be used for mathematical analysis of dynamic flight performance and for flight simulation. Note that the propulsion constraints, the stall constraint, and the load constraint only constrain the aircraft input variables, namely the thrust magnitude, the angle of attack, and the bank angle about the velocity vector.

The above differential equations can be written in several different forms. As one example, it is possible to introduce the aircraft climb rate as a state variable in place of the flight path angle; these two flight variables are related by a simple transformation involving the airspeed.

As in our analysis of steady flight in previous chapters, it is often justified to make the assumption that the flight path angle and the angle of attack are small angles in radians. In such cases, the approximations for the sine of a small angle and the cosine of a small angle can be made to further simplify the above differential equations. This simplification is not made in the subsequent development.

13.3 Including Engine Characteristics and Fuel Consumption

The prior differential equations are based on the assumption of a constant aircraft mass or weight and hence are most suitable for relatively short time periods where this approximation is justified. For longer flight times, the change in aircraft mass or weight should be taken into account, reflecting the engine consumption of fuel. In particular, the time rate of change of the aircraft weight, including fuel, is the negative of the product of a specific fuel consumption rate and the engine thrust (for an ideal jet engine) or the engine power (for an ideal propeller and internal combustion engine); see chapter 4. This differential equation can be adjoined to the prior differential equations, which must also be slightly modified to include the time varying aircraft mass in the expression for the time derivative of the linear momentum of the aircraft in the derivation.

The differential equations for the aircraft translational dynamics hold for any type of propulsion system. For an aircraft whose propulsion system is an ideal jet engine, it is natural to view the thrust produced by the engine as an aircraft input. For an aircraft whose propulsion system consists of an ideal internal combustion engine and a propeller, it is natural to view the power produced by the engine and propeller as an input; assuming the engine is in steady operation, the thrust produced is the ratio of the power produced and the airspeed of the aircraft. Two different sets of differential equations for the aircraft translational dynamics are now presented that incorporate fuel consumption and engine characteristics.

In the case of an aircraft with an ideal jet engine the differential equations, taking into account that the aircraft weight is not constant in time, are

$$\frac{W}{g}\frac{dV}{dt} + V\frac{dW}{dt} = -W\sin\gamma - D + T\cos\alpha, \tag{13.15}$$

$$\frac{W}{g}V\cos\gamma\frac{d\sigma}{dt} = L\sin\mu + T\sin\alpha\sin\mu, \tag{13.16}$$

$$\frac{W}{g}V\frac{d\gamma}{dt} = -W\cos\gamma + L\cos\mu + T\sin\alpha\cos\mu, \tag{13.17}$$

$$\frac{dW}{dt} = -cT. \tag{13.18}$$

The constant c is the thrust specific fuel consumption rate for the jet engine; T denotes the thrust developed by the jet engine.

In the case of an aircraft powered by an ideal internal combustion engine and propeller the differential equations, taking into account that the aircraft weight is not constant in time, are

$$\frac{W}{g}\frac{dV}{dt} + V\frac{dW}{dt} = -W\sin\gamma - D + \frac{P}{V}\cos\alpha, \tag{13.19}$$

$$\frac{W}{g}V\cos\gamma\frac{d\sigma}{dt} = L\sin\mu + \frac{P}{V}\sin\alpha\sin\mu, \tag{13.20}$$

$$\frac{W}{g}V\frac{d\gamma}{dt} = -W\cos\gamma + L\cos\mu + \frac{P}{V}\sin\alpha\cos\mu, \tag{13.21}$$

$$\frac{dW}{dt} = -c\frac{P}{\eta}. \tag{13.22}$$

The constant c is the power specific fuel consumption rate for the internal combustion engine and η is the propeller efficiency; P denotes the net power developed by the engine and the propeller that is available for flight.

The differential equations for the aircraft translational kinematics can be added to these sets of equations. The following comments can be made about these differential equations:

- There are seven state variables for the above differential equations: the three translational position components of the aircraft in the ground frame, the aircraft airspeed, the heading angle, and the flight path angle that describe the direction of the velocity vector, and the weight of the aircraft.

- There are three aircraft input variables, namely the thrust magnitude (or the power) developed by the propulsion system, the angle of attack, and the bank angle about the velocity vector.

- If the lift coefficient is assumed to depend on the time rate of change of the angle of attack, it is natural to view the angle of attack as an additional state variable and to view the time rate of change of the angle of attack as an input variable.

- Assuming that the atmosphere is stationary, then the time rate of change of the airspeed, the time rate of change of the heading angle, the time rate of change of the flight path angle, and the time rate of change of the aircraft weight do not depend on the horizontal position variables x and y and depend on the altitude variable z only through the air density term. This implies that the five differential equations, namely equations (13.15), (13.16), (13.17), (13.18), and (13.14), or (13.19), (13.20), (13.21), (13.22), and (13.14), for the aircraft airspeed, the heading angle, the flight path angle, the aircraft weight, and the altitude can be considered alone. If the additional assumption is made that the altitude change is small, then the air density is essentially constant, and the four differential equations, namely equations (13.15), (13.16), (13.17), and (13.18), or (13.19), (13.20), (13.21), and (13.22), for the airspeed, the heading angle, the flight path angle, and the aircraft weight can be considered alone.

- Important constraints are imposed as inequality constraints:
  - propulsion constraints on the thrust or power magnitude that arise due to engine limitations;
  - stall constraint that reflects the aerodynamics of the aircraft;
  - load constraint that reflects the structural limitations of the aircraft wing.

The aircraft translational differential equations, which include the effects of varying aircraft weight, are most important for studying flight dynamics over relatively long flight times where the aircraft weight can change significantly. In particular, such models are useful in evaluating aircraft range and aircraft endurance for dynamic flight.

13.4 Differential Equations for Longitudinal Translational Flight Dynamics

Assuming the weight of the aircraft is constant, the differential equations of motion for an aircraft that flies in a fixed vertical plane can be described by $y = 0$. Assume that the heading angle is identically zero and the roll angle about the velocity vector is identically zero; this guarantees longitudinal flight of the aircraft in the vertical plane. The differential equations (13.15)–(13.18) simplify to the following longitudinal form.

The translational differential equations that characterize longitudinal flight are

$$\frac{W}{g}\frac{dV}{dt} = -W\sin\gamma - D + T\cos\alpha, \tag{13.23}$$

$$\frac{W}{g}V\frac{d\gamma}{dt} = -W\cos\gamma + L + T\sin\alpha. \tag{13.24}$$

The lift and drag forces in the translational dynamics equations are given by

$$L = \frac{1}{2}\rho V^2 S C_L, \tag{13.25}$$

$$D = \frac{1}{2}\rho V^2 S C_D. \tag{13.26}$$

The lift coefficient and drag coefficient are functions of the angle of attack according to

$$C_L = C_{L_0} + C_{L_\alpha}\alpha, \tag{13.27}$$

and the quadratic drag polar expression is

$$C_D = C_{D_0} + K C_L^2. \tag{13.28}$$

These expressions for the lift and drag are essentially the same as those used for steady flight analysis. As previously, the lift coefficient and the drag coefficient may have a significant dependence on the rate of change of the angle of attack and on the rate of change of the pitch angle.

The translational kinematics equations express the components of the velocity vector of the aircraft in the ground frame:

$$\frac{dx}{dt} = V\cos\gamma, \tag{13.29}$$

$$\frac{dz}{dt} = V\sin\gamma. \tag{13.30}$$

The above longitudinal equations of motion for three-dimensional flight of an aircraft are expressed as four first order differential equations, namely equations (13.23), (13.24), (13.29), and (13.30), for the two aircraft translational degrees of freedom. The following observations can be made:

- There are four state variables for the above equations: the two translational position components of the aircraft in the ground frame, the airspeed, and the flight path angle that describes the direction of the velocity vector.

- There are two aircraft input variables, namely the thrust magnitude and the angle of attack.

- If the lift coefficient is assumed to depend on the time rate of change of the angle of attack, it is natural to view the angle of attack as an additional state variable and to view the time rate of change of the angle of attack as an input variable.

- Since the atmosphere is stationary, the time rate of change of the airspeed and the time rate of change of the flight path angle do not depend on the horizontal position variable x and depend on the altitude variable z only through the air density term. This implies that the three differential equations, namely equations (13.23), (13.24), and (13.30), for the airspeed, the flight path angle, and the altitude can be considered alone. If one makes the additional assumption that the altitude change is small so that the air density is essentially constant, then the two differential equations, namely equations (13.23) and (13.24), for the airspeed and the flight path angle can be considered alone.

- Important constraints are imposed as inequality constraints:
  - propulsion constraints on the thrust or power magnitude that arise due to engine limitations;
  - stall constraint that reflects the aerodynamics of the aircraft.

The above four differential equations and the associated two constraints define the longitudinal differential equations that characterize flight performance in a fixed vertical plane. These are the simplest such differential equations that can be used for mathematical analysis of longitudinal flight performance and for flight simulation of the longitudinal dynamics. Note that the propulsion constraints and the stall constraint constrain the input variables only, namely the thrust magnitude and the angle of attack.

13.5 Differential Equations for Takeoff and Landing

Aircraft takeoff and landing maneuvers have not been previously examined in this book since such maneuvers necessarily involve nontrivial (positive or negative) acceleration; they are not steady flight maneuvers. Such maneuvers, strictly speaking, are not even flight maneuvers since the aircraft maintains rolling contact with the ground.

It is an easy matter to modify the translational differential equations that characterize longitudinal translational flight to obtain differential equations that describe the takeoff and landing dynamics during the ground roll of the aircraft. Assuming that the runway is horizontal, it follows that takeoffs and landings occur when the flight path angle $\gamma = 0$. A vertical ground reaction force, denoted by N, and a horizontal friction force, denoted by $\mu_F N$, characterize the forces of the runway on the aircraft and must be included in the description of the takeoff and landing dynamics. Here μ_F is the constant coefficient of friction with the runway. It is easy to modify the longitudinal translational equations given by equations (13.23) and (13.24) to include these assumptions; this leads to the translational differential equations that characterize takeoffs and landings, namely

$$\frac{W}{g}\frac{dV}{dt} = -D - \mu_F N + T\cos\alpha, \tag{13.31}$$

$$0 = -W + L + N + T\sin\alpha. \tag{13.32}$$

The vertical ground reaction force can be determined from equation (13.32) and substituted into equation (13.31) to obtain

$$\frac{W}{g}\frac{dV}{dt} = -D - \mu_F (W - L) + T\cos\alpha\,(1 + \mu_F). \tag{13.33}$$

As usual, the lift and drag forces in the translational dynamics equations are given by

$$L = \frac{1}{2}\rho V^2 S C_L, \tag{13.34}$$

$$D = \frac{1}{2}\rho V^2 S C_D. \tag{13.35}$$

The lift coefficient and drag coefficient are functions of the angle of attack according to

$$C_L = C_{L_0} + C_{L_\alpha}\alpha, \tag{13.36}$$

and the quadratic drag polar expression is

$$C_D = C_{D_0} + K C_L^2. \tag{13.37}$$

The translational kinematics equation expresses the horizontal component of the velocity vector of the aircraft in the ground frame during a takeoff or landing as

$$\frac{dx}{dt} = V. \tag{13.38}$$

The above differential equations of motion for one-dimensional motion of an aircraft during a takeoff or landing are expressed as two first order differential equations, namely equations (13.33) and (13.38). The following observations can be made:

- There are two state variables for the above equations: the horizontal position of the aircraft in the ground frame and the airspeed.

- There are two aircraft input variables, namely the thrust magnitude and the angle of attack.

- Several terms in the differential equation (13.33) can be significantly influenced by special aircraft features. On landing, thrust reversers, if the engine has this capability, and braking are often used. Many aircraft have special flaps that, if deployed, can significantly increase the lift on takeoff and increase the drag on landing. These effects are easily included in the differential equations for takeoffs and landings.

13.6 Steady Flight and the Translational Flight Dynamics

In this section, suppose that the input variables, namely the thrust magnitude, the angle of attack, and the bank angle about the velocity vector, are constant in time; we also assume that the weight of the aircraft is effectively constant. The most general form of steady flight, as developed in the prior chapters, occurs when the center of mass of the aircraft flies along a helical path with vertical axis. In this case, the aircraft airspeed is constant and the flight path angle is constant. The heading angle is not constant, but the time rate of change of the heading angle in steady flight is constant and is given by

$$\frac{d\sigma}{dt} = \frac{V \cos \gamma}{r}, \tag{13.39}$$

where r denotes the constant radius of the helical path. Thus for steady flight

$$\frac{dV}{dt} = 0, \tag{13.40}$$

$$\frac{d\gamma}{dt} = 0, \tag{13.41}$$

$$\frac{d\sigma}{dt} = \frac{V \cos \gamma}{r}. \tag{13.42}$$

Substituting these conditions into the previous differential equations for dynamic flight gives the algebraic equations

$$0 = -W \sin \gamma - D + T \cos \alpha, \tag{13.43}$$

$$\frac{W}{g} \frac{V^2 \cos^2 \gamma}{r} = L \sin \mu + T \sin \alpha \sin \mu, \tag{13.44}$$

$$0 = -W \cos \gamma + L \cos \mu + T \sin \alpha \cos \mu, \tag{13.45}$$

which characterize the most general form of steady flight. These equations are identical to the steady climbing and turning flight equations (10.1), (10.2), and (10.3).

In summary, the most general form of steady aircraft flight corresponds to solutions of the differential equations for the translational flight dynamics for which the aircraft airspeed, the flight path angle, and the rate of change of the heading angle are constants in time. Such solutions are referred to as equilibrium solutions of the differential equations for the translational flight dynamics.

This feature has many important applications. As one illustration, these differential equations can be approximated, for small perturbations from a specific steady flight condition, by linear differential equations. Such linear approximations can be utilized as a basis for analytical studies of the flight dynamics for small perturbations from the steady flight condition.

For the special case of dynamic longitudinal flight, it is also true that steady longitudinal flight corresponds to solutions of the longitudinal differential equations for which the magnitude of the aircraft velocity and the flight path angle are constants in time. Setting the derivatives of these constant solutions to zero leads to the steady longitudinal flight equations that have been previously developed and used in the prior chapters of this book. The steady longitudinal flight solutions are equilibrium solutions of the differential equations for the longitudinal translational flight dynamics.

13.7 Dynamic Flight Stability

The concept of static flight stability was introduced in section 12.1. The conditions given there are based on a purely static analysis of the forces and moments that act on an aircraft when perturbed from a steady flight condition. A more detailed analysis is required to obtain conditions for flight stability of a steady flight condition that incorporates the aircraft dynamics. Dynamic flight stability is beyond the scope of the present book, but a few comments are made and readers are referred to more extensive accounts of flight stability in Stengel (2004) and references therein.

Dynamic flight stability of a steady flight condition of an aircraft is the following property: arbitrary small initial perturbations in the state variables from steady flight cause subsequent perturbations of the aircraft from steady flight that remain small thereafter. Since the initial flight perturbations can be arbitrary (but small), the resulting aircraft flight need not be steady; that is, the effect of small initial perturbations requires the analysis of flight dynamics, as described by differential equations, to assess the subsequent motion. Dynamic flight stability is a property of the particular aircraft and of a particular steady flight condition.

In addition to dynamic flight stability, it is often desirable that the effects of initial state perturbations decay to zero. If this is true for arbitrary small initial state perturbations, the steady flight condition is said to be asymptotically stable. Practical aircraft can have steady flight conditions that are either stable or unstable, depending on the aircraft design. An unstable steady flight condition is often difficult for a pilot to fly, since the effects of the perturbations must be compensated by manual control of the aircraft inputs. In many cases where high performance flight is desired, flight control systems are introduced to automatically adjust the aircraft inputs to attenuate the effects of flight perturbations. In this way, an unstable aircraft that incorporates a control system, sometimes referred to as an automatic pilot, may be made asymptotically stable with suitable handling qualities.

Based on differential equations that describe the aircraft translational dynamics, as described previously in this chapter, mathematical approaches are available to assess the stability properties of a particular steady flight condition. The most common approach is to approximate the differential equations by linear differential equations that are valid for small perturbations from the steady flight condition. Then flight stability of the steady flight condition is directly related to the eigenvalues of these linear differential equations.

In the subsequent flight simulation examples presented in this chapter, it is easy to assess dynamic flight stability directly from the flight simulation results.

13.8 Computing Dynamic Flight Performance Measures and Flight Envelopes

Dynamic flight performance is a subject of substantial importance for many modern aircraft. Although a detailed study of dynamic performance is beyond the scope of this book, several comments can be made about the way in which the differential equations for the translational flight dynamics can be used to obtain analytical flight performance relationships and computational flight performance results.

Typically, the differential equations are viewed as defining an initial value problem. This means that the inputs, namely the thrust magnitude (or power) provided by the propulsion system, the angle of attack, and the bank angle about the velocity vector, are specified as functions of time that satisfy the aircraft constraints; the initial values of the state variables, such as the aircraft airspeed, the flight path angle, and the heading angle, and the three variables that describe the location of the center of mass of the aircraft, are also specified. It follows that the differential equations, for the specified input functions, have a unique solution that satisfies the specified initial values.

Although it is usually not possible to obtain such solutions to an initial value problem in simple analytical terms, there are special cases where analytical solutions can be obtained.

- If the aircraft input functions are constant in time and satisfy the flight constraints, then there may be steady flight solutions, corresponding to constant values of the airspeed, the flight path angle, and the rate of change of the heading angle. As seen in previous chapters of this book, there may be no, one, or several steady flight solutions. The relevant initial values are determined by the values of the steady flight solutions.

- If the aircraft input functions are constant in time and satisfy the flight constraints, and the initial values are small perturbations of a steady flight solution, then the differential equations for the translational flight dynamics can be approximated by linear differential equations. Standard methods are available for obtaining analytical solutions for these linear differential equations; the solutions of these approximating linear differential equations are approximations of the solutions of the differential equations for the translational flight dynamics, at least over sufficiently short time periods.

More generally, numerical methods can be used to obtain solution approximations of initial value problems for the differential equations. Various computational environments, including Matlab, can be used to obtain such approximate solutions for an initial value problem. This process is referred to as flight simulation.

There are many important dynamic performance measures that are used to quantify dynamic flight performance. Some of these are natural modifications of performance measures used for steady flight performance. In other cases, dynamic performance measures have no counterpart in steady flight. Examples of dynamic performance measures include the following:

- maximum airspeed of an aircraft after a fixed time of flight, assuming specified initial flight conditions;

- minimum airspeed of an aircraft after a fixed time of flight, assuming specified initial flight conditions;

- maximum flight path angle of an aircraft after a fixed time of flight, assuming specified initial flight conditions;

- maximum flight altitude of an aircraft after a fixed time of flight, assuming specified initial flight conditions;

- minimum turn radius of an aircraft after a fixed time of flight, assuming specified initial flight conditions;

- maximum aircraft range for a fixed amount of fuel, assuming specified initial flight conditions;

- maximum endurance for a fixed amount of fuel, assuming specified initial flight conditions;

- minimum ground run distance to achieve takeoff, assuming airspeed is initially zero;

- minimum ground run distance to stop during landing, assuming specified initial airspeed at touchdown.

Many different dynamic flight envelopes can be described based on the differential equations for the translational dynamics of an aircraft. Examples of dynamic flight envelopes include:

- the flight envelope that defines the set of all possible altitudes and airspeeds that the aircraft can achieve after a fixed time of flight, assuming specified initial flight conditions and flight along a fixed horizontal straight line;

- the flight envelope that defines the set of all possible altitudes, flight path angles (or climb rates), and airspeeds that the aircraft can achieve after a fixed time of flight, assuming specified initial flight conditions and flight in a fixed vertical plane;

- the flight envelope that defines the set of all possible altitudes, turn radii, and airspeeds that the aircraft can achieve after a fixed time of flight, assuming specified initial flight conditions and flight in a fixed horizontal plane;

- the flight envelope that defines the set of all possible altitudes, flight path angles (or climb rates), turn radii, and airspeeds that the aircraft can achieve after a fixed time of flight, assuming specified initial flight conditions;

- the flight envelope that defines the set of all possible altitudes, flight path angles (or climb rates), turn radii, and airspeeds that the aircraft can achieve for any time of flight, assuming specified initial flight conditions.

Computation of a specific flight performance metric or a specific dynamic flight envelope almost always involves the use of advanced computational algorithms and numerical software. In particular, the theoretical and computational methods of optimal control provide the basis for such computations. Although the details of these methods are beyond the scope of this book, the methods make important use of the differential equations for the translational dynamics of the aircraft that have been developed in this chapter. Determination of dynamic flight envelopes and dynamic flight performance measures, which generalize the steady flight envelopes and steady flight performance measures that have been the primary emphasis of this book, requires the use of more advanced theoretical and computational methods.

Although it is beyond the scope of the present book to develop, in detail, dynamic flight envelopes or to evaluate dynamic performance measures, it is possible to simulate the differential equations for the translational flight dynamics and to compare the dynamic simulation results with the results for steady flight developed previously in the book. The steps in a typical flight simulation are: select the relevant flight dynamics model, specify the values of the parameters in the differential equations, specify the aircraft input functions (for example, the thrust or power delivered by the propulsion system, the angle of attack, and the bank angle about the velocity vector as functions of time), and specify initial state values (for example, the initial value of the aircraft airspeed, the initial value of the heading angle, the initial value of the flight path angle, and initial values for any other state variables incorporated in the simulation model). Then Matlab can numerically integrate the initial value problem and provide graphical output of the resulting solution data for the initial value problem.

13.9 Flight Simulations: Executive Jet Aircraft

Several flight simulations of the executive jet aircraft are developed in this section. The data for the executive jet aircraft are the same as in the prior chapters. For each simulation, we specify the differential equations that define the initial value problem, we specify input variables as functions of time, and we specify the initial values of the state variables; then we use Matlab to numerically integrate the initial value problem. Plots of the solutions of the initial value problem are provided. The physical meaning of the development is emphasized throughout.

Simulation of a perturbation from a maximum speed steady climbing flight condition

Assuming the aircraft weight is constant, a longitudinal flight model, based on the longitudinal performance differential equations (13.23) and (13.24), is used to study the aircraft dynamics. The input functions are the thrust magnitude, assumed to be a constant 10,409.0 lbs, the angle of attack, assumed to be a constant 1.37 degrees, and the bank angle about the velocity vector, assumed to be identically zero. These constant input functions satisfy the engine constraints, the stall constraint, and the load constraint; the thrust value corresponds to full throttle at an altitude of 10, 000 ft. The flight simulation is for a sufficiently short duration that the air density has the essentially constant value of $0.0017553 \, \text{slug/ft}^3$.

The results in chapter 8 show that the executive jet aircraft is in steady climbing flight if the aircraft airspeed is 688.25 ft/s and the flight path angle is 3 degrees. This flight condition corresponds to maximum airspeed for the executive jet aircraft in steady climbing flight. This flight condition motivates the flight simulation results that are subsequently developed. This steady climbing flight condition is based on approximations made in chapter 8.

The flight simulation results for the executive jet aircraft in this section are based on the differential equations for longitudinal flight dynamics given by equations (13.15) and (13.17), with μ and σ identically zero. In this flight simulation, we are not directly interested in the displacement of the aircraft as it varies with time; hence no aircraft kinematics equations are included in the simulation.

The flight simulation results are obtained for the initial value of the airspeed given by 620 ft/s and the initial value of the flight path angle given by 1.07 degrees. The initial value of the aircraft airspeed is smaller than the value required for steady climbing flight; the initial value of the flight path angle is slightly smaller than the value required for steady climbing flight.

The simulation of this initial value problem provides the time responses that describe the resulting motion of the aircraft. This resulting motion does not correspond to steady flight: the resulting airspeed and flight path angle are not constants but vary with time. Although the specified values of the thrust magnitude and the angle of attack correspond to the maximum aircraft airspeed in steady flight, these constant values do not maximize the aircraft airspeed for arbitrary initial values of the airspeed and the flight path angle.

Numerical flight simulation is easily carried out in Matlab; details on numerical integration in Matlab are available through numerous Internet sources and in many Matlab tutorials. It is necessary to specify the total simulation time; here we simulate the aircraft flight for 300 s. A specific numerical integration algorithm can be specified; here we use a standard Runge-Kutta algorithm; for simplicity, no additional integration options are specified.

The differential equations are specified in the following Matlab m-file.

Matlab function 13.1 FSEJ1.m

```
function dX = FSEJ1(t,X)
% Input: values of magnitude of velocity vector V (ft/s)
% and flight path angle gamma (rad) stored in the vector X
% Output: values of time derivative of magnitude of velocity
% vector dV (ft/s^2) and time derivative of flight path angle
% dgamma (rad/s) stored in the vector dX
% Organize function input data into scalar components
V=X(1); gamma=X(2);
% Enter parameter data for executive jet aircraft
W=73000.0; CD0=0.015; K=0.05; CL0=0.02; CLa=0.12; S=950.0;
% Enter air density (slugs/ft^3), assumed constant depending
% on the approximate flight altitude
rho=1.7553E-3;
% Specify input functions: thrust magnitude (lbs), angle of
% attack (rad)
```

```
T=7961.7;
alpha=1.37*3.14159/180;
% Compute lift and drag on executive jet aircraft; angle of
% attack in degrees here
CL=CL0+CLa*180*alpha/3.14159;
CD=CD0+K*CL*CL;
L=0.5*rho*V*V*S*CL;
D=0..5*rho*V*V*S*CD;
% Compute time derivatives of magnitude of aircraft velocity
% vector and flight path angle based on performance
% differential equations
dV=-32.2*sin(gamma)-D*32.2/W+T*32.2*cos(alpha)/W;
dgamma=-32.2*cos(gamma)/V+L*32.2/(W*V)+T*32.2*sin(alpha)/(W*V);
% Organize function output time derivative data into a column
% vector
dX =[dV;dgamma];
```

The numerical integration of these differential equations for the executive jet aircraft, with the specified input functions and the specified initial values, is executed over the simulation time by the following Matlab commands: the discretized values of the simulation time are stored in the data vector TT and discretized "solution" values of the two state variables are stored in the columns of the data array Xsol. Several typical plot commands are included to obtain graphics output.

```
Tsim=300;
V0=620.0; gamma0=0.0187;
[TT,Xsol] = ode45(@FSEJ1,[0,Tsim],[V0;gamma0]);
plot(TT,Xsol(:,1))
axis([0 300 0 800])
xlabel('Time (s)')
ylabel('Magnitude of aircraft velocity vector (ft/s)')
plot(TT,Xsol(:,2)*180/3.14159)
axis([0 240 -8 8])
xlabel('Time (s)');
ylabel('Flight path angle (degrees)');
```

The graphics output plots shown in figure 13.1 have the following interpretation. The perturbations in the initial values of the airspeed and the flight path angle cause perturbations in the aircraft motions from the maximum speed steady climbing flight solution; since the perturbations lie within the fixed vertical flight plane, the resulting aircraft flight remains within that fixed vertical plane. These perturbations exhibit a long period, slowly decaying oscillation, sometimes referred to as a phugoid oscillation, toward the steady climbing flight values. The oscillation frequency and the decay rate can be quantified, but we do not provide that analysis. The slow decay of the perturbations in time is a consequence of the flight stability property of this steady climbing flight condition for the executive jet aircraft.

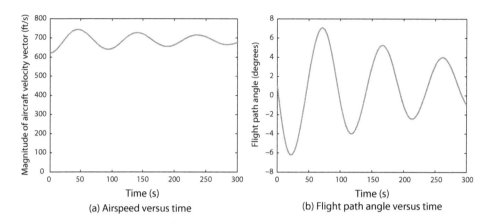

Figure 13.1. Time responses of aircraft airspeed and flight path angle in longitudinal flight: executive jet aircraft.

Simulation of a Perturbation from a Steady Turning and Climbing Flight Condition

Assuming the aircraft weight is constant, a 3D flight model, based on the differential equations (13.5)–(13.7), is used to study the aircraft dynamics. The input functions are the thrust magnitude, assumed to be a constant 7,961.7 lbs, the angle of attack, assumed to be a constant 4.4 degrees, and the bank angle about the velocity vector, assumed to be a constant 15 degrees. These constant input functions satisfy the engine constraints, the stall constraint, and the load constraint. As before, the flight simulation is for a sufficiently short duration that the air density is essentially constant at the value of 0.0017553 slug/ft^3, which corresponds to a flight altitude of 10,000 ft.

The results obtained in chapter 10 show that the executive jet aircraft is in steady climbing and turning flight if the aircraft airspeed is 406.87 ft/s, the turn radius is 19,186.0 ft, and the flight path angle is 3 degrees. The path of the aircraft in this steady flight condition is along the arc of a helix with a vertical axis. The constant rate of change of the heading angle in this steady climbing turn is

$$\frac{d\sigma}{dt} = \frac{V \cos \gamma}{r} = 0.02118 \, \text{rad/s}.$$

This steady flight condition for the executive jet aircraft motivates the flight simulation results that are subsequently developed. This steady climbing and turning flight condition is based on approximations made in chapter 10.

The flight simulation results for the executive jet aircraft in this section are based on the differential equations given by equations (13.5), (13.6), and (13.7). In this flight simulation, we are not directly interested in the displacement of the aircraft as it varies with time; hence no aircraft kinematics equations are included in the simulation. The flight simulation results are obtained for the initial value of the airspeed given by 380 ft/s and the initial value of the flight path angle given by 3 degrees. The initial value of the aircraft airspeed is smaller than the value required for steady climbing and turning flight; the initial value of the flight path angle is the value required for steady climbing flight.

The simulation of this initial value problem provides the time responses that describe the resulting motion of the aircraft. This resulting motion does not correspond to steady flight: the resulting airspeed and flight path angle are not constants but vary with time.

Numerical flight simulation is carried out in Matlab. We choose to simulate the aircraft flight for 300 s, which is the time required for approximately three-fourths of one revolution for the steady climbing turn solution.

The differential equations are specified in the following Matlab m-file.

Matlab function 13.2 FSEJ2.m
```
function dX = FSEJ2(t,X)
% Input: values of magnitude of velocity vector V (ft/s),
% heading angle sigma (rad), and flight path angle gamma (rad)
% stored in the vector X
% Output: values of time derivative of magnitude of velocity
% vector dV (ft/s^2), time derivative of heading angle dsigma
% (rad/s), and time derivative of flight path angle dgamma
% (rad/s) stored in the vector dX
% Organize function input data into scalar components
V=X(1); sigma=X(2); gamma=X(3);
%  Enter parameter data for executive jet aircraft
W=73000.0; CD0=0.015; K=0.05; CL0=0.02; CLa=0.12; S=950.0;
% Enter air density (slugs/ft^3), assumed constant depending
% on the approximate flight altitude
rho=1.7553E-3;
% Specify input functions: thrust magnitude (lbs), angle of
% attack (rad) and bank angle about velocity vector (rad)
T=7961.7;
alpha=4.40*3.14159/180;
mu=15*3.14159/180;
% Compute lift and drag on executive jet aircraft;
% angle of attack in degrees here
CL=CL0+CLa*180*alpha/3.14159;
CD=CD0+K*CL*CL;
L=0.5*rho*V*V*S*CL;
D=0..5*rho*V*V*S*CD;
% Compute time derivatives of magnitude of aircraft velocity
% vector, heading angle, and flight path angle based on
% performance differential equations
dV=-32.2*sin(gamma)-D*32.2/W+T*32.2*cos(alpha)/W;
dsigma=L*32.2*sin(mu)*sec(gamma)/(W*V)+T*32.2*sin(alpha)*sin(mu)
      *sec(gamma)/(W*V);
```

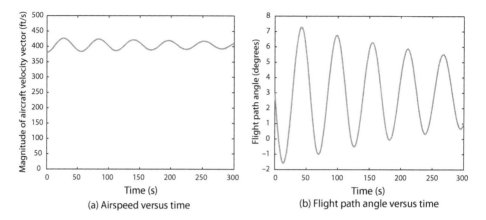

Figure 13.2. *Time responses of aircraft airspeed and flight path angle in a turn: executive jet aircraft.*

```
dgamma=-32.2*cos(gamma)/V+L*32.2*cos(mu)/(W*V)+T*32.2*sin(alpha)
     *cos(mu)/(W*V);
% Organize function output time derivative data into a
% column vector
dX =[dV;dsigma;dgamma];
```

The numerical integration of the differential equations for the executive jet aircraft, with the specified input functions and the specified initial values, is executed over the simulation time by the following Matlab commands; the discretized values of the simulation time are stored in the data vector TT and discretized "solution" values of the three state variables are stored in the columns of the data array Xsol. Several typical plot commands are included to obtain graphics output.

```
Tsim=300;
V0=380.0; sigma0=0.0; gamma0=0.0526;
[TT,Xsol] = ode45(@FSEJ2,[0,Tsim],[V0;sigma0;gamma0]);
plot(TT,Xsol(:,1))
axis([0 240 0 500])
xlabel('Time (s)')
ylabel('Air speed (ft/s)')
plot(TT,Xsol(:,3)*180/3.14159)
axis([0 240 -10 10])
xlabel('Time (s)');
ylabel('Flight path angle (degrees)');
```

The graphics output plots shown in figure 13.2 have the following interpretation. The perturbations in the initial value of the airspeed cause perturbations in the aircraft motions from the steady climbing turn solution; the resulting perturbations to the steady climbing turn necessarily occur in three dimensions. As in the previous simulation, these perturbations in the airspeed and in the flight path angle exhibit a long period phugoid oscillation with a slow decay toward the steady climbing turn flight values. The slow decay of the perturbations in time is a consequence of the flight stability property of this steady climbing turn flight condition for the executive jet aircraft.

13.10 Flight Simulations: General Aviation Aircraft

Several flight simulations of the general aviation aircraft are developed in this section. The data for the general aviation aircraft are the same as in the prior chapters. For each simulation, we specify the differential equations that define the initial value problem, we specify input variables as functions of time, and we specify the initial values of the state variables; then we use Matlab to numerically integrate the initial value problem. Plots of the solutions of the initial value problem are provided. The physical meaning of the development is emphasized throughout.

Simulation of a Perturbation from a Minimum Power Steady Level Flight Condition

Assuming the aircraft weight is constant, a flight model, based on the differential equations (13.23), (13.24), (13.12), and (13.14), is used to study the aircraft longitudinal flight dynamics. The input functions are the net engine power, taking into account the propeller efficiency, assumed to be a constant $31,454\,\text{ft lbs/s}$, the angle of attack, assumed to be a constant $9.85\,\text{degrees}$, and the bank angle about the velocity vector, assumed to be identically zero. These constant input functions satisfy the engine constraints, the stall constraint, and the load constraint; the engine power is the minimum value required for steady level flight at an altitude of $10,000\,\text{ft}$. The air density has the constant value of $0.0017553\,\text{slug/ft}^3$.

The results obtained in chapter 7 show that the general aviation aircraft is in steady level flight, with minimum required power from the propulsion system if the required power from the propulsion system is $31,454\,\text{ft lbs/s}$, the angle of attack is $9.85\,\text{degrees}$, the bank angle about the velocity vector is $0\,\text{degrees}$, the airspeed is $125.34\,\text{ft/s}$, and the flight path angle is $0\,\text{degrees}$. This steady level flight condition is introduced to motivate the flight simulation results that are subsequently developed. This steady flight condition is based on approximations made in chapter 7.

The flight simulation results for the general aviation aircraft in this section are based on the longitudinal differential equations given by equations (13.19) and (13.21), with bank angle and heading angle identically zero. In this flight simulation, we are interested in the displacement of the aircraft as it varies with time; hence we include the aircraft kinematics equations (13.12) and (13.14) in the simulation.

The flight simulation results are obtained for the initial value of the airspeed given by $124.4\,\text{ft/s}$ and the initial value of the flight path angle given by $5\,\text{degrees}$. The initial value of the airspeed is approximately the value required for steady climbing flight; the initial value of the flight path angle is positive, corresponding to an initial climb. Physically, the initial flight condition is perturbation from the minimum power steady level flight condition just described.

The simulation of this initial value problem provides the time responses that describe the resulting motion of the aircraft. This resulting motion does not correspond to steady flight: the resulting airspeed and flight path angle are not constants but vary with time. Although the specified values of the power and the angle of attack are the values for minimum power steady level flight, these constant values do not exactly minimize the power required for arbitrary initial values of the airspeed and the flight path angle.

Numerical flight simulation is easily carried out in Matlab; we simulate the aircraft flight for 150 s, using a standard Runge-Kutta algorithm.

The differential equations are specified in the following Matlab m-file.

Matlab function 13.3 FSGA1.m
```
function dX = FSGA1(t,X)
% Input: values of magnitude of velocity vector V (ft/s),
% flight path angle gamma (rad), horizontal displacement (ft),
% and vertical displacement (ft) stored in the vector X
% Output: values of time derivative of magnitude of velocity
% vector dV (ft/s^2), time derivative of flight path angle
% dgamma (rad/s), time derivative of horizontal displacement
% dx (ft), and time derivative of vertical displacement dz (ft)
% stored in the vector dX
% Organize function input data into scalar components
V=X(1); gamma=X(2);
% Enter parameter data for general aviation aircraft
W=2900.0; CD0=0.026; K=0.054; CL0=0.02; CLa=0.12; S=175.0;
% Enter air density (slugs/ft^3), assumed constant depending on
% the approximate flight altitude
rho=1.7553E-3;
% Specify input functions: engine power (ft lbs/s) and angle of
% attack (rad)
P=31454;
alpha=9.849*3.14159/180;
% Compute lift and drag on general aviation aircraft; angle of
% attack in degrees here
CL=CL0+CLa*180*alpha/3.14159;
CD=CD0+K*CL*CL;
L=0.5*rho*V*V*S*CL;
D=0..5*rho*V*V*S*CD;
% Compute time derivatives of magnitude of aircraft velocity
% vector and flight path angle based on performance differential
% equations
dV=-32.2*sin(gamma)-D*32.2/W+P*32.2*cos(alpha)/(W*V);
dgamma=-32.2*cos(gamma)/V+L*32.2/(W*V)+P*32.2*sin(alpha)
        /(W*V^2);
dx=V*cos(gamma);
dz=V*sin(gamma);
```

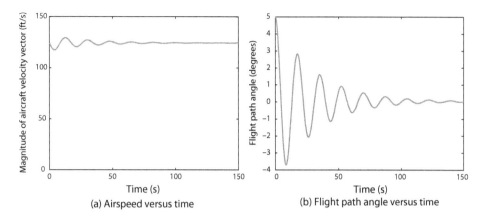

Figure 13.3. Time responses of aircraft airspeed and flight path angle in longitudinal flight: general aviation aircraft.

```
% Organize function output time derivative data into a column
% vector
dX =[dV;dgamma;dx;dz];
```

The numerical integration of the differential equations for the general aviation aircraft, with the specified input functions and the specified initial values, is executed over the simulation time by the following Matlab commands: the discretized values of the simulation time are stored in the data vector TT and discretized "solution" values of the three state variables are stored in the columns of the data array Xsol. Several typical plot commands are included to obtain graphics output.

```
Tsim=150;
V0=124.4; gamma0=5.0*3.14159/180;
[TT,Xsol] = ode45(@FSGA1,[0,Tsim],[V0;gamma0]);
plot(TT,Xsol(:,1))
axis([0 150 0 150])
xlabel('Time (s)')
ylabel('Aircraft airspeed (ft/s)')
plot(TT,Xsol(:,2)*180/3.14159)
axis([0 150 -4  5])
xlabel('Time (s)');
ylabel('Flight path angle (degrees)');
plot(Xsol(:,3),Xsol(:,4))
xlabel('Horizontal displacement of aircraft (ft)');
ylabel('Vertical displacement of aircraft (ft)');
```

The graphics output plots shown in figures 13.3(a) and 13.3(b) have the following interpretation. The perturbation in the initial value of the flight path angle (the aircraft

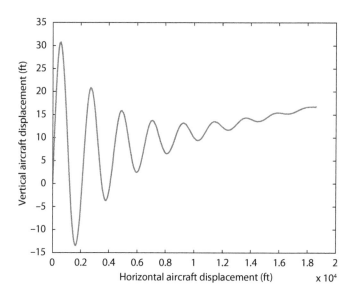

Figure 13.4. Path of aircraft in the vertical plane in longitudinal flight: general aviation aircraft.

is initially climbing) causes perturbations in the airspeed of the aircraft and the flight path angle; each of these flight variables exhibits a phugoid oscillation with decay toward the steady flight value. The decay of the perturbations in the airspeed and the flight path angle with time is a consequence of the flight stability property of this steady level flight condition for the general aviation aircraft.

Figure 13.4 shows the path of the center of mass of the aircraft in the vertical plane of its motion. The altitude response, relative to the horizontal distance traveled, is oscillatory with the aircraft moving above its initial attitude at some times and moving below its initial attitude at other times. As the time from the initial perturbation increases, the horizontal displacement of the aircraft increases and the vertical displacement of the aircraft shows a very slow increase. After the horizontal displacement of the aircraft has increased by 18,000 ft, the altitude of the aircraft has increased by approximately 15 ft.

Simulation of a Perturbation from a Minimum Radius Steady Level Turn

Assuming the aircraft weight is constant and flight occurs in three dimensions, a flight model, based on the differential equations (13.5)–(13.7), (13.2), and (13.4), is used to study the aircraft translational dynamics. The input functions are the net engine power, taking into account the propeller efficiency, assumed to be a constant 102,170 ft lbs/s, the angle of attack, assumed to be a constant 19.8 degrees, and the bank angle about the velocity vector, assumed to be a constant 60 degrees. These constant input functions satisfy the engine constraints, the stall constraint, and the load constraint; the stall and load constraints are satisfied as equalities. We assume that the air density is essentially constant at the value of 0.0017553 slug/ft^3, which corresponds to an approximate flight altitude of 10,000 ft.

The results obtained in chapter 9 show that the minimum radius steady level turn of a general aviation aircraft occurs if the power from the propulsion system is 102,170 ft lbs/s, the angle of attack is 19.8 degrees, the bank angle about the velocity vector is 60 degrees, the airspeed is 125.4 ft/s, and the flight path angle is 0 degrees. The minimum steady turn radius is 282.12 ft. The path of the aircraft in this steady level turn is the arc of a circle in a horizontal plane. The constant rate of change of the heading angle in this steady level turn is

$$\frac{d\sigma}{dt} = \frac{V \cos \gamma}{r} = 0.444 \, \text{rad/s}.$$

This minimum radius turn is a very aggressive maneuver for the general aviation aircraft, since it is operating exactly at stall and exactly at its maximum load factor. This steady level turning flight condition for the general aviation aircraft is introduced to motivate the flight simulation results that are subsequently developed. This steady flight condition is based on approximations made in chapter 9.

The flight simulation results for the general aviation aircraft in this section are based on the differential equations for three-dimensional flight given by equations (13.19), (13.20), and (13.21). In this flight simulation, we are interested in the displacement of the aircraft as it varies with time; hence we include the three-dimensional aircraft kinematics equations (13.12), (13.13), and (13.14) in the simulation.

Flight simulation results are obtained for the initial value of the airspeed given by 135.0 ft/s and the initial value of the flight path angle given by 0 degrees. The initial value of the aircraft airspeed exceeds the value required for steady level turning flight. Initialization of the heading angle is irrelevant, since the heading angle does not appear in the differential equations. The initial values of the position components of the center of mass of the aircraft can be selected arbitrarily; for graphical purposes, the initial horizontal position components are $x = 0$ ft, $y = -200$ ft, and the initial vertical position is $z = 0$ ft. Physically, the initial flight condition is a perturbation from the minimum radius steady level flight conditions just described; in particular, the initial value of the aircraft airspeed is larger than the value required for a steady level turn.

The simulation of this initial value problem provides the time responses that describe the resulting motion of the aircraft; this resulting motion does not correspond to steady flight, and the resulting airspeed and flight path angle are not constants but vary with time. Although the specified values of the power and the angle of attack are the values for a minimum radius steady level turn, these constant values do not exactly minimize the turn radius for arbitrary initial values of the airspeed and the flight path angle.

We simulate the aircraft flight for 30 s, which is the time required for approximately two complete revolutions for the minimum radius steady level turn solution. We use a standard Runge-Kutta algorithm.

The differential equations are specified in the following Matlab m-file.

Matlab function 13.4 FSGA2.m

```
function dX = FSGA2(t,X)
% Input: values of magnitude of velocity vector V (ft/s),
% heading angle sigma (rad), and flight path angle gamma (rad)
```

```
% stored in the vector X
% Output: values of time derivative of magnitude of velocity
% vector dV (ft/s^2), time derivative of heading angle dsigma
% (rad/s), time derivative of flight path angle dgamma (rad/s),
% and time derivatives of x, y, and z displacements of the
% aircraft stored in the vector dX
% Organize function input data into scalar components
V=X(1); sigma=X(2); gamma=X(3);
% Enter parameter data for general aviation aircraft
W=2900.0; CD0=0.026; K=0.054; CL0=0.02; CLa=0.12; S=175.0;
% Enter air density (slugs/ft^3), assumed constant depending
% on the approximate flight altitude
rho=1.7553E-3;
% Specify input functions: engine power (ft lbs/s), angle of
% attack (rad) and bank angle about velocity vector (rad)
P=102170.0;
alpha=19.8*3.14159/180;
mu=60*3.14159/180;
% Compute lift and drag on general aviation aircraft; angle of
% attack in degrees here
CL=CL0+CLa*180*alpha/3.14159;
CD=CD0+K*CL*CL;
L=0.5*rho*V*V*S*CL;
D=0..5*rho*V*V*S*CD;
% Compute time derivatives of magnitude of aircraft velocity
% vector, heading angle, flight path angle and x, y, z
% displacements based on performance differential equations
dV=-32.2*sin(gamma)-D*32.2/W+P*32.2*cos(alpha)/(W*V);
dsigma=L*32.2*sin(mu)*sec(gamma)/(W*V)+P*32.2*sin(alpha)*sin(mu)
        *sec(gamma)/(W*V^2);
dgamma=-32.2*cos(gamma)/V+L*32.2*cos(mu)/(W*V)+P*32.2*sin(alpha)
        *cos(mu)/(W*V^2);
dx=V*cos(sigma)*cos(gamma);
dy=V*sin(sigma)*cos(gamma);
dz=V*sin(gamma);
% Organize function output time derivative data into a column
% vector
dX =[dV;dsigma;dgamma;dx,dy,dz];
```

The numerical integration of the differential equations for the general aviation aircraft, with the specified input functions and the specified initial values, is executed over the simulation time by the following Matlab commands: the discretized values of the simulation time are stored in the data vector TT and discretized "solution" values of the three state variables are stored in the columns of the data array Xsol. Several typical plot commands are included to obtain graphics output.

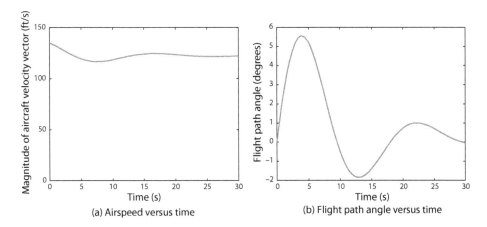

(a) Airspeed versus time　(b) Flight path angle versus time

Figure 13.5. Time responses of aircraft airspeed and flight path angle in a climbing turn: general aviation aircraft.

```
Tsim=30;
V0=135.0; sigma0=0.0; gamma0=0.0; x0=0.0; y0=-200.0; z0=0;
[T,X] = ode45(@FSGA2,[0,Tsim],[V0;sigma0;gamma0,x0;y0;z0]);
plot(TT,Xsol(:,1))
axis([0 30 0 150])
xlabel('Time (s)')
ylabel('Aircraft airspeed (ft/s)')
plot(TT,Xsol(:,3)*180/3.14159)
axis([0 30 -2  6])
xlabel('Time (s)');
ylabel('Flight path angle (degrees)');
plot(Xsol(:,4),Xsol(:,5))
xlabel('Horizontal aircraft displacement along ground x axis (ft)')
ylabel('Horizontal aircraft displacement along ground y axis (ft)')
plot(TT,Xsol(:,6))
xlabel('Time (s))
ylabel('Vertical displacement along ground z axis (ft)')
```

The graphics output plots in Figures 13.5(a) and 13.5(b) have the following interpretation. The airspeed of the aircraft decreases from its initial value and oscillates with decay toward its steady value. In addition, the flight path angle exhibits a decaying oscillation toward its steady value. This necessarily implies that the instantaneous turn radius oscillates with decay about its steady flight value corresponding to the steady level turn. This decaying phugoid oscillation of the perturbations in the airspeed and the flight path angle with time is a consequence of the flight stability property of this steady flight condition for the general aviation aircraft.

Figures 13.6(a) and 13.6(b) show the path of the center of mass of the general aviation aircraft in three dimensions. At the scale shown, the perturbations in the horizontal path

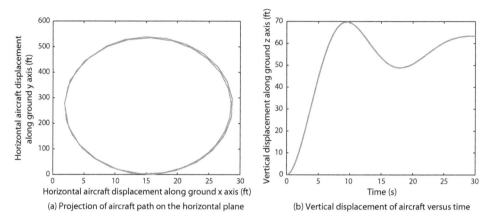

(a) Projection of aircraft path on the horizontal plane   (b) Vertical displacement of aircraft versus time

Figure 13.6. *Path of aircraft in a climbing turn: general aviation aircraft.*

of the aircraft from its steady circular path are small. At the same time, the excess initial speed of the aircraft causes the altitude of the aircraft to increase with time; after 30 s the altitude increases by approximately 60 ft as a consequence of the initial perturbation in the airspeed of the aircraft.

13.11 Conclusions

This chapter has provided a brief introduction to dynamic flight by developing a set of differential equations for the aircraft translational dynamics and demonstrating how these differential equations can be used for computing various dynamic flight envelopes and dynamic performance measures. These differential equations have been used to illustrate flight simulations for the executive jet aircraft and the general aviation aircraft.

As mentioned earlier, this chapter is intended as a bridge between steady flight, which is based on algebraic equations, and dynamic flight, which is based on differential equations. This connection is easy and direct: the algebraic equations for steady flight are simply the equilibrium equations of the differential equations for the aircraft translational flight dynamics. This simplicity is obtained due to the choice of flight variables and the fact that the differential equations do not include aircraft rotational effects.

13.12 Problems

13.1. Consider the executive jet aircraft in flight at an altitude of 33,000 ft. Assume the aircraft weight is a constant 73,000 lbs. Assume the aircraft is operating at a constant 90 percent of full throttle with a constant angle of attack of 2.5 degrees and a constant bank angle of 0 degrees.
    (a) Describe a steady longitudinal flight condition for these aircraft inputs. What is the constant airspeed of the aircraft? What is the constant flight path angle of the aircraft?

(b) Develop a flight simulation of the longitudinal flight dynamics assuming that the initial aircraft airspeed exceeds its steady flight value by 15 ft/s; assume the initial flight path angle is exactly the steady flight value. Develop time response plots for the airspeed, the flight path angle, and the change in altitude of the aircraft.

(c) Develop a flight simulation of the longitudinal flight dynamics assuming that the initial flight path angle exceeds its steady flight value by 2 degrees; assume the aircraft airspeed is exactly the steady flight value. Develop time response plots for the airspeed, the flight path angle, and the change in altitude of the aircraft.

13.2. Consider the executive jet aircraft in flight at an altitude of 33,000 ft. Assume the aircraft weight is a constant 73,000 lbs. Assume the aircraft is operating at a constant 90 percent of full throttle with a constant angle of attack of 2.5 degrees and a constant bank angle of 15 degrees.

(a) Describe a steady turning flight condition for these aircraft inputs. What is the constant aircraft airspeed? What is the constant flight path angle of the aircraft? What is the radius of the turn?

(b) Develop a flight simulation of the turning flight dynamics assuming that the initial aircraft airspeed exceeds its steady flight value by 15 ft/s; assume the initial flight path angle is exactly the steady flight value. Develop time response plots for the airspeed, the flight path angle, and the change in altitude of the aircraft. Develop a plot of the projection of the aircraft path onto a horizontal plane.

(c) Develop a flight simulation of the turning flight dynamics assuming that the initial flight path angle exceeds its steady flight value by 2 degrees; assume the aircraft airspeed is exactly the steady flight value. Develop time response plots for the airspeed, the flight path angle, and the change in altitude of the aircraft. Develop a plot of the projection of the aircraft path on to a horizontal plane.

13.3. Consider the propeller-driven general aviation aircraft in flight at an altitude of 18,000 ft. Assume the aircraft weight is a constant 2,900 lbs. Assume the aircraft is operating at a constant 90 percent of full throttle with a constant angle of attack of 2.5 degrees and a constant bank angle of 0 degrees.

(a) Describe a steady longitudinal flight condition for these aircraft inputs. What is the constant aircraft airspeed? What is the constant flight path angle of the aircraft?

(b) Develop a flight simulation of the longitudinal flight dynamics assuming that the initial aircraft airspeed exceeds its steady flight value by 10 ft/s; assume the initial flight path angle is exactly the steady flight value. Develop time response plots for the airspeed, the flight path angle, and the change in altitude of the aircraft.

(c) Develop a flight simulation of the longitudinal flight dynamics assuming that the initial flight path angle exceeds its steady flight value by 3 degrees; assume the initial aircraft airspeed is exactly the steady flight value. Develop time response plots for the airspeed, the flight path angle, and the change in altitude of the aircraft.

13.4. Consider the propeller-driven general aviation aircraft in flight at an altitude of 18,000 ft. Assume the aircraft weight is a constant 2,900 lbs. Assume the aircraft is operating at a constant 90 percent of full throttle with a constant angle of attack of 2.5 degrees and a constant bank angle of 20 degrees.

   (a) Describe a steady turning flight condition for these aircraft inputs. What is the constant aircraft airspeed? What is the constant flight path angle of the aircraft? What is the radius of the turn?

   (b) Develop a flight simulation of the turning flight dynamics assuming that the initial aircraft airspeed exceeds its steady flight value by 10 ft/s; assume the initial flight path angle is exactly the steady flight value. Develop time response plots for the airspeed, the flight path angle, and the change in altitude of the aircraft. Develop a plot of the projection of the aircraft path onto a horizontal plane.

   (c) Develop a flight simulation of the turning flight dynamics assuming that the initial flight path angle exceeds its steady flight value by 3 degrees; assume the aircraft airspeed is exactly the steady flight value. Develop time response plots for the airspeed, the flight path angle, and the change in altitude of the aircraft. Develop a plot of the projection of the aircraft path onto a horizontal plane.

13.5. Consider the UAV in flight at an altitude of 2,000 ft. Assume the aircraft weight is a constant 45 lbs. Assume the aircraft is operating at a constant 90 percent of full throttle with a constant angle of attack of 3 degrees and a constant bank angle of 0 degrees.

   (a) Describe a steady longitudinal flight condition for these aircraft inputs. What is the constant aircraft airspeed? What is the constant flight path angle of the aircraft?

   (b) Develop a flight simulation of the longitudinal flight dynamics assuming that the initial aircraft airspeed exceeds its steady flight value by 6 ft/s; assume the initial flight path angle is exactly the steady flight value. Develop time response plots for the airspeed, the flight path angle, and the change in altitude of the aircraft.

   (c) Develop a flight simulation of the longitudinal flight dynamics assuming that the initial flight path angle exceeds its steady flight value by 1.5 degrees; assume the initial aircraft airspeed is exactly the steady flight value. Develop time response plots for the airspeed, the flight path angle, and the change in altitude of the aircraft.

13.6. Consider the UAV in flight at an altitude of 2,000 ft. Assume the aircraft weight is a constant 45 lbs. Assume the aircraft is operating at a constant

90 percent of full throttle with a constant angle of attack of 3 degrees and a constant bank angle of 30 degrees.

(a) Describe a steady turning flight condition for these aircraft inputs. What is the constant aircraft airspeed? What is the constant flight path angle of the aircraft? What is the radius of the turn?

(b) Develop a flight simulation of the turning flight dynamics assuming that the initial aircraft airspeed exceeds its steady flight value by 6 ft/s; assume the initial flight path angle is exactly the steady flight value. Develop time response plots for the airspeed, the flight path angle, and the change in altitude of the aircraft. Develop a plot of the projection of the aircraft path onto a horizontal plane.

(c) Develop a flight simulation of the turning flight dynamics assuming that the initial flight path angle exceeds its steady flight value by 1.5 degrees; assume the initial aircraft airspeed is exactly the steady flight value. Develop time response plots for the airspeed, the flight path angle, and the change in altitude of the aircraft. Develop a plot of the projection of the aircraft path onto a horizontal plane.

13.7. Consider the executive jet aircraft in longitudinal flight from just after takeoff, at an altitude of 2,000 ft, until it reaches its maximum possible flight altitude. Assume the aircraft inputs are constant in time: full throttle, angle of attack of 2.5 degrees, and bank angle of 0 degrees. Take into account the important translational dynamics of the executive jet aircraft, the decrease in weight of the aircraft due to fuel consumption, and the change in air density of the atmosphere as a function of altitude.

(a) Give differential equations that describe the flight dynamics based on the above assumptions. Include all parameter values and the constant input functions in describing your differential equations. Assume the initial weight of the aircraft, including fuel, is 73,000 lbs, the initial airspeed of the aircraft is 200 ft/s and the initial flight path angle is 1.5 degrees.

(b) Develop a flight simulation based on the above initial value problem. Develop time response plots for the airspeed, the flight path angle, and the altitude of the aircraft.

(c) What is the maximum possible altitude of the aircraft based on your flight simulation results? How does this maximum dynamic altitude compare with the flight ceiling of the aircraft in steady level flight? How long does it take for the aircraft to reach its maximum possible altitude? How much fuel is burned?

13.8. Consider the propeller-driven general aviation aircraft in longitudinal flight from just after takeoff, at an altitude of 1,000 ft, until it reaches its maximum possible flight altitude. Assume the aircraft inputs are constant in time: full throttle, angle of attack of 2.5 degrees, and bank angle of 0 degrees. Take into account the important translational dynamics of the general aviation aircraft, the decrease in weight of the aircraft due to fuel consumption, and the change in air density of the atmosphere as a function of altitude.

   (a) Give differential equations that describe the flight dynamics based on the above assumptions. Include all parameter values and the constant input functions in describing your differential equations. Assume the initial weight of the aircraft, including fuel, is 2,900 lbs, the initial airspeed of the aircraft is 150 ft/s, and the initial flight path angle is 1.5 degrees.

   (b) Develop a flight simulation based on the above initial value problem. Develop time response plots for the airspeed, the flight path angle, and the altitude of the aircraft.

   (c) What is the maximum possible altitude of the aircraft based on your flight simulation results? How does this maximum dynamic altitude compare with the flight ceiling of the aircraft in steady level flight? How long does it take for the aircraft to reach its maximum possible altitude? How much fuel is burned?

13.9. Consider the UAV in longitudinal flight from just after takeoff, at sea level, until it reaches its maximum possible flight altitude. Assume the aircraft inputs are constant in time: full throttle, angle of attack of 2.5 degrees, and bank angle of 0 degrees. Take into account the important translational dynamics of the UAV, the decrease in weight of the aircraft due to fuel consumption, and the change in air density of the atmosphere as a function of altitude.

   (a) Give differential equations that describe the flight dynamics based on the above assumptions. Include all parameter values and the constant input functions in describing your differential equations. Assume the initial weight of the aircraft, including fuel, is 45 lbs, the initial airspeed of the aircraft is 75 ft/s, and the initial flight path angle is 1.5 degrees.

   (b) Develop a flight simulation based on the above initial value problem. Develop time response plots for the airspeed, the flight path angle, and the altitude of the aircraft.

   (c) What is the maximum possible altitude of the aircraft based on your flight simulation results? How does this maximum dynamic altitude compare with the flight ceiling of the aircraft in steady level flight? How long does it take for the aircraft to reach its maximum possible altitude? How much fuel is burned?

13.10. Consider the ground run of an executive jet aircraft during takeoff at an altitude of 3,000 ft. Assume the aircraft weight is a constant 73,000 lbs. Assume the aircraft is operating at full throttle with a constant angle of attack of 21 degrees and a constant bank angle of 0 degrees. The coefficient of friction between the aircraft tires and the runway is 0.025.

   (a) Give differential equations that describe the ground roll dynamics based on the above assumptions. Include all parameter values and the constant input functions in describing your differential equations. Assume the aircraft begins at rest.

(b) Develop a flight simulation of this takeoff maneuver from initiation until takeoff, which is assumed to occur when the contact force of the aircraft with the ground $N = 0$. Develop a time response plot for the the airspeed. What is the distance traveled by the aircraft from initialization to takeoff?

13.11. Consider the ground run of the propeller-driven general aviation aircraft during takeoff at an altitude of 1,000 ft. Assume the aircraft weight is a constant 2,900 lbs. Assume the aircraft is operating at 90 percent of full throttle with a constant angle of attack of 18 degrees and a constant bank angle of 0 degrees. The coefficient of friction between the aircraft tires and the runway is 0.04.

(a) Give differential equations that describe the ground roll dynamics based on the above assumptions. Include all parameter values and the constant input functions in describing your differential equations. Assume the initial aircraft ground roll speed is 5 ft/s.

(b) Develop a flight simulation of this takeoff maneuver from initiation until takeoff, which is assumed to occur when the contact force of the aircraft with the ground $N = 0$. Develop a time response plot for the the airspeed. What is the distance traveled by the aircraft from initialization to takeoff?

13.12. Consider the ground run of an executive jet aircraft while landing at an altitude of 3,000 ft. Assume the aircraft weight is a constant 65,000 lbs. Assume the aircraft is operating at zero throttle with a constant angle of attack of 8 degrees and a constant bank angle of 0 degrees. The coefficient of friction between the aircraft tires and the runway, taking into account steady braking, is 0.18.

(a) Give differential equations that describe the ground roll dynamics based on the above assumptions. Include all parameter values and the constant input functions in describing your differential equations. Assume the aircraft airspeed at touchdown is 185 ft/s.

(b) Develop a flight simulation of this landing maneuver from touchdown until the aircraft comes to rest. Develop a time response plot for the the airspeed. What is the distance traveled by the aircraft from touchdown to rest?

13.13. Consider the ground run of the propeller-driven general aviation aircraft during landing at an altitude of 1,000 ft. Assume the aircraft weight is a constant 2,700 lbs. Assume the aircraft is operating at zero throttle with a constant angle of attack of 6 degrees and a constant bank angle of 0 degrees. The coefficient of friction between the aircraft tires and the runway, taking into account steady braking, is 0.17.

(a) Give differential equations that describe the ground roll dynamics based on the above assumptions. Include all parameter values and the constant input functions in describing your differential equations. Assume the aircraft airspeed at touchdown is 98 ft/s.

(b) Develop a flight simulation of this landing maneuver from touchdown until the aircraft comes to rest. Develop a time response plot for the the airspeed. What is the distance traveled by the aircraft from touchdown to rest?

Appendix A

The Standard Atmosphere Model

The standard atmospheric model is described in chapter 2. This model can be used to generate tables of data in British or U.S. units (slugs-lbs-ft) for the standard atmosphere using the Matlab m-file FigStdAtpUS given in chapter 2. The following data table lists the altitude, temperature, pressure, density, and speed of sound for the standard atmosphere.

Further information about the historical development of the standard atmospheric model and more extensive data tables are available through the Web site http://nssdc.gsfc.nasa.gov/space/model/atmos/us_standard.html.

Table A.1. Standard Atmosphere (US units)

Altitude h (ft)	Temperature T (R)	Pressure P (lb/ft^2)	Density ρ (slug/ft^3)	Speed of Sound a (ft/s)
−2000	525.80	2.2737E+003	2.5191E − 003	1124.10
−1000	522.24	2.1938E+003	2.4472E − 003	1120.28
0	**518.67**	**2.1162E+003**	**2.3769E − 003**	**1116.45**
1000	515.10	2.0409E+003	2.3081E − 003	1112.61
2000	511.54	1.9677E+003	2.2409E − 003	1108.75
3000	507.97	1.8966E+003	2.1751E − 003	1104.88
4000	504.41	1.8277E+003	2.1109E − 003	1100.99
5000	500.84	1.7608E+003	2.0481E − 003	1097.09
6000	497.27	1.6959E+003	1.9867E − 003	1093.18
7000	493.71	1.6329E+003	1.9268E − 003	1089.25
8000	490.14	1.5719E+003	1.8683E − 003	1085.31
9000	486.57	1.5127E+003	1.8111E − 003	1081.36
10000	**483.01**	**1.4553E+003**	**1.7553E − 003**	**1077.39**
11000	479.44	1.3997E+003	1.7008E − 003	1073.40
12000	475.88	1.3459E+003	1.6476E − 003	1069.40
13000	472.31	1.2937E+003	1.5957E − 003	1065.39
14000	468.74	1.2432E+003	1.5450E − 003	1061.36
15000	465.18	1.1943E+003	1.4956E − 003	1057.31
16000	461.61	1.1469E+003	1.4474E − 003	1053.25
17000	458.05	1.1011E+003	1.4004E − 003	1049.18
18000	454.48	1.0568E+003	1.3546E − 003	1045.08
19000	450.91	1.0139E+003	1.3100E − 003	1040.97
20000	**447.35**	**9.7249E+002**	**1.2664E − 003**	**1036.85**
21000	443.78	9.3243E+002	1.2240E − 003	1032.71
22000	440.21	8.9372E+002	1.1827E − 003	1028.55

Table A.1. Continued.

Altitude h (ft)	Temperature T (R)	Pressure P (lb/ft^2)	Density ρ (slug/ft^3)	Speed of Sound a (ft/s)
23000	436.65	8.5632E+002	1.1425E − 003	1024.38
24000	433.08	8.2019E+002	1.1033E − 003	1020.19
25000	429.52	7.8531E+002	1.0651E − 003	1015.98
26000	425.95	7.5164E+002	1.0280E − 003	1011.75
27000	422.38	7.1915E+002	9.9187E − 004	1007.51
28000	418.82	6.8781E+002	9.5671E − 004	1003.24
29000	415.25	6.5758E+002	9.2252E − 004	998.96
30000	**411.69**	**6.2843E+002**	**8.8927**E − 004	**994.66**
31000	408.12	6.0035E+002	8.5695E − 004	990.35
32000	404.55	5.7328E+002	8.2553E − 004	986.01
33000	400.99	5.4721E+002	7.9500E − 004	981.66
34000	397.42	5.2212E+002	7.6534E − 004	977.28
35000	393.85	4.9796E+002	7.3654E − 004	972.89
36000	390.29	4.7471E+002	7.0857E − 004	968.47
37000	389.97	4.5244E+002	6.7587E − 004	968.08
38000	389.97	4.3120E+002	6.4416E − 004	968.08
39000	389.97	4.1097E+002	6.1393E − 004	968.08
40000	**389.97**	**3.9168E+002**	**5.8512**E − 004	**968.08**
41000	389.97	3.7330E+002	5.5766E − 004	968.08
42000	389.97	3.5579E+002	5.3149E − 004	968.08
43000	389.97	3.3909E+002	5.0655E − 004	968.08
44000	389.97	3.2318E+002	4.8278E − 004	968.08
45000	389.97	3.0801E+002	4.6013E − 004	968.08
46000	389.97	2.9356E+002	4.3853E − 004	968.08
47000	389.97	2.7978E+002	4.1795E − 004	968.08
48000	389.97	2.6665E+002	3.9834E − 004	968.08
49000	389.97	2.5414E+002	3.7965E − 004	968.08
50000	**389.97**	**2.4221E+002**	**3.6183**E − 004	**968.08**
51000	389.97	2.3085E+002	3.4485E − 004	968.08
52000	389.97	2.2001E+002	3.2867E − 004	968.08
53000	389.97	2.0969E+002	3.1325E − 004	968.08
54000	389.97	1.9985E+002	2.9855E − 004	968.08
55000	389.97	1.9047E+002	2.8454E − 004	968.08
56000	389.97	1.8153E+002	2.7119E − 004	968.08
57000	389.97	1.7301E+002	2.5846E − 004	968.08
58000	389.97	1.6490E+002	2.4633E − 004	968.08
59000	389.97	1.5716E+002	2.3477E − 004	968.08
60000	**389.97**	**1.4978E+002**	**2.2375**E − 004	**968.08**
61000	389.97	1.4275E+002	2.1325E − 004	968.08
62000	389.97	1.3606E+002	2.0325E − 004	968.08
63000	389.97	1.2967E+002	1.9371E − 004	968.08
64000	389.97	1.2359E+002	1.8462E − 004	968.08
65000	389.97	1.1779E+002	1.7596E − 004	968.08
66000	390.18	1.1226E+002	1.6761E − 004	968.34
67000	390.73	1.0700E+002	1.5953E − 004	969.02

Table A.1. Continued.

Altitude h (ft)	Temperature T (R)	Pressure P (lb/ft^2)	Density ρ (slug/ft^3)	Speed of Sound a (ft/s)
68000	391.28	1.0199E+002	1.5185E − 004	969.70
69000	391.83	9.7222E+001	1.4455E − 004	970.38
70000	**392.37**	**9.2684E+001**	**1.3761E − 004**	**971.06**
71000	392.92	8.8364E+001	1.3101E − 004	971.74
72000	393.47	8.4251E+001	1.2474E − 004	972.41
73000	394.02	8.0334E+001	1.1877E − 004	973.09
74000	394.57	7.6605E+001	1.1310E − 004	973.77
75000	395.12	7.3053E+001	1.0771E − 004	974.44
76000	395.67	6.9671E+001	1.0258E − 004	975.12
77000	396.22	6.6450E+001	9.7702E − 005	975.80
78000	396.76	6.3381E+001	9.3061E − 005	976.47
79000	397.31	6.0459E+001	8.8648E − 005	977.15
80000	**397.86**	**5.7675E+001**	**8.4449E − 005**	**977.82**
81000	398.41	5.5022E+001	8.0455E − 005	978.50
82000	398.96	5.2496E+001	7.6654E − 005	979.17
83000	399.51	5.0088E+001	7.3038E − 005	979.84
84000	400.06	4.7794E+001	6.9597E − 005	980.51
85000	400.60	4.5608E+001	6.6323E − 005	981.19
86000	401.15	4.3524E+001	6.3207E − 005	981.86
87000	401.70	4.1539E+001	6.0241E − 005	982.53
88000	402.25	3.9646E+001	5.7418E − 005	983.20
89000	402.80	3.7843E+001	5.4731E − 005	983.87
90000	**403.35**	**3.6123E+001**	**5.2173E − 005**	**984.54**
91000	403.90	3.4484E+001	4.9738E − 005	985.21
92000	404.44	3.2921E+001	4.7420E − 005	985.88
93000	404.99	3.1432E+001	4.5212E − 005	986.55
94000	405.54	3.0011E+001	4.3111E − 005	987.22
95000	406.09	2.8656E+001	4.1109E − 005	987.88
96000	406.64	2.7365E+001	3.9203E − 005	988.55
97000	407.19	2.6133E+001	3.7388E − 005	989.22
98000	407.74	2.4958E+001	3.5659E − 005	989.88
99000	408.29	2.3837E+001	3.4012E − 005	990.55
100000	**408.83**	**2.2768E+001**	**3.2443E − 005**	**991.21**

Appendix B

End-of-Chapter Problems

Problems are included at the end of most of the chapters, many of which require detailed analysis of steady flight and steady performance for one of three aircraft categories. The three categories of aircraft are: an ideal executive jet aircraft described in section 5.7, a single engine ideal propeller-driven general aviation aircraft described in section 5.8, and an uninhabited aerial vehicle described in section 5.9. The relevant data for these three aircraft are provided here for use in those problems. The atmosphere is assumed to be the standard atmosphere and it is assumed to be stationary.

B.1 Executive Jet Aircraft

This executive jet aircraft, as described in section 5.7, is powered by two jet engines. The weight of the aircraft, with a full fuel tank, is 73,000 lbs. The wing surface area is 950 ft^2; the aspect ratio is 5.9. The aerodynamic drag polar is given by

$$C_D = 0.015 + 0.05C_L^2.$$

The lift coefficient, expressed in terms of the angle of attack, is

$$C_L = 0.02 + 0.12\alpha,$$

where the angle of attack is measured in degrees. The maximum lift coefficient at stall is 2.8, which occurs at an angle of attack of 23.2 degrees. The pitch moment coefficient, expressed in terms of the angle of attack and the elevator deflection, is given by

$$C_M = 0.24 - 0.18\alpha + 0.28\delta_e,$$

where the angle of attack α and the elevator deflection δ_e are measured in degrees.

The jet aircraft is powered by two jet engines, each of which can provide a maximum sea-level thrust of 6,250 lbs. The engines are configured in the aircraft so that they do not generate any pitch moment on the aircraft. The thrust produced by the propulsion system depends on the flight altitude and the throttle setting according to

$$T = \delta_t \left(\frac{\rho}{2.3769 \times 10^{-3}} \right)^{0.6} 12,500 \text{ lbs,}$$

where δ_t is the throttle setting and ρ is the air density in slug/ft^3 at the flight altitude. The maximum fuel that can be carried in the aircraft is 28,000 lbs. The thrust specific fuel consumption rate for the jet engines is 0.69 lb fuel/hr/lb.

The maximum load factor for the aircraft is 2.

B.2 Single Engine Propeller-Driven General Aviation Aircraft

This single engine propeller-driven general aviation aircraft, as described in section 5.8, is powered by an internal combustion engine. The weight of the aircraft, with a full fuel tank, is 2,900 lbs. The wing surface area is 175 ft^2; the aspect ratio is 7.4. The aerodynamic drag polar is given by

$$C_D = 0.026 + 0.054C_L^2.$$

The lift coefficient, expressed in terms of the angle of attack, is

$$C_L = 0.02 + 0.12\alpha,$$

where the angle of attack is measured in degrees. The maximum lift coefficient at stall is 2.4, which occurs at an angle of attack of 19.8 degrees. The pitch moment coefficient, in terms of the angle of attack and the elevator deflection, expressed in terms of the angle of attack and the elevator deflection, is given by

$$C_M = 0.12 - 0.08\alpha + 0.075\delta_e,$$

where the angle of attack α and the elevator deflection δ_e are measured in degrees.

A single propeller generates the thrust. The propeller is driven by an internal combustion engine that produces a maximum of 290 hp at sea level. The propulsion system is configured in the aircraft so that the thrust does not generate any pitch moment on the aircraft. The power produced by the propulsion system that is available for flight depends on the propeller efficiency, the flight altitude, and the throttle setting according to

$$P = \delta_t \eta \left(\frac{\rho}{2.3769 \times 10^{-3}} \right)^{0.6} 290 \text{ hp},$$

where $\eta = 0.8$ is the propeller efficiency, δ_t is the throttle setting, and ρ is the air density in slug/ft^3 at the flight altitude. The maximum fuel that can be carried in the aircraft is 370 lbs. The power specific fuel consumption rate for the internal combustion engine is 0.45 lb fuel/hr/hp.

The maximum load factor for the aircraft is 2.

B.3 Uninhabited Aerial Vehicle (UAV)

A small uninhabited aerial vehicle is described in section 5.9. The weight of the UAV, with a full fuel tank, is 45 lbs. The wing surface area is 10.2 ft^2; the wing aspect ratio is 10. The aerodynamic drag polar is given by

$$C_D = 0.025 + 0.05 \, (C_L - 0.7)^2 \, .$$

Note that this drag polar differs slightly from the previously assumed form. The lift coefficient, expressed in terms of the angle of attack, is

$$C_L = 0.18 + 0.06\alpha,$$

where the angle of attack is measured in degrees. The maximum lift coefficient at stall is 1.7, which occurs at an angle of attack of 25.3 degrees. The pitch moment coefficient, expressed in terms of the angle of attack and the elevator deflection, is given by

$$C_M = 0.22 - 0.15\alpha + 0.23\delta_e,$$

where the angle of attack α and the elevator deflection δ_e are measured in degrees.

The UAV is powered by a propeller and engine that is configured in the aircraft so that it does not generate any pitch moment on the aircraft. The thrust produced by the propulsion system, accounting for the propeller efficiency, depends on the airspeed, flight altitude, and the throttle setting according to

$$T = \delta_t \left(\frac{\rho}{2.3769 \times 10^{-3}} \right)^{0.6} (22 - 0.126V) \text{ lbs,}$$

where δ_t is the throttle setting, V is the airspeed in ft/s, and ρ is the air density in slug/ft^3 at the flight altitude. This expression for the thrust is valid for airspeeds from zero up to a maximum of 170 ft/s. Note that this expression for the thrust provided by the propulsion system depends explicitly on the airspeed of the aircraft; this expression differs from the previously assumed form. The maximum fuel that can be carried in the aircraft is 18 lbs. The thrust specific fuel consumption rate for the engine is 0.6 lb fuel/hr/lb.

The maximum load factor for the aircraft is 2.37.

The UAV is described by a drag polar expression and an engine thrust expression that differ from the standard assumptions that form the basis for the analytical developments in the prior chapters.

REFERENCES

Anderson, J. D. 1998. *Aircraft Performance and Design*. McGraw-Hill.

———. 2000. *Introduction to Flight*. 4th ed. McGraw-Hill.

Asselin, M. 1997. *An Introduction to Aircraft Performance*. AIAA Education Series.

Barnard, R. H., and D. R. Philpott. 1995. *Aircraft Flight: A Description of the Physical Principles of Aircraft Flight*. 2nd ed. Longman Scientific and Technical.

Collinson, R.P.G. 1996. *Introduction to Avionics*. Chapman and Hall.

Eshelby, M. E. 2000. *Aircraft Performance: Theory and Practice*. AIAA.

Etkin, B. 1972. *Dynamics of Atmospheric Flight*. John Wiley.

Layton, D. 1988. *Aircraft Performance*. Matrix Publishers.

Lowry, J. T. 1999. *Performance of Light Aircraft*. AIAA.

Mair, W. A., and D. L. Birdsall. 1996. *Aircraft Performance*. Cambridge University Press.

McCormick, B. 1994. *Aerodynamics, Aeronautics, and Flight Mechanics*. 2nd ed. Wiley.

Pamadi, B. N. 1998. *Performance, Stability, Dynamics and Control of Airplanes*. AIAA.

Roskam, J., and C.T.E. Lan. 1997. *Airplane Aerodynamics and Performance*. DARcorporation.

Saarias, M. 2006. *Aircraft Performance*. Wiley.

Shevell, R. S. 1988. *Fundamentals of Flight*. Prentice Hall.

Stengel, R. F. 2004. *Flight Dynamics*. Princeton University Press.

Torenbeek, E., and H. Wittenberg. 2009. *Flight Physics: Essentials of Aeronautical Disciplines and Technology, with Historical Notes*. Springer.

Vinh, N. X. 1995. *Flight Mechanics of High-Performance Aircraft*. Cambridge University Press.

Wegener, P. P. 1991. *What Makes Airplanes Fly? History, Science, and Applications of Aerodynamics*. Springer-Verlag.

Yechout, T. R., and D. E. Bossert. 2003. *Introduction to Aircraft Flight Mechanics: Performance, Static Stability, Dynamic Stability, and Classical Feedback Control*. AIAA Education Series.

INDEX

www.ingramcontent.com/pod-product-compliance
Ingram Content Group UK Ltd.
Pitfield, Milton Keynes, MK11 3LW, UK
UKHW010828161224
452264UK00002B/35